GLOBAL ISSUES IN FOOD SCIENCE AND TECHNOLOGY

GLOBAL ISSUES IN FOOD SCIENCE AND TECHNOLOGY

Edited by

**GUSTAVO BARBOSA-CÁNOVAS,
ALAN MORTIMER, DAVID LINEBACK,
WALTER SPIESS, KEN BUCKLE** *AND*
PAUL COLONNA

AMSTERDAM ● BOSTON ● HEIDELBERG ● LONDON
NEW YORK ● OXFORD ● PARIS ● SAN DIEGO
SAN FRANCISCO ● SINGAPORE ● SYDNEY ● TOKYO

Academic Press is an imprint of Elsevier

Academic Press is an imprint of Elsevier
30 Corporate Drive, Suite 400, Burlington, MA 01803, USA
525 B Street, Suite 1900, San Diego, CA 92101-4495, USA
32 Jamestown Road, London, NW1 7BY, UK
360 Park Avenue South, New York, NY 10010-1710, USA

First edition 2009

Library of Congress Cataloging-in-Publication Data
A catalog record for this book is available from the Library of Congress

British Library Cataloguing in Publication Data
A catalogue record for this book is available from the British Library

ISBN: 978-0-12-374124-0

For information on all Academic Press publications
visit our website at www.elsevierdriect.com

Typeset by TNQ Books and Journals, Chennai, India

Printed and bound in the United States of America

09 10 11 12 13   10 9 8 7 6 5 4 3 2 1

CONTENTS

CONTRIBUTORS

Yves Bertheau
Institut National de la Recherche Agronomique, (INRA), route de Saint Cyr, F-78026 Versailles, France

Patrick Borel
INSERM, U476, Nutrition Humaine et Lipides, Marseille, F-13385, France; INRA, UMR1260, Marseille, F-13385, France; Univ Méditerranée Aix-Marseille 2, Faculté de Médecine, IPHM-IFR 125, Marseille, F-13385, France

Frank Busta
The National Center for Food Protection and Defense, University of Minnesota, St. Paul, MN, USA

Nigel Cook
Central Science Laboratory, Sand Hutton, York YO41 1LZ, UK

Martin D'Agostino
Central Science Laboratory, Sand Hutton, York, UK YO41 1LZ

Michael Davidson
Department of Food Science and Technology, University of Tennessee, Knoxville, TN 37996-4591, USA

John Davison
INRA, Route de Versailles, F-78026 Versailles, France

Dennis Degeneffe
The Food Industry Center, University of Minnesota, St. Paul, MN, USA

Hulya Dogan
Department of Grain Science, Kansas State University, Manhattan, KS 66506, USA

Francesco Donsì
Department of Chemical and Food Engineering University of Salerno, Fisciano (SA), 84084 – Italy

Robert Engel
University of Karlsruhe, Institute of Bio and Food Process Engineering, Fritz-Haber-Weg 2, D-76131 Karlsruhe, Germany

Margaret Everitt
Director Sensory & Consumer Research, Sensory Dimensions Ltd, Science & Technology Centre, Earley Gate, Whiteknights Road, Reading, RG6 6BZ, UK

Bita Farhang
University of Guelph, Department of Food Science, 50 Stone Road East, Guelph, Ontario, Canada, NIG2W1

Giovanna Ferrari
ProdAl Scarl and Department of Chemical and Food Engineering, University of Salerno, Fisciano (SA), 84084 – Italy

Daniel Fung
Kansas State University, Animal Science and Industry, 207 Call Hall, Manhattan, KS 66506, USA

Sylvia Gaysinsky
Department of Food Science, University of Massachusetts, Amherst, MA 01003, USA

Koel Ghosh
The Food Industry Center, University of Minnesota, St. Paul, MN, USA

Beatriz Guerra
Federal Institute for Risk Assessment (BfR), Department of Biological Safety, Unit of Antimicrobial Resistance and Resistance Determinants, National Reference Laboratory for Antimicrobial Resistance, Diedersdorfer Weg 1, D-12277 Berlin, Germany

Shigeru Hayakawa
Department of Biochemistry and Food Science Faculty of Agriculture, Kagawa University, 2393 Ikenobe, Miki, Kagawa, Japan 761-0795

Reiner Helmuth
Federal Institute for Risk Assessment (BfR), Department of Biological Safety, Unit of Antimicrobial Resistance and Resistance Determinants, National Salmonella Reference Laboratory, Diedersdorfer Weg 1, D-12277 Berlin, Germany

Marta Hernández
Laboratorio de Biologia Molecular, Instituto Tecnológico Agrario de Castilla y León (ITACyL), Valladolid, Spain

Qingrong Huang
Center for Advanced Food Technology and Department of Food Science, Rutgers University, New Brunswick, NJ, 08901, USA

John Jenkins
Institute of Food Research, Norwich Research Park, Colney Norwich, NR4 7UA, UK

Lidia Kempa
University of Karlsruhe, Institute of Bio and Food Process Engineering,
Fritz-Haber-Weg 2, D-76131 Karlsruhe, Germany

Mark Kerslake
Numsight, 80–82 rue Galliéni 92100, Boulogne-Billancourt, France

Jean Kinsey
Department of Applied Economics and The Food Industry Center, University of
Minnesota, St. Paul, MN, USA

Jozef Kokini
Center for Advanced Food Technology and Department of Food Science, Rutgers
University, New Brunswick, NJ, 08901, USA;
Department of Food Science and Human Nutrition, University of Illinois Urbana
Champaign – Urbana, IL 61801, USA

Alexandre Leclercq
Institut Pasteur - National Reference Centre and Collaborative Centre for World
Health Organization for Listeria, 28 rue du Docteur Roux – 75724 Paris cedex
15, France

Bertrand Lombard
Coordinator of Scientific & Technical Support Community Reference
Laboratories & International Relations, AFSSA-LERQAP (French Agency for
Food Safety – Laboratory for Study & Research on Food Quality & Food
Processes), 23 avenue du Général De Gaulle – 94706 Maisons-Alfort, France

Julian McClements
Department of Food Science, University of Massachusetts, Amherst, MA 01003,
USA

Albert McGill
James Martin Institute for Science and Civilization, Said Business School, Oxford
University, Park End Street, Oxford OX1 1HP, UK

Alan Mackie
Institute of Food Research, Norwich Research Park, Colney Norwich, NR4
7UA, UK

Burkhard Malorny
Federal Institute for Risk Assessment (BfR), Department of Biological Safety, Unit
of Antimicrobial Resistance and Resistance Determinants, National Salmonella
Reference Laboratory, Diedersdorfer Weg 1, D-12277 Berlin, Germany

Paola Maresca
ProdAl Scarl, University of Salerno, Fisciano (SA), 84084 – Italy

Klaus Menrad
Chair of Marketing and Management, University of Applied Sciences of
Weihenstephan, Straubing Center of Science, Straubing, Germany

Angelika Miko
Federal Institute for Risk Assessment (BfR), Department of Biological Safety, Unit of Microbial Toxins, National Reference Laboratory for Escherichia coli, Diedersdorfer Weg 1, D-12277 Berlin, Germany

Clare Mills
Institute of Food Research, Norwich Research Park, Colney, Norwich, NR4 7UA, UK

Carmen Moraru
Department of Food Science, Cornell University, Ithaca, NY, USA

Supaporn Naknukool
Department of Biochemistry and Food Science, Faculty of Agriculture, Kagawa University, Ikenobe, Miki, Kagawa, Japan 761-0795

Masahiro Ogawa
Department of Biochemistry and Food Science, Faculty of Agriculture, Kagawa University, Ikenobe, Miki, Kagawa, Japan 761-0795

Faustine Régnier
INRA, UR 1303 ALISS, F-94205 Ivry-sur-Seine, France

Neil Rigby
Institute of Food Research, Norwich Research Park, Colney Norwich, NR4 7UA, UK

David Rodríguez-Lázaro
Food Safety and Technology Group, Instituto Tecnológico Agrario de Castilla y León (ITACyL), Carretera de Burgos, Km.119, 47071 Valladolid, Spain

Artur Rzeżutka
National Veterinary Research Institute, Department of Food & Environmental Virology, Al. Partyzantów 57, 24-100 Puławy, Poland

Ana Sancho
Institute of Food Research, Norwich Research Park, Colney Norwich, NR4 7UA, UK

Andreas Schroeter
Federal Institute for Risk Assessment (BfR), Department of Biological Safety, Unit of Antimicrobial Resistance and Resistance Determinants, National Reference Laboratory for Antimicrobial Resistance; National Salmonella Reference Laboratory, Diedersdorfer Weg 1, D-12277 Berlin, Germany

Helmar Schubert
Institute of Process Engineering in Life Sciences, Section I: Food Process Engineering, Kaiserstr. 12, D - 76131 Karlsruhe, Germany

Eyal Shimoni
Faculty of Biotechnology and Food Engineering, The Russel Berrie
Nanotechnology Institute, Technion – Israel Institute of Technology, Haifa 32000,
Israel

Joel Sidel
Tragon Corporation, 350 Bridge Parkway, Redwood Shores, CA 94065-1061,
USA

H.V. Smith
Scottish Parasite Diagnostic Laboratory, Stobhill Hospital, 133 Balornock Road,
Glasgow G21 3UW, UK

Kai Sparke
TNS Infratest GmbH, Landsberger Str. 338, 80637 Munich, Germany

Tom Stinson
Department of Applied Economics, University of Minnesota, St. Paul, MN, USA

Herbert Stone
Senior Advisor, Tragon Corporation, 350 Bridge Parkway, Redwood Shores, CA
94065-1061, USA

Paul Takhistov
Center for Advanced Food Technology and Department of Food Science, Rutgers
University, New Brunswick, NJ, 08901, USA

Takahiro Uno
Department of Biochemistry and Food Science, Faculty of Agriculture, Kagawa
University, Ikenobe, Miki, Kagawa, Japan 761-0795

Jochen Weiss
Dept. of Food Physiochemistry and Biophysics, Institute of Food Science and
Biotechnology, Garbenstrasse 21, 70599 Stuttgart, Germany

The International Union of Food Science and Technology (IUFoST) is a not-for-profit, country-membership organization with a global voice on food science and technology. IUFoST undertakes a variety of international activities promoting the advancement of food science and technology through its education programs, workshops, and regional symposia, as well as through the International Academy of Food Science and Technology (IAFoST). IAFoST is an organization of elected distinguished food scientists and technologists who collectively form a pool of non-aligned expert advice on scientific matters.

IUFoST serves to link the world's top food scientists and technologists together as one scientific body. IUFoST was created in 1970 during the 3rd International Congress of Food Science and Technology held in Washington, DC, USA. Previous and subsequent International Congresses were held in London, UK (1962); Warsaw, Poland (1966); Madrid, Spain (1974); Kyoto, Japan (1979); Dublin, Ireland (1983); Singapore (1987); Toronto, Canada (1991); Budapest, Hungary (1995); Sydney, Australia (1999); Seoul, South Korea (2001); Chicago, USA (2003); and Nantes, France (2006). Future congresses will be held in Shanghai, China (2008); Cape Town, South Africa (2010); and Salvador, Brazil (2012).

The 13th IUFoST World Congress was held in Nantes, France in 2006 where more than a thousand delegates representing 72 countries attended plenary and scientific lectures, workshops, technical sessions and oral, and poster presentations on issues of global importance to those in the field. The theme of the congress was "Food is Life."

While World Congresses have been an integral part of IUFoST's activities since its inception, there has been little systematic effort by congress organizers to collect the papers presented. Thus, the IUFoST Governing Council decided to publish a book containing selected key papers from the Nantes Congress, and are delighted that Elsevier – Academic Press agreed to publish this important volume.

This book contains 24 chapters representing key selections from those presentations, organized into four sections: Contemporary Topics, Consumer Trends, Food Safety, and Nanotechnology in Food Applications. Twenty-two papers were selected from presentations at the Nantes Congress and two nanotechnology papers in Section 4 were commissioned by the editors. The chapters were selected based on their scientific merit and relevance to current global issues in food science and technology. Most of the chapters are authored by colleagues that were invited speakers at the World Congress, and the selection was made after listening to many presentations and reviewing all available supporting materials. The editors are confident that this book gathers the important highlights of this successful congress led by Professors Pierre Feillet and Paul Colonna. At the same time, the editors recognize that there were many other presentations of great relevance. The fact that they are not included in this book should not imply a lack of superior quality on their part.

Gustavo Barbosa-Cánovas, Alan Mortimer,
David Lineback, Walter Spiess,
Ken Buckle and Paul Colonna

ACKNOWLEDGMENT

The Editors want to express their gratitude and appreciation to Sharon Himsl, Publications Coordinator, Washington State University, for her professionalism and dedication in facilitating so many of the steps needed to complete this challenging project. She edited and kept track of all manuscripts, and effectively interacted with the editors, authors and Elsevier Publishing. Her performance was remarkable at all times, and there are not enough words to praise her fantastic job.

SECTION *One*

Contemporary Topics

CHAPTER 1

Principles of Structured Food Emulsions: Novel Formulations and Trends

Helmar Schubert, Robert Engel, *and* Lidia Kempa

Contents

Abstract

Most properties of food emulsions depend on emulsion microstructure, which is largely influenced by mean droplet diameter and droplet size distribution. The correlation between the properties and the microstructure is called the *property function*, and the relationship between the microstructure and the process is called the *process function*. If both functions are known, the properties of an emulsion may be derived directly from the process parameters.

The droplet size of an emulsion depends on the intensity and mechanism of droplet disruption and on the extent of superimposed or subsequent droplet coalescence. Recent experiments have shown that the emulsifier, by reducing the interfacial tension, does not improve droplet disruption because only the interfacial tension of the unoccupied interface is relevant for droplet comminution.

Several methods suitable for the production of emulsions with the desired microstructure will be presented. Among these, newly developed valves inducing elongational flow have allowed for greatly increased homogenization efficiency in high-pressure homogenizers. Advances and new processes in emulsification with membranes and microstructured systems as well as their potential will also be discussed.

In an example it is demonstrated how nanoemulsions can be used to design functional foods. Phytosterols may significantly reduce cholesterol levels in humans. Due to their poor solubility in water and oil, product engineering is required to achieve satisfactory dose responses. With

Global Issues in Food Science and Technology
© 2009 Elsevier Inc.

ISBN 9780123741240

reference to cell culture studies on the bioavailability, a property function for carotenoid-loaded emulsions can be derived showing how the concept can be applied for formulation and quality improvement of a product.

I. PRODUCTION OF EMULSIONS

Emulsions are systems made up of at least two, practically immiscible liquids (e.g. oil and water), in which one liquid is finely dispersed in the other liquid. The dispersed liquid is also referred to as the disperse phase, whereas the other liquid is referred to as the continuous phase. The two basic types of emulsions are dispersions of a lipophilic or oil phase in a hydrophilic or watery phase or vice versa. With oil and water being the most common liquids for the preparation of food emulsions, these basic types of emulsions are referred to as oil-in-water- (o/w-) emulsions and water-in-oil- (w/o-) emulsions, respectively. More complex types consist of three or more phases, which can be achieved, for example, by dispersing a w/o-emulsion into a second watery phase, leading to a water-in-oil-in-water- (w/o/w-) emulsion. Due to the interfacial tension between the immiscible liquids, emulsions containing only oil and water are thermo-dynamically unstable. By using surface-active molecules called emulsifiers, emulsions can be kinetically stabilized. Furthermore, a difference in the densities of the two liquids may cause undesired creaming or sedimentation of the dispersed droplets. The occurrence of this physical instability can be delayed or prevented by increasing the viscosity of the continuous phase by so-called stabilizers (e.g. macromolecular substances). The basic types of emulsions as well as the role of emulsifiers and stabilizers are demonstrated in Figure 1.1.

Most of the properties of emulsions depend on the emulsion micro-structure, the emulsifiers used, and the viscosity of the continuous phase. The microstructure is mainly a function of droplet size and droplet size distribution. Droplet size is of essential importance because of its great influence on physical and microbiological stability, rheological and optical characteristics, bioavailability or dose response, taste and many other properties (Schubert, 2005a, b). In many cases, the aim of emulsification is to produce droplets as fine as possible in such a way that the resulting emulsion is stable.

Emulsions with fine-dispersed droplets, so-called 'fine emulsions,' can be produced in many different ways (Schubert, 2005b) (Figure 1.2). Mechanical processes are most frequently applied. They have been the subject of recent publications (Schubert, 2003; Schubert and Ax, 2001). Widely applicable rotor-stator systems are capable of producing emulsions both continuously

Production of fine-dispersed emulsions

Mechanical processes			Non-mechanical processes
Droplet disruption in rotor-stator systems o/w (fine emulsion) o/w (pre-mix)	Droplet disruption in high-pressure systems o/w (fine emulsion) o/w (pre-mix)		**Precipitation** of the dispersed phase, previously dissolved in the continuous phase
Droplet disruption by ultrasound Ultrasound probe o/w (fine emulsion) o/w (pre-mix)	Droplet formation at micropores Membrane emulsification w o/w ↓O	Premix membrane emulsification o/w o/w	Microchannel emulsification o/w w O	**Phase inversion**
			Phase inversion temperature method (PIT-method)

Figure 1.1 Principle types of emulsions and emulsifying agents.

and discontinuously. High–pressure systems, frequently referred to as high-pressure homogenizers, are used to continuously produce fine-dispersed emulsions. As in rotor-stator systems, mechanical energy is the driving force for droplet disruption. In high-pressure systems the mechanical energy is applied in the form of a pressure difference. A simple orifice valve has been found to be very efficient (Stang *et al.*, 2001; Tesch *et al.*, 2002). Its design is being modified and optimized at present. Recently obtained results show that this system is very promising (Freudig *et al.*, 2002).

Fine-dispersed emulsions can also be produced by ultrasound (Behrend *et al.*, 2000; Behrend and Schubert, 2001; Behrend, 2002). This method has

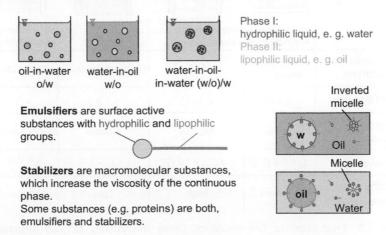

oil-in-water o/w   water-in-oil w/o   water-in-oil-in-water (w/o)/w

Phase I: hydrophilic liquid, e. g. water
Phase II: lipophilic liquid, e. g. oil

Emulsifiers are surface active substances with hydrophilic and lipophilic groups.

Stabilizers are macromolecular substances, which increase the viscosity of the continuous phase.
Some substances (e.g. proteins) are both, emulsifiers and stabilizers.

Inverted micelle
w   Oil

Micelle
oil   Water

Figure 1.2 Overview of processes suitable for fine-emulsion formation.

mainly been employed in laboratories because wide droplet size distributions raise problems in continuous operations in industrial processes.

Besides the processes mentioned above, which are based on droplet disruption, emulsions may also be produced by droplet formation at membranes and microstructured systems (Figure 1.3). In this case, the disperse phase is forced through the micropores of a membrane. The droplets forming at the pore outlets are detached by the flow of the continuous phase parallel to the membrane surface. The method developed by Nakashima (1994) in Japan in the 1990s offers many new possibilities for gently producing very fine droplets and narrow droplet size distributions. Recent studies have shown that the process of forcing a coarse emulsion through membranes, known as premix membrane emulsification, is a means for achieving high throughputs, inducing phase inversions where desired and producing multiple emulsions (Suzuki, 2000; Vladisavljevic et al., 2005). At the Institute of Process Engineering in Life Sciences, Section of Food Engineering, University of Karlsruhe, membrane emulsification has been a subject of intense studies (Schröder, 1999; Schröder and Schubert, 1999; Altenach-Rehm et al., 2002; Vladisavljevic et al., 2002; Lambrich et al., 2005). Microchannel emulsification is another novel process studied recently to produce monodisperse emulsions (Nakajima, 2000; Nakajima and Kobayashi, 2001) (Figure 1.3).

Except for high-pressure and membrane or microchannel emulsification, emulsions may be produced either batch-wise or continuously. In the case of continuous emulsification, the ingredients are usually dosed separately and

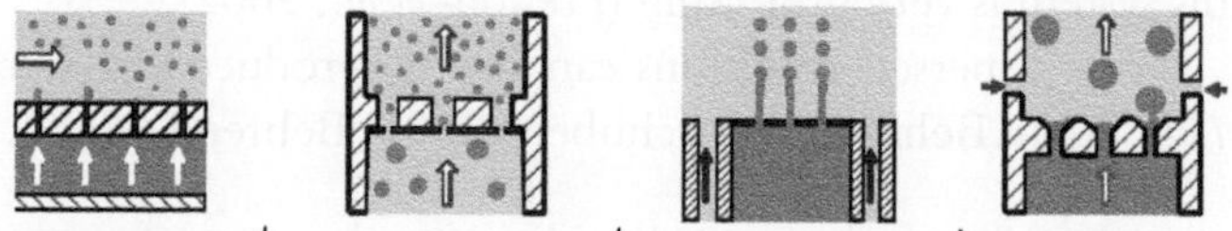

Process	Membrane emulsification	Premix-membrane-emulsification	Membrane jetting emulsification	Microchannel emulsification
Droplet detachment (formation) by	Wall shear stress	Flux	Rayleigh break-up	Instability
Droplet size (μm)	0.1–10	0.1–10	> 0.2 ?	> 3
Droplet size distribution	Narrow	Narrow	Narrow	Monodisperse
Flux (m^3/m^{2}h)	$J_d < 0.4$ (0.2)	$1 < J_{em} < 10$ (20)	$J_d > 1$	$J_b < 0.01$
Danger of membrane fouling	Medium	High	Low	Low

Figure 1.3 Production of fine-emulsions using membranes or microstructured systems.

premixed in a blender. The resulting coarse-disperse raw emulsion is then fed into the droplet disruption machine for fine emulsification. Energy input required for the formation of the raw emulsion is negligible compared to that required for fine emulsification. Single-step continuous processes excel at energy efficiency, while discontinuous or multi-stage continuous ones usually allow the production of emulsions with narrower droplet size distributions.

Besides these mechanical processes, there are several non-mechanical emulsifying processes applied to produce specific products, for example in the chemical industry. A typical example of such processes is based on the precipitation of the disperse phase previously dissolved in the external phase. Changes in the phase behavior of the substances to be emulsified, prompted by variation in temperature or composition, or mechanical stress, are used to achieve the desirable dispersed state of the system. Another process of interest is the phase-inversion temperature (PIT) method discussed by von Rybinski (2005).

Most emulsions are produced in a continuous mode by mechanical droplet disruption. However, the prevailing principles differ depending on the type of mechanism used for droplet disruption. Fundamental work in this field has been done by Walstra and Smulders (1998). A differentiation among the various rotor-stator systems, ultrasound systems, and high-pressure homogenizers results in different Reynolds numbers, Re_{flow} of the continuous phase and $Re_{droplet}$ of the flow around droplets in $Re_{droplet}$; the droplet diameter is considered the characteristic length (Figure 1.4). It becomes

	TURBULENT inertia forces	TURBULENT shear forces	LAMINAR shear or stretch forces
Re_{flow}	> 2000	> 2000	< 1000
$Re_{droplet}$	> 1	< 1	< 1
Rotor-stator mill, ultrasound-homogenizer	$x_{max} \propto P_v^{-0,4}$ $x_{max} \propto t^{-0,3 *)}$ $x_{max} \cong A_1 \cdot E_v^{-0,35 **)}$	$x_{max} \propto P_v^{-0,5}$ $x_{max} \propto t^{-0,3 *)}$ $x_{max} \cong A_2 \cdot E_v^{-0,4 **)}$	$x_{max} \cong A_3 \cdot E_v^{-1}$
High-pressure homogenizer	$x_{max} = A_4 \cdot E_v^{-0,6}$	$x_{max} = A_5 \cdot E_v^{-0,75}$	$x_{max} = A_6 \cdot E_v^{-1}$

In general:

$$x_{3,2} = A_i E_v^{-b}$$

$x_{3,2}$ = sauter diameter [μm]
E_v = energy density [J/m^3]
$A_i = f(\eta_d,\eta_d / \eta_c,\rho_c,\gamma_L,...)$

*) exponent from experiments
**) approximated exponent

Figure 1.4 Mechanisms of droplet disruption in continuous emulsification (see also Walstra and Smulders, 1998).

obvious that the maximum droplet size depends on both power density P_v and residence time t. The concept of energy density is based on these functions (Karbstein, 1994; Schubert and Karbstein, 1994). Experiments have shown that a simple relationship exists between maximum droplet size and the Sauter diameter $x_{3,2}$ (mean droplet size of droplet collective) (Armbruster, 1990). The maximum droplet size may hence be replaced by the Sauter diameter, if the respective proportionality factor is adequately adjusted. In high-pressure homogenizers the Sauter diameter is a direct function of energy density E_v, which equals the effective pressure difference Δp at the homogenizing valve. In rotor-stator systems and ultrasound homogenizers, Sauter diameter and energy density are only linked by approximations on the basis of experimental data (Figure 1.4).

II. THE CONCEPT OF ENERGY DENSITY OF EMULSIONS

In general, the Sauter diameter in continuous emulsification with droplet disruption may be described by the relationship:

$$x_{3,2} = a_i \cdot E_v^{-b} \tag{1.1}$$

with the constants a_i and b depending on the emulsifying agents, the properties of the continuous and the disperse phase, and the emulsification equipment applied.

As mentioned above (Karbstein, 1994), some secondary conditions must be fulfilled to make sure this relationship applies. First of all, the forces at the droplet surface must exceed a critical threshold above which droplets are disrupted. Furthermore, the time allowed for droplet deformation must exceed a critical deformation time in order to deform droplets to an extent sufficient for break up. This means that a critical power density P_v must be exceeded for droplets to be disrupted.

The concept of energy density (Karbstein, 1994) according to Eq. (1.2) developed at the Institute of Process Engineering in Life Sciences, Section of Food Engineering (University of Karlsruhe, Germany) has been found to be a useful approximation for mechanical emulsification practice. It is applied to design, scale-up, and control of continuously operating emulsifying apparatus.

$$E_v = \frac{P}{\dot{V}} = \Delta p \tag{1.2}$$

Energy density is easy to measure because in an emulsifying system, power input P into the machine, volume flow rate $\dot{V} = dV/dt$ of the emulsion, and pressure difference Δp, in case of a high-pressure homogenizer, are all easily accessible. The concept of energy density may also be applied to compare different mechanical emulsification processes. In membrane emulsification, mean droplet sizes (Sauter diameters) not only depend on energy density, but also on volume proportion φ of the disperse phase (Schröder, 1999). A comparison, where droplet coalescence, which was minimized by choosing adequate emulsifiers in the experiments, has been ignored, illustrates that for equal energy densities, different types of emulsification equipment yield a wide range of different droplet sizes (Figure 1.5).

Another important factor for homogenization of emulsions is the ratio of the viscosity of the disperse phase η_d to that of the continuous phase η_c. A comparison of the results of homogenizing an o/w-emulsion with a high ratio η_d/η_c in different high-pressure homogenization valves shows a great advantage in terms of energy required for homogenization as well as the result regarding droplet sizes for a simple orifice valve (Figure 1.6). These results can be explained by the three-dimensional elongational flow in front of an orifice valve resulting in deformation of the droplets to filaments, which will then be broken up by inertial forces and instabilities caused by the turbulent flow regime behind the orifice valve.

Besides the viscosity of the disperse phase, the interfacial tension between the continuous and the disperse phase plays an important role in hindering droplet disruption. The minimal required energy for droplet disruption is directly proportional to the resulting increase in interfacial area as well as the interfacial tension between the two phases. Therefore, reducing the interfacial tension by the addition of an emulsifier, and its absorption to the

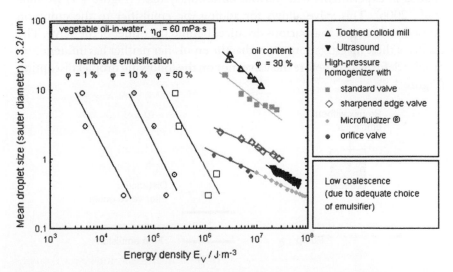

Figure 1.5 Comparison of various homogenization processes by means of energy density.

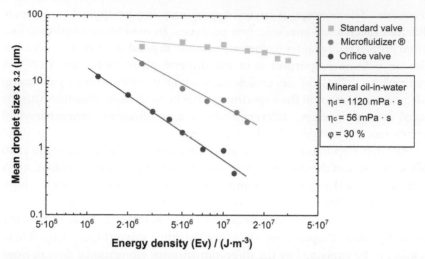

Figure 1.6 Comparison of homogenization efficiency of various homogenization valves for an o/w-emulsion with a high viscosity ratio η_d/η_c.

interface before homogenization, has long been considered a means for improving homogenization efficiency by facilitating droplet disruption. In order to investigate the emulsifier's effect on droplet disruption and homogenization efficiency, it was the aim to allow for investigating droplet disruption separately from droplet coalescence. For this purpose, a special homogenization valve for high-pressure homogenization was designed and applied in experiments with various emulsion systems (Figure 1.7) (Kempa et al., 2006). This valve allows for adding the emulsifier to the emulsion immediately as well as at various distances behind the outlet of the valve. The results of these investigations show that the emulsifier neither has influence on droplet deformation before the valve, nor on the following droplet disruption (Figure 1.8).

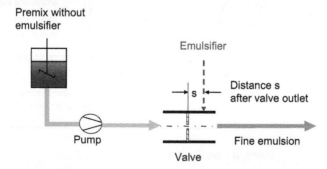

Figure 1.7 Schematic drawing of the high-pressure homogenization valve allowing for addition of the emulsifier at various distances behind the valve.

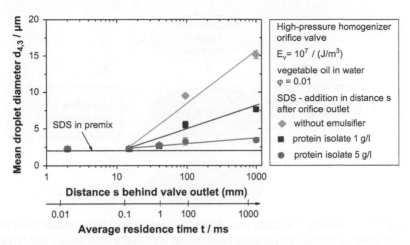

Figure 1.8 Mean droplet diameters in emulsions homogenized with and without the presence of an emulsifier and different localization for addition of emulsifier behind the homogenizing valve.

It also becomes evident that adding the emulsifier too far behind the valve causes droplet coalescence due to the interfacial area not being covered by emulsifier molecules and the droplets thus being unprotected against coalescence (Kempa *et al.*, 2006). The same holds for emulsifiers (e.g. protein isolates), which do not absorb to the interface quickly enough to protect the droplets from coalescence, although parts of this effect may be compensated by increasing the emulsifier concentration.

III. ADJUSTMENT OF EMULSION PROPERTIES

The results discussed so far provide a basis for product design and the formulation of emulsions. In this context, product design means creating a product with certain desired properties that are adjusted by engineering methods. As it is very expensive to vary all essential parameters of a technical apparatus in order to ultimately achieve the desired property of a product, an interim attribute is introduced. In the case of emulsions, their microstructure is taken for the interim attribute, which is mainly determined by mean droplet size, droplet size distribution, and type of emulsifier and stabilizer. For most problems, it may be sufficient to simply characterize the microstructure by the mean droplet size, a physical value easily measurable and adjustable by means of emulsifying equipment.

The link between the emulsion's microstructure and the process or emulsifying apparatus, respectively, is called the *process function* (Krekel and Polke, 1992). In the present simplified case, the process function can be

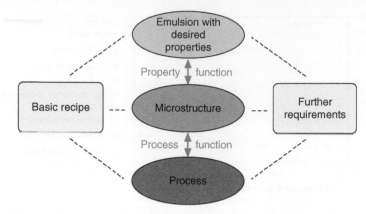

Figure 1.9 Achieving a product with desired properties by taking into account the microstructure, process, and property functions, as well as the basic recipe and certain boundary conditions.

described by Eq. (1.1). The mean droplet size, accordingly, depends on the energy density, which can either be directly determined or taken from Figure 1.5 for the different apparatus. Linking the microstructure and emulsion properties has been found to be much more complex. According to Rumpf (1967), this link is called the *property function*. In many cases it is difficult to define this property function, i.e. to link product properties and physical characteristics (microstructure). Therefore, in most cases, the property function is a far more complex equation and stands for a complex connection between properties and microstructure (Figure 1.9). The concept of product design and formulation is currently in the process of further improvement. In the case of emulsions, the process function provides a useful approximation, while property functions for most applications remain to be developed. If in the present cases of carotenoid or phytosterol formulations dose response was merely a function of droplet size, it would be easy to derive a property function. Process and property function would then provide the possibility to achieve the desired properties by selecting the appropriate process. Investigations discussed in Section V will show if, and to what extent, this is a practicable way.

IV. STABILITY OF EMULSIONS

Without doubt, stability is the most important property of emulsions. However, it has to be differentiated between physical, chemical, and microbiological stability, of which only the first will be subject to detailed discussion here. An emulsion is called physically stable if its dispersed state does not change, i.e. if its droplet size distribution remains constant regardless of time or volume element observed. First of all, droplets must not sediment, or aggregate or coalesce; changes in droplet sizes due to Ostwald ripening, i.e. the growth of large droplets at the expense of small ones due to different capillary pressures, or phase inversion, are not admissible either. Emulsions with sufficiently small droplets, which are prevented from aggregating or coalescing by the use of suitable emulsifiers, exhibit great physical stability. In many cases, Ostwald ripening is controlled by diffusion. Because the various liquid phases of an emulsion are usually poorly soluble in each other, Ostwald ripening mostly is of minor importance. However, if Ostwald ripening is to be minimized, this undesirable process is slowed down by producing droplets as equal in size as possible, thereby reducing the differences in capillary pressures.

The chemical stability of an emulsion reflects its resistance against chemical changes. Mostly oxidation of fats and oils is the critical reaction for chemical deterioration of emulsions. The addition of antioxidants and protection against external influences such as light or excessive heat, as well as suitable diffusion barriers on the surfaces between the different liquid phases, improve the chemical stability.

The physical stability of emulsions during and immediately after emulsification, which we have called short-term stability, presents specific problems too. The short-term stability of an emulsion indicates whether the newly formed droplets of an emulsion are sufficiently protected against coalescence during or directly after emulsification. Therefore, this protection determines the success or failure of the emulsification process because mechanical emulsification processes are expected to not only size-reduce, but also to immediately stabilize these smaller droplets. Droplets successfully protected against coalescence remain small, resulting in the desired fine-dispersed emulsion. If coalescence cannot be successfully prevented, droplets will flow together and form larger droplets. In a critical case, this could completely reverse the previous size reduction. In unfavorable cases, it may even cause emulsions to break. Therefore, avoiding droplet coalescence during and immediately following the formation of new droplets, i.e. achieving sufficient short-term stability, is of utmost importance for any emulsification process.

Despite many indications for droplet coalescence in emulsifying apparatus, this question is discussed controversially in the pertinent literature. Based on the Gibbs-Marangoni effect, some authors are of the opinion that newly formed droplets in emulsions immediately following the formation cannot coalesce with an emulsifier present. For immediate detection of droplet coalescence during emulsification, a method has been developed at the Institute of Process Engineering in Life Sciences, Section of Food Process Engineering by Danner (2001a) called the coloring method.

According to his work, the droplets of two identical raw emulsions are stained with different colors. Thereafter, a fine emulsion is produced from the mixture of these two raw emulsions involving disruption of the large stained droplets. The colors have been selected in such a way that a third color can only result from the coalescence of differently colored droplets and the subsequent mixture of their colorants, but not from diffusion. The presence of droplets of this third color after emulsification indicates that coalescence has occurred. Assays of the fraction of the droplets of the third color at different times of a batch–wise emulsification process allow for quantification of the extent of coalescence (Figure 1.10). In the specific case studied the coalescence of white and black droplets results in droplets of the third color, gray. After a short emulsification time t_1, only a few droplets have coalesced and therefore are gray, whereas after an extended emulsification time t_2, nearly all droplets have been subject to coalescence at least once. By the application of an image-evaluating software, reliable and reproducible data of droplet coalescence during emulsification are obtained (Danner, 2001b).

Danner (2001b) developed a stochastic model in order to describe the processes of coalescence in emulsions in greater detail. Provided the frequency of droplet collision is known, both probability and rate of coalescence may be calculated. In this way, coalescence in different types of emulsifying equipment can be monitored quantitatively (Figure 1.11). In the case studied, the droplets of two raw emulsions were stained, one blue and the other yellow, with coalesced droplets turning green. It becomes evident from the study that coalescence depends on the application of Tween® 80 or egg yolk as emulsifier. Tween® 80 provides a much better short-term stability than egg yolk. The coloring method

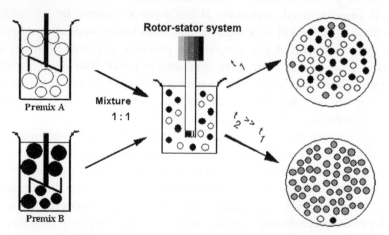

Figure 1.10 Principle of the coloring method used for determining the occurrence and rate of coalescence.

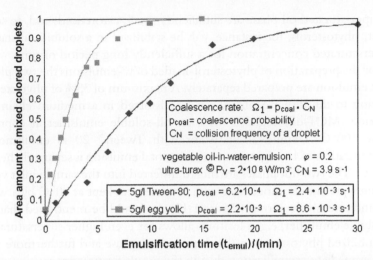

Figure 1.11 Increase of coalescence with emulsification time measured by the coloring method used for emulsions prepared with different emulsifiers.

hence is a very useful and practice-oriented test to judge the stabilizing efficiency of different emulsifiers and stabilizers. It may also serve for studying further essential parameters influencing droplet coalescence during and immediately after emulsification. Finally, the method has also been found helpful by those investigating possibilities of stabilizing emulsions against droplet coalescence hydrodynamically by trying to develop stabilizing zones (Stang, 1998) immediately following the area of size reduction (Danner, 2001b).

V. FORMULATION OF EMULSIONS CONTAINING POORLY SOLUBLE COMPOUNDS

Some compounds with interesting properties for food and feed products or pharmaceuticals are insoluble in water and insufficiently soluble in edible oils at room temperature. Among these are some carotenoids with health-promoting properties, phytosterols with their cholesterol-lowering and anticarcinogenic effects, as well as the majority of novel pharmaceutically active substances. Because of insolubility or poor solubility of their usually crystalline form, they are barely bioavailable and have poor dose responses.

In order to improve bioavailability and dose response, products containing these compounds may be formulated as emulsions (Ax *et al.*, 2001) in which the poorly soluble compound is present in the dispersed lipid phase at a highly supersaturated concentration. Provided the droplets of the supersaturated oil phase are sufficiently small, carotenoids and, to some extent, phytosterols, for instance, can be stabilized in a solubilized state at a supersaturated concentration for a sufficiently long period of time.

For the preparation of phytosterol-loaded o/w-emulsions the two phases of the emulsion are prepared separately. A maximum of 35% of phytosterols, referring to the disperse phase, can be dissolved in a medium chain tri-glyceride (MCT-oil) together with an oil-soluble emulsifier at approximately 100°C. Demineralized water with Tween® 20 as emulsifier at a concentration of 1 weight percent of the total emulsion is separately heated to 90°C. The two hot phases are then transferred into the sample inlet vessel of a high-pressure homogenizer with both devices kept at 90°C by a water bath and intensively stirred (Figure 1.12). It should be mentioned that the oil-soluble emulsifier, e.g. a lecithin, allows for even higher supersaturation of solubilized phytosterols in the dispersed oil phase and furthermore pre-vents immediate crystallization during the emulsification process.

The resulting coarse raw emulsion was homogenized at a pressure of 1000 bar at 90°C. After emulsification, the stable fine-dispersed emulsion with a Sauter diameter well below 1 μm could cool down to room temperature.

Since phytosterols, due to their molecular structure, are interface-active components, they tend to migrate to and crystallize at the oil–water in-terface of the oil droplets in o/w-emulsions (Engel and Schubert, 2005). In order to suspend this effect on the emulsion's short-term stability, an emulsifier system of one oil- and one water-soluble and fast-stabilizing emulsifier such as Tween® 20 had to be employed.

Regarding their physical stability, the small Sauter diameter of approx-imately ten times the phytosterol-supersaturated oil droplets, together with the crystallization inhibitor, makes sure that no crystallization or creaming occurs.

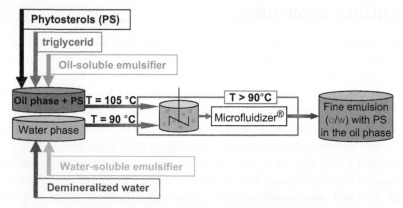

Figure 1.12 Process for production of phytosterol loaded o/w-emulsions.

The formulation of emulsions enriched with poorly soluble compounds such as phytosterols or carotenoids may – in a greatly simplified way – be demonstrated by means of the process and property functions. Following Eq. (1.1) the process function for emulsions produced in a high-pressure homogenizer with an orifice valve, or in a Microfluidizer® with a homogenizing pressure of $\Delta p = E_v$, is (Ax, 2004)

$$X_{3,2} = 4, 5 \cdot 10^{-3} \Delta p^{-0.5} \tag{1.3}$$

with a resulting Sauter diameter of $X_{3,2}$ in m and a given effective pressure difference of Δp in Pa.

Above that, it can be assumed that the bioavailability B of the added compound (e.g. carotenoids) increases with decreasing Sauter diameter. The validity of this interrelationship could be demonstrated for example by Ax et al. (2001) for carotenoids. As possible values of B are in a range between B_{min} and B_{max}, B and $X_{3,2}$ may be linked by the simple equation,

$$B = \frac{B_{max} + B_{min} \cdot k \cdot X_{3,2}}{1 + k \cdot X_{3,2}} \tag{1.4}$$

For the border cases $B_{min} = 0$ and $B_{max} = 1$, Eq. (1.4) reduces to

$$B = \frac{1}{1 + k \cdot X_{3,2}} \tag{1.5}$$

For this case Eqs. (1.4) and (1.5), respectively, represent the applicable property function. The constant values B_{min}, B_{max} and k must be derived by experiment, for instance, from human studies. Until now, the equations for this property function have only been verified in in vitro tests for carotenoids. Therefore, precise values for the constant required cannot yet be derived from appropriate experiments. However, to illustrate the procedure of developing emulsions by making use of the process and property functions, it is assumed that $k = 1 \ \mu m^{-1} = 10^6 \ m^{-1}$. By introducing Eq. (1.3) into Eq. (1.5), the following equation can be derived:

$$\frac{\Delta p}{Pa} = \left(\frac{4, 5 \cdot 10^{-3} \cdot k \cdot B \cdot m}{1 - B} \right)^2 \tag{1.6}$$

If, for instance, the bioavailability of the incorporated compound is 60% ($B = 0.6$), a homogenizing pressure of $\Delta p \approx 450$ bar would be needed to produce the desired emulsion. The example elucidates – even if some assumptions and substantial simplifications have been made – how useful the process and property functions are, which may be much more complex in many cases. Usually, some secondary conditions have to be considered as well. For example, in the present case, the aforementioned stability of the resulting emulsion and the chemical stability of the incorporated compound would have to be taken into account, too (Ax et al., 2001).

ACKNOWLEDGMENTS

These studies were financially supported by Deutsche Forschungsgemeinschaft (DFG) within the joint research project 'Lipide und Phytosterole in der Ernährung' and within the framework of the DFG Graduiertenkolleg 366 'Grenzflächenphänomene in aquatischen Systemen und wässrigen Phasen.' The authors would also like to thank the 'Bundes-ministerium für Bildung und Forschung' for the financial support of the subproject 0312248A/7 'Synthese und Formulierung von Carotinoiden.'

REFERENCES

Altenach-Rehm, J., Schubert, H., & Suzuki, K. (2002). Premix Membranemulgieren mittels hydrophiler und hydrophober PFTE-Membranen zur Herstellung von O/W-Emulsionen. Chemie Ingenieur Technik, 74, 587–588.

Armbruster, H. (1990). Untersuchungen zum kontinuierlichen Emulgierprozess in Kolloidmühlen. Germany: Dissertation University of Karlsruhe.

Ax, K. (2004). Emulsionen und Liposomen als Trägersysteme für Carotinoide. Shaker Verlag Aachen, Germany: Dissertation University of Karlsruhe.

Ax, K., Schubert, H., Briviba, K., Rechkemmer, G., & Tevini, M. (2001). Oil-in-water emulsions as carriers of bioavailable carotenoids. Proceedings PARTEC, 27–29. March, Nuremberg, Germany, pp. 1–8.

Behrend, O., Ax, K., & Schubert, H. (2000). Influence of continuous phase viscosity on emulsification by ultrasound. Ultrasonics Sonochemistry, 7, 77–85.

Behrend, O., & Schubert, H. (2001). Influence of hydrostatic pressure and gas content on continuous ultrasound emulsification. Ultrasonics Sonochemistry, 8, 271–276.

Behrend, O. (2002). Mechanisches Emulgieren mit Ultraschall. GCA-Verlag Herdecke, Germany: Dissertation University of Karlsruhe.

Danner, T. (2001). Tropfenkoaleszenz in Emulsionen. GCA-Verlag Herdecke, Germany: Dissertation University of Karlsruhe.

Engel, R., & Schubert, H. (2005). Formulation of phytosterols in emulsions for increased dose response in functional foods. Innovative Food Science and Emerging Technologies, 6, 233–237.

Freudig, B., Tesch, S., & Schubert, H. (2002). Herstellen von Emulsionen in Hochdruckhomogenisatoren, Teil 2. Chemie Ingenieur Technik, 74, 880–884.

Karbstein, H. (1994). Untersuchungen zum Herstellen und Stabilisieren von O/W Emulsionen. Germany: Dissertation University of Karlsruhe.

Kempa, L., Schuchmann, H. P., & Schubert, H. (2006). Tropfenzerkleinerung und Tropfenkoaleszenz beim mechanischen Emulgieren mit Hochdruckhomogenisatoren. Chemie Ingenieur Technik, 78, 765–768.

Krekel, J., & Polke, R. (1992). Qualitätssicherung bei der Verfahrensentwicklung. Chemie Ingenieur Technik, 64, 528–535.

Lambrich, U., Schröder, V., Vladisavljevic, G.T. (2005). Emulgieren mit mikroporösen Systemen. In: Schubert, H. (ed.) Emulgiertechnik – Grundlagen, Verfahren und Anwendungen, pp. 369–431. B. Behr's Verlag, Hamburg, Germany.

Nakajima, M. (2000). Microchannel emulsification for monodispersed microspheres. *Proc.* NFRI-BRAIN Workshop, Oct. 16-17, Tsukuba, Japan, pp. 626-628.

Nakajima, M., & Kobayashi, I. (2001). *Emulsification with newly fabricated through-type microchannel. 11th World Congress of Food Science and Technology, April 22–27.* Korea: Seoul.

Nakashima, T., Nakamura, K., Kochi, M., Iwasaki, Y., & Tomita, M. (1994). Development of membrane emulsification and its applications to food industries. *Shokuhin Kogyo Gakkaishi, 41*, 70–76.

Rumpf, H. (1967). Über die Eigenschaften von Nutzstäuben. *Staub-Reinhaltung der Luft, 27*, 70–76.

Schröder, V. (1999). *Herstellen von Öl-in-Wasser-Emulsionen mit mikroporösen Membranen.* Shaker Verlag Aachen, Germany: Dissertation University of Karlsruhe.

Schröder, V., & Schubert, H. (1999). *Influence of emulsifier and pore size on membrane emulsification. In: Food Emulsions and Foams.* Cambridge, UK: Royal Society of Chemistry.

Schubert, H. (2003). Neue Entwicklungen auf dem Gebiet der Emulgiertechnik. In: Kraume, M. (ed.) Mischen und Rühren, pp. 313–342. Wiley-VCH Verlag, Weinheim, Germany.

Schubert, H. (ed.) (2005a). Emulgiertechnik – Grundlagen, Verfahren und Anwendungen. B. Behr's Verlag, Hamburg, Germany.

Schubert, H. (2005b). Einführung in die Emulgiertechnik. In: Schubert, H. (ed.) Emulgiertechnik - Grundlagen, Verfahren und Anwendungen, pp. 1-15. B. Behr's Verlag, Hamburg, Germany.

Schubert, H., & Ax, K. (2001). Verbesserung der gesundheitlichen Qualität von Lebensmitteln durch Erhöhung und Modifikation des Carotinoidgehaltes. In: Innovationsfeld Lebensmittel, Vol. 59, pp. 106–137. Diskussionstagung, Forschungskreis der Ernährungsindustrie, e.V. (FEI). Bonn, Germany.

Schubert, H., & Karbstein, H. (1994). *Mechanical emulsification. In: Development in Food Engineering, Part 1.* London: Blacky Academic & Professional.

Stang, M. (1998). *Zerkleinern und Stabilisieren von Tropfen beim mechanischen Emulgieren.* Fortschrittberichte VDI, Reihe: Dissertation University of Karlsruhe 3, Nr. 527.

Stang, M., Schuchmann, H., & Schubert, H. (2001). Emulsification in high-pressure homogenizers. *Eng. Life Sci., 1*(4), 151–157.

Suzuki, K. (2000). Membrane emulsification for food processing. *Proc.* NFRI-BRAIN Workshop, Oct. 16–17, Tsukuba, Japan, pp. 635–636.

Tesch, S., Freudig, B., & Schubert, H. (2002). Herstellen von Emulsionen in Hochdruck-homogenisatoren, Teil 1. *Chemie Ingenieur Technik, 74*, 875–880.

Vladisavljevic, G.T., Shimizu, M., & Nakashima, T. (2005). Preparation of uniform multiple emulsions using multi-stage premix membrane emulsification technique. In: Schubert, H. (ed.) Emulgiertechnik – Grundlagen, Verfahren und Anwendungen, pp. 433–468. B. Behr's Verlag, Hamburg, Germany.

Vladisavljevic, G. T., & Tesch, S. (2002). Preparation of water-in-oil emulsions using microporous polypropylene hollow fibers. *Chemical Engineering and Processing, 41*, 231–238.

von Rybinski, W. (2005). Herstellen von Emulsionen nach der Phaseninversions-Methode. In: Schubert, H. (ed.) Emulgiertechnik – Grundlagen, Verfahren und Anwendungen, pp. 469–485. B. Behr's Verlag, Hamburg, Germany.

Walstra, P., & Smulders, P. E. A. (1998). *Emulsion formation. In: Modern Aspects of Emulsion Science.* Cambridge, UK: Royal Society of Chemistry.

CHAPTER 2

The Effect of Processing and the Food Matrix on Allergenicity of Foods

Clare Mills, Ana Sancho, Neil Rigby, John Jenkins, *and* Alan Mackie

Contents

Abstract

The agents that cause IgE-mediated allergies are known as allergens and are almost always proteins. It is emerging that the way in which food proteins elicit an allergic reaction can be modified by food processing procedures. This is because food processing alters the structure of food proteins through either unfolding and aggregation, or covalent modification by other food components such as sugars. In this way the IgE-recognition sites on an allergen, known as epitopes, can either be destroyed or new epitopes formed. Processing can destroy the allergenicity of some proteins, notably the Bet v 1 homologs, which both unfold and become modified with plant polyphenols. Others, such as the prolamin superfamily members, the nsLTP and 2S albumin allergens, have stable protein scaffolds and either do not unfold or refold on cooling, retaining their allergenicity. Some allergens are highly thermostable because of their mobile structures, which are not disrupted on heating, such as the caseins and seed storage prolamins of wheat. Others, such as seed storage globulins, only partially unfold and can retain much of their allergenicity. The structure of the food matrix may also affect the release and stability of allergens impacting on the elicitation of reactions in food-allergic individuals. Such complexity makes it a challenge to develop generic food processing procedures capable of removing or reducing allergenicity that are effective for all allergic consumers. A better knowledge of how

Global Issues in Food Science and Technology
© 2009 Elsevier Inc.

processing affects the allergenicity of food is also important for risk assessors and managers involved in managing allergenic food hazards.

I. INTRODUCTION

Food allergies are generally held to be adverse reactions to foods that have an immunological basis. They include both IgE-mediated allergies usually classified as type I hypersensitivity reactions and the gluten intolerance syndrome, celiac's disease. With regards to the former, IgE is produced as part of the normal functioning of the immune system in response to parasitic infections. For reasons not fully understood, some individuals begin to make IgE to various environmental agents, including dust, pollens, and foods. IgE-mediated allergies develop in two phases: (1) sensitization when IgE production is stimulated, and (2) elicitation when an individual experiences an adverse reaction upon re-exposure to an allergen. Both stages are triggered by allergens, which are almost always proteins. In an allergic reaction allergen is recognized by IgE bound to the surface of histamine-containing mast cells, cross-linking the IgE in the process and triggering the release of inflammatory mediators such as histamine. These mediators cause the acute inflammatory reactions that become manifested as respiratory (asthma, rhinitis), cutaneous (eczema, urticaria), or gastrointestinal (vomiting, diarrhea) symptoms, which may occur alone or in combination in an allergic reaction. A rare but very severe reaction is anaphylactic shock characterized by respiratory symptoms, fainting, itching, urticaria, swelling of the throat or other mucous membranes, and a dramatic loss of blood pressure.

In contrast the gluten intolerance syndrome celiac's disease is manifested in a much slower manner than IgE-mediated allergies, an individual taking hours or days rather than seconds or minutes to react. Thought to affect around 1% of the population, this disease afflicts more women than men and arises as a consequence of deamidation of the glutamine residues in gluten peptides by the gut mucosal transglutaminase. The modified peptides are able to bind to class II human histocompatibility leukocyte antigen (HLA) molecules DQ2 and DQ8. This recognition event appears to orchestrate an abnormal cellular-mediated immune response that triggers an inflammatory reaction resulting in the flattened mucosa characteristic of celiac's disease (Hischenhuber et al., 2006).

There are two major questions frequently asked in food allergy research, particularly in relation to IgE-mediated allergies. What makes one person, and not another, become allergic? What are the attributes of some foods and food proteins that make them more allergenic than others? Seeking answers to these questions is more difficult with food allergies than inhalant allergies, partly because we lack effective animal models for oral sensitization. Many animal models require co-administration of adjuvants, such as cholera toxin or polysaccharides (e.g. carageenen), before an IgE response can be elicited (Knippels and Penninks, 2003). Studies are more complex in general because the food proteins involved in sensitizing or eliciting allergic reactions are altered by food processing procedures. For example the food proteins often become an insoluble mass not amenable to extraction in the simple salt solutions routinely employed for serological or clinical analyses. Furthermore the structure of a food itself may affect the way in which allergens are broken down during digestion and presented to the immune system during the sensitization and elicitation phases of allergic responses. It has been shown that the structure of the food matrix can affect the elicitation of allergic reactions in food-allergic individuals; for example fat-rich matrices such as chocolate, affect the kinetics of allergen release, potentiating the severity of allergic reactions (Grimshaw et al., 2003).

As there is no cure for IgE-mediated allergies or celiac's disease, individuals with these conditions have to exclude known problem foods and any derived ingredients from their diet. Consequently food-labeling legislation has been modified around the world in response to the Codex Alimentarius Commission amendment to the Codex General Standard for the Labeling of Prepackaged Foods (Codex General Standard for the Labeling of Prepackaged Foods, 1999) (CODEX STAN 1-1985; Rev. 1, 1991), which aims to ensure that allergic consumers have good-quality information on the content of allergens in pre-packaged foods. In addition, any risk assessment of a novel food (including GMOs) or novel food processing method has to include an explicit allergenic risk assessment. Understanding how food processing affects allergenic potential is therefore important if allergenic risks are to be managed effectively and the risks of adverse reactions to foods minimized. Since we do not understand the molecular events involved when an individual becomes sensitized to a particular protein, and we do not have adequate animal models for food allergy, it is difficult to study how food processing can affect the ability of an allergen or a type of food to sensitize an individual. Consequently this chapter is focused on knowledge of how processing affects elicitation of IgE-mediated allergic reactions to food.

II. ALLERGENS AND EPITOPES IN IgE-MEDIATED ALLERGIES

The World Health Organization and the International Union of Immunological Societies produce an official list of allergens, designated by the Allergen Nomenclature sub-committee. Allergens included in this list must induce an IgE-mediated (atopic) allergy in humans with a prevalence of IgE reactivity above 5%. An allergen is termed 'major' if recognized by IgE in at least 50% of a cohort of allergic individuals, but it does not carry any connotation of allergenic potency; allergens are otherwise termed 'minor.' The allergen designation is then based on the Latin name of the species from which it originates and is composed of the first three letters of the genus, followed by the first letter of the species, finishing with an Arabic number, for example Ara h 1, an allergen from *Arachis hypogea* (peanuts).

The sites on a protein recognized by an antibody are known as epitopes and can be either linear or conformational (Van Regenmortel, 1992). In the former only the primary sequence of a polypeptide is involved in antibody recognition, but in conformational epitopes the three-dimensional structure of a protein is also important. Thus, the latter epitopes are formed from a number of segments of the polypeptide chain, which may be quite distant in the amino acid sequence of a protein but are brought together spatially as a consequence of the tertiary and quaternary structure of the protein. Most epitopes are thought to be conformational in nature and are particularly difficult to define in relation to food allergens, where processing can have such a disruptive effect on native protein structure. The ability of an allergen to elicit a reaction can be modified by disrupting the IgE epitopes. It is also possible that food processing could introduce new epitopes, sometimes termed neo-epitopes. This may occur through protein unfolding, revealing inner portions of a protein's structure not generally available for antibody binding or the covalent modification of a protein by sugars or other food components.

II.A. Processing labile allergens

The Bet v 1 superfamily of plant food allergens, which are involved in the pollen-fruit/vegetable cross-reactivity syndrome, is generally unstable to common food processing technologies. This is probably because the main cross-reactive IgE-binding sites on these proteins are primarily directed towards conformational epitopes (Gajhede *et al.*, 1996; Neudecker *et al.*, 2001). As a general rule, both the IgE-reactivity and ability of Bet v 1 homologs to trigger a reaction in sensitized individuals are reduced by food processing, thought to result from disruption of these conformational epitopes. However, the extent to which they unfold may depend on the form of food processing and the properties of the matrix concerned. Thus, heating in low-water systems, such as roasting of hazelnuts, while reducing the allergenicity of Bet v 1 homologs, does not abolish reactions in all patients (Hansen *et al.*, 2003). In contrast, cooking of fruits, such as apple, which have a high water activity, reduces their allergenicity in birch-pollen-allergic individuals. These observations can be explained by the general observation that protein thermostability is dependent on water activity (Gekko and Timasheff, 1981).

II.B. Processing stable allergens

Allergens that do not lose their IgE-epitopes following processing generally fall into one of three different categories based on their properties and characteristics, as summarized next.

II.B.1. Thermostable proteins that refold upon cooling

Depending on the conditions of heating many of the protein scaffolds characteristic of food allergens either resist thermal denaturation and/or are able to refold upon cooling to the native folded conformation. One group of proteins in this category is the prolamin superfamily. These proteins are characterized by a conserved pattern of cysteine residues, with six or eight such residues forming three or four intra-chain disulfide bonds, which constrain the folded structure of the proteins. Consequently, they resist thermal denaturation and may refold to the original native conformation. In general, the allergenic properties of these proteins are retained following processing as has been found for the ns LTP allergens, which retain their allergenic properties after heating, including cooking of foods such as polenta (Pastorello et al., 2000) or fermentation to produce beer (Asero et al., 2001) or wine (Pastorello et al., 2003).

II.B.2. Inherently mobile proteins

There are a number of proteins, which, instead of adopting compact globular structures, appear to comprise regions of ill-defined, disordered, mobile structures (Dunker et al., 2001). Such structures are dynamic, possessing a range of rapidly interchanging conformations, and consequently do not show the co-operative transition on heating which is characteristic of a globular protein as it moves from a folded to an unfolded or partially folded structure. This class of protein is represented by two examples of food allergens, caseins from milk and the prolamin seed storage proteins from cereals. The absence of a co-operative transition for caseins as determined by DSC and their lack of secondary structure has led to their being termed rheomorphic (rheo meaning 'to flow'; morphe meaning 'shape'), a property shared by seed storage prolamins (Paulsson and Dejmek, 1990; Holt and Sawyer, 1993). As a result of their dynamic nature these proteins possess many linear and, hence, potentially thermo-stable IgE-epitopes, and probably explains why the IgE-binding capacity of caseins and wheat prolamins is largely unaltered by thermal processing (Kohno et al., 1994; Simonato et al., 2001b).

II.B.3. Proteins that only partially unfold

Many food proteins unfold to some extent during processing. The partially unfolded proteins then interact to form aggregates or aggregated protein networks such as those found in heat-set gels or interfacial layers in foams and emulsions. This group probably represents the majority of globular food proteins and includes, for example, cow's milk whey proteins, β-lactoglobulin and α-lactalbumin, and the seed storage globulins such as the 11S and 7S seed storage globulins of soya (Mills *et al.*, 2002).

Both 11S and 7S globulins, in common with other members of the cupin superfamily, are thermo-stable proteins, the 7S globulins having their major thermal transition at around 70–75°C, while 11S globulins unfold at temperatures above 94°C, as determined by differential scanning calorimetry. However, while these proteins form large aggregates following heating and unfolding, it is accompanied by little change in the native-like β-sheet structures (Mills *et al.*, 2003). Such observations suggest that some local unfolding occurs following heating and that the β-sheet structure of the cupin fold is largely unaltered. Processes such as the adsorption to interfaces can also cause proteins to unfold and form aggregated networks (Mackie *et al.*, 2000). However, indications are that this process also results in only limited unfolding in several protein systems, including the whey protein β-lactoglobulin (β-Lg) (Husband *et al.*, 2001) or plant seed proteins such as the 2S albumins (Burnett *et al.*, 2002) and α-amylase inhibitors of cereals (Gilbert *et al.*, 2003).

II.C. Processing induced chemical modification of allergens

The reaction between free amino groups on proteins and the aldehyde or ketone groups of sugars known as Maillard's reaction is one of the major chemical reactions that take place in foods during processing. As a result of these non-enzymatic glycation reactions, food proteins can become modified in a complex, diverse way with Amadori products or advanced glycation end products (AGEs). The Maillard adduct rearrangement products can also cross-link food proteins, and studies on the IgE-reactivity of bread in a panel of wheat-allergic individuals suggested that some of the IgE-reactive protein was extensively cross-linked by Maillard adducts (Simonato *et al.*, 2001a). This cross-linking ability may enhance the IgE-binding capacity of allergens, and indications are that some, such as Ara h 1 and Ara h 2 from peanut, are able to form high-molecular-weight aggregates, which bind IgE more effectively than unmodified allergens (Maleki *et al.*, 2000; Chung and Champagne, 2001). Such results are consistent with the observation that the serum IgE-binding in peanut-allergic individuals was greater in roasted, compared to boiled or fried peanuts (Beyer *et al.*, 2001). These data indicate that certain types of thermal processing can introduce additional IgE-binding sites. However, these observations may be complicated by the fact that peanut allergens leach out of peanuts during boiling, lowering the residual allergen content in the boiled nuts

(Mondoulet *et al.*, 2005). Maillard modification has also been found to increase the IgE-binding capacity of the allergenic shellfish tropomyosin (Nakamura *et al.*, 2005). Individuals in this study may have become sensitized to glycated tropomyosin itself through consumption of dried fish products, especially in Oriental cuisine.

In contrast to observations on peanuts and shellfish, glycation of fruit allergens does not appear to increase their allergenicity. For example, glycation of Pru av 1, the allergenic Bet v 1 homolog of cherry, significantly reduces its IgE reactivity, while modification with carbonyl compounds formed during carbohydrate breakdown, such as glyoxal and glycoaldehyde, almost completely abolished IgE binding (Gruber *et al.*, 2005). In addition, glycation of the ns LTP allergen from apple, Mal d 3, also protected the IgE-binding capacity of the protein following harsh thermal treatment (Sancho *et al.*, 2005).

Other types of processing-induced modification that may affect allergenicity include interactions with oxidized lipids (Doke *et al.*, 1989) and enzymatic modification with polyphenols catalyzed by the polyphenol oxidase. Modification with epichatechin and caffeic acid was found to reduce the IgE-binding capacity of Pru av 1. However, the extent to which it was reduced was highly dependent on the polyphenol involved, quercetin, quercetin glycoside and rutin having a lesser effect (Gruber *et al.*, 2004). Such enzymatic modifications may be responsible for the highly labile nature of many fruit Bet v 1 type allergens.

III. CONCLUSION

While understanding the impact of food processing and food structure on allergenic potential is central to managing allergen risks in the food chain, there do not appear to be any clear rules regarding how different allergens within the same food respond to food processing. In fruits and vegetables, for example, the allergenicity members of the Bet v 1 family of allergens found in fruits is generally destroyed by cooking. However, the allergenicity of another type of allergen belonging to the ns LTP family is largely unaltered by food processing. As a consequence the efficacy of food processing in removing allergenicity is a function of both the processing procedure and the type of allergy an individual suffers from. Such complexity makes it a challenge to develop generic food processing procedures capable of removing or reducing allergenicity effective for all allergic consumers. Our lack of understanding of the impact of conventional food processing procedures on allergenicity also makes the assessment of novel processes, such as high pressure, or novel thermal processing procedures, such as pulsed electric field, less certain.

Two areas that remain neglected with regards to investigating the impact of food processing on allergenicity, relate to measures of sensitization potential, and the way it may alter thresholds for elicitation of allergic reactions in sensitized individuals. The latter is essential information for managing allergens in a factory environment, particularly in relation to cross-contact allergens, which find their way into foods otherwise free from allergens through parallel or common processing lines. It may be that certain types of food structures, for example fat-continuous versus aqueous-continuous matrices, may raise or lower the threshold doses for important allergens such as those from peanuts. It is also known that food processing can affect the responsiveness of the immunoassay methods used to monitor allergens in foods and equipment clean-down (Poms *et al.*, 2004). A better understanding of how processing affects allergen structure, and hence allergen-screening assays, would help support interpretation of immunoassay results especially when used to monitor highly processed ingredients.

ACKNOWLEDGMENTS

This work was supported by the competitive strategic grant to IFR from Biological and Biotechnological Sciences Research Council.

REFERENCES

Asero, R., Mistrello, G., Roncarolo, D., Amato, S., & van Ree, R. (2001). A case of allergy to beer showing cross-reactivity between lipid transfer proteins. *Ann. Allergy Asthma Immunol., 87*, 65–67.

Beyer, K., Morrow, E., Li, X. M., Bardina, L., Bannon, G. A., Burks, A. W., & Sampson, H. A. (2001). Effects of cooking methods on peanut allergenicity. *J. Allergy Clin. Immunol., 107*, 1077–1081.

Burnett, G. R., Rigby, N. M., Mills, E. N. C., Belton, P. S., Fido, R. J., Tatham, A. S., & Shewry, P. R. (2002). Characterization of the emulsification properties of 2S albumins from sunflower seed. *J. Colloid Interface Sci., 247*, 177–185.

Chung, S. Y., & Champagne, E. T. (2001). Association of end-product adducts with increased IgE binding of roasted peanuts. *J. Agric. Food Chem., 49*, 3911–3916.

Doke, S., Nakamura, R., & Torii, S. (1989). Allergenicity of food proteins interacted with oxidized lipids in soybean-sensitive individuals. *Agri. Biol. Chem., 53*, 1231–1235.

Dunker, A. K., Lawson, J. D., Brown, C. J., et al. (2001). Intrinsically disordered protein. *J. Mol. Graph. Model., 19*, 26–59.

Gajhede, M., Osmark, P., Poulsen, F. M., et al. (1996). X-ray and NMR structure of Bet v 1, the origin of birch pollen allergy. *Nature Struct. Biol., 3*, 1040–1045.

Gekko, K., & Timasheff, S. N. (1981). Mechanism of protein stabilization by glycerol – preferential hydration in glycerol–water mixtures. *Biochemistry, 20*, 4667–4676.

Gilbert, S. M., Burnett, G. R., Mills, E. N. C., Belton, P. S., Shewry, P. R., & Tatham, A. S. (2003). Identification of the wheat seed protein CM3 as a highly active emulsifier using a novel functional screen. *J. Agri. Food Chem., 51*, 2019–2025.

Grimshaw, K. E. C., King, R. M., Nordlee, J. A., Hefle, S. L., Warner, J. O., & Hourihane, J. O. B. (2003). Presentation of allergen in different food preparations affects the nature of the allergic reaction – a case series. *Clin. Exp. Allergy, 33*, 1581–1585.

Gruber, P., Becker, W. M., & Hofmann, T. (2005). Influence of the Maillard reaction on the allergenicity of rAra h 2, a recombinant major allergen from peanut (*Arachis hypogaea*), its major epitopes, and peanut agglutinin. *J. Agri. Food Chem., 53*, 2289–2296.

Gruber, P., Vieths, S., Wangorsch, A., Nerkamp, J., & Hofmann, T. (2004). Maillard reaction and enzymatic browning affect the allergenicity of Pru av 1, the major allergen from cherry (*Prunus avium*). *J. Agri. Food Chem., 52*, 4002–4007.

Hansen, K. S., Ballmer-Weber, B. K., et al. (2003). Roasted hazelnuts – allergenic activity evaluated by double-blind, placebo-controlled food challenge. *Allergy, 58*, 132–138.

Hischenhuber, C., Crevel, R., Jarry, B., et al. (2006). Review article: safe amounts of gluten for patients with wheat allergy or coeliac disease. *Alim. Pharmacol. Therapeutics, 23*, 559–575.

Holt, C., & Sawyer, L. (1993). Caseins as rheomorphic proteins – interpretation of primary and secondary structures of the alpha-S1-caseins, beta-caseins and kappa-caseins. *J. Chem. Soci. – Faraday Transact, 89*, 2683–2692.

Husband, F. A., Garrood, M. J., Mackie, A. R., Burnett, G. R., & Wilde, P. J. (2001). Adsorbed protein secondary and tertiary structures by circular dichroism and infrared spectroscopy with refractive index matched emulsions. *J. Agri. Food Chem., 49*, 859–866.

Knippels, L. M., & Penninks, A. H. (2003). Assessment of the allergic potential of food protein extracts and proteins on oral application using the brown Norway rat model. *Environ. Health Perspect., 111*, 233–238.

Kohno, Y., Honma, K., Saito, K., Shimojo, N., Tsunoo, H., Kaminogawa, S., & Niimi, H. (1994). Preferential recognition of primary-protein structures of alpha-casein by IgG and IgE antibodies of patients with milk allergy. *Ann. Allergy, 73*, 419–422.

Mackie, A. R., Gunning, A. P., Wilde, P. J., & Morris, V. J. (2000). Orogenic displacement of protein from the oil/water interface. *Langmuir, 16*, 2242–2247.

Maleki, S. J., Chung, S. Y., Champagne, E. T., & Raufman, J. P. (2000). The effects of roasting on the allergenic properties of peanut proteins. *J. Allergy Clin. Immunol., 106*, 763–768.

Mills, E. N., Jenkins, J., Marigheto, N., Belton, P. S., Gunning, A. P., & Morris, V. J. (2002). Allergens of the cupin superfamily. *Biochem. Soc. Trans., 30*, 925–929.

Mills, E. N. C., Marigheto, N. A., Wellner, N., et al. (2003). Thermally induced structural changes in glycinin, the 11S globulin of soya bean (*Glycine max*) – an in situ spectroscopic study. *Biochim. Biophys. Acta – Proteins Proteom., 1648*, 105–114.

Mondoulet, L., Paty, E., Drumare, M. F., et al. (2005). Influence of thermal processing on the allergenicity of peanut proteins. *J. Agri. Food Chem., 53*, 4547–4553.

Nakamura, A., Watanabe, K., Ojima, T., Ahn, D. H., & Saeki, H. (2005). Effect of Maillard reaction on allergenicity of scallop tropomyosin. *J. Agri. Food Chem., 53*, 7559–7564.

Neudecker, P., Schweimer, K., Nerkamp, J., et al. (2001). Allergic cross-reactivity made visible: solution structure of the major cherry allergen Pru av 1. *J. Biol. Chem., 276*, 22756–22763.

Pastorello, E. A., Farioli, L., Pravettoni, V., et al. (2000). The maize major allergen, which is responsible for food-induced allergic reactions, is a lipid transfer protein. *J. Allergy Clin. Immunol., 106*, 744–751.

Pastorello, E. A., Farioli, L., Pravettoni, V., et al. (2003). Identification of grape and wine allergens as an endochitinase 4, a lipid-transfer protein, and a thaumatin. *J. Allergy Clin. Immunol., 111*, 350–359.

Paulsson, M., & Dejmek, P. (1990). Thermal-denaturation of whey proteins in mixtures with caseins studied by differential scanning calorimetry. *J. Dairy Sci., 73*, 590–600.

Poms, R. E., Klein, C. L., & Anklam, E. (2004). Methods for allergen analysis in food: a review. *Food Add. Contam., 21*, 1–31.

Sancho, A. I., Rigby, N. M., Zuidmeer, L., et al. (2005). The effect of thermal processing on the IgE reactivity of the non-specific lipid transfer protein from apple, Mal d 3. *Allergy, 60*, 1262–1268.

Simonato, B., Pasini, G., Giannattasio, M., Peruffo, A.D., De Lazzari, F., & Curioni, A. (2001a). Food allergy to wheat products: the effect of bread baking and in vitro digestion on wheat allergenic proteins. A study with bread dough, crumb, and crust. *J. Agric. Food Chem.*, *49*, 5668–5673.

Simonato, B., Pasini, G., Giannattasio, M., Peruffo, A.D., De Lazzari, F., & Curioni, A. (2001b). Food allergy to wheat products: the effect of bread baking and in vitro digestion on wheat allergenic proteins. A study with bread dough, crumb, and crust. *J. Agric. Food Chem.*, *49*, 5668–5673.

Van Regenmortel, M. H. V. (1992). Molecular dissection of protein antigens. In Van Regenmortel. (Ed.), *Structure of Antigens, Vol. 1* (pp. 1–28). CRC Press Inc.

CHAPTER 3

Nutrigenetic Effect on Intestinal Absorption of Fat-Soluble Microconstituents (Vitamins A, E, D and K, Carotenoids and Phytosterols)

Patrick Borel

Contents

Abstract

The absorption efficiency of fat-soluble microconstituents (FSMs), i.e. vitamins A, E, D, and K, carotenoids and phytosterols, is quite variable and depends on numerous factors. The mnemonic 'SLAMENGHI' is used as a reminder of these factors (Species of FSM, molecular Linkage, Amount, etc.). Recent studies have shown that membrane transporters are involved in absorption of several FSMs by the intestinal cell. It is likely that mutations, or genetic polymorphisms in gene coding for these transporters, lead to modifications in absorption efficiency of FSM. This probably explains the high inter-individual variability in FSM absorption. This genetic-related absorption efficiency can in turn lead to modification of the plasma and tissue status of FSM and, finally, their protective effect on some diseases. Ongoing studies are being carried out to verify this hypothesis, assessing whether the absorption efficiency or plasma status of FSM is related to genetic polymorphisms in transporters of FSMs. Since most FSMs are beneficial at low doses and can be hazardous at higher doses, it is probable that recommended dietary allowances of FSMs or doses in

Global Issues in Food Science and Technology
© 2009 Elsevier Inc.

functional foods/supplements will be adapted to groups in the population carrying common genetic variants in genes known to affect FSM bioavailability. Knowledge of the nutrigenetic factors that affect FSM bioavailability would allow food companies to state that the dose of FSM incorporated into their manufactured food is not harmful, even for higher responders to the micronutrient incorporated. It would also be possible to elaborate different doses of FSM for low and high responders to FSM.

I. INTRODUCTION

Fat-soluble vitamins (A, E, D, and K), carotenoids, and phytosterols are the main fat-soluble microconstituents (FSMs) in the human diet (see Table 3.1 for more details on FSMs). There is a renewed interest in fat-soluble vitamins absorption and metabolism because they are apparently implicated in the prevention of several diseases (Booth *et al.*, 2003; Holick, 2005; Traber, 2007). The plant pigment carotenoids are also suspected to participate in the prevention of age-related diseases (Jiménez-Jiménez *et al.*, 1993; Rao and Agarwal, 1999; Gale *et al.*, 2003; Osganian *et al.*, 2003), and the xanthophyll subclass may play a role in eye function (Stahl, 2005) as well. Finally, phytosterols, due to their beneficial effect on blood cholesterol, are used worldwide in the prevention of cardiovascular diseases (Quilez *et al.*, 2003). Recent findings showing that some proteins are involved in intestinal absorption of these compounds have potential important consequences on recommended dietary allowances (RDAs) of these molecules. Indeed these RDA may be fitted to groups of individuals bearing common genetic variants in gene coding for these proteins.

II. FACTORS AFFECTING THE ABSORPTION OF FSM

The absorption efficiency of FSM is very variable and depends on several factors (Borel, 2003), and thus numerous studies have been dedicated to understand the role of these factors. These studies have shown that absorption efficiency depends on (1) the state of food (raw, processed, or cooked), (2) the nature of dishes consumed with the FSM, (3) the activities of digestive enzymes, and (4) the efficiency of transport across the enterocyte, etc. The mnemonic 'SLAMENGHI,' proposed to list the factors thought to govern carotenoid bioavailability (West and Castenmiller, 1998), can be used to list those suspected to affect FSM bioavailability. Each letter of the acronym stands for one factor: Species of FSM, molecular Linkage, Amount of FSM consumed in a meal, Matrix in which FSM is incorporated, Effectors of absorption and bioconversion, Nutrient status of host, Genetic factors, Host-related factors, and mathematical Interactions.

Table 3.1 Characteristics of fat-soluble microconstituents

Common name	Main molecular species found in the Western diet	RDA[1]	Mean/median daily intake[2]	Main biological activity
Pre-formed vitamin A	Retinyl palmitate	900 REA[3]	598–682 µg	Vitamin A activities
Provitamin A carotenoids	β-carotene α-carotene β-cryptoxanthin	– – –	2.15–2.62 mg 0.39 mg 0.12–0.14	Vitamin A and antioxidant activities
Vitamin E	d-α-tocopherol d-γ-tocopherol	15 mg –	9.8–10.3 mg –	Vitamin E activity, i.e. mainly antioxidant
Vitamin D	Cholecalciferol	5 µg	2.9 µg	Vitamin D activities
Vitamin K	Phylloquinone Menaquinone	120 µg –	70–80 µg 21 µg	Vitamin K activities
Non-provitamin A carotenoids	Lycopene Lutein/Zeaxanthin	– –	6.6–12.7 mg 2.0–2.3[4]	Antioxidant activity
Phytosterols	Sitosterol Stigmasterol Campesterol	– – –	167–437 mg	Inhibitors of cholesterol absorption

1 US RDA (recommended dietary allowances), or adequate intake (vitamin D), for adult male population.
2 Mean/median daily intake for adult male population.
3 REA: Retinol Activity Equivalents.
4 Lutein + zeaxanthin.

Table 3.2 Available data on factors suspected to affect the absorption of fat-soluble microconstituents (FSM)

FSM species	S¹	L	A	M	E	N	G	H	I
Vitamin A	0	+	0	++	++	+	0	++	0
Provitamin A carotenoids	++	+	0	++	++	+	+	+	0
Vitamin E	++	+	+	+	++	0	+	++	0
Vitamin D	+	+	0	+	++	+	0	++	0
Vitamin K	++	0	+	++	++	0	0	++	0
Non-provitamin A carotenoids	++	+	0	++	0	0	+	+	0
Phytosterols	+	0	+	+	0	0	++	0	0

1 The mnemonic SLAMENGHI enumerates the factors suspected to govern FSM bioavailability. Each letter represents one factor: Species of FSM, molecular Linkage (mainly esterification), Amount of FSM consumed in a meal, Matrix in which FSM is incorporated, Effectors of absorption and bioconversion (mainly disease, age, sex), Nutrient status of host (in corresponding FSM), Genetic factors, Host-related factors, and mathematical Interactions.
++: indicates there were more than three references on factor suspected to affect absorption;
+: indicates there were only some references (between 1 and 3) on this factor;
0: indicates there was apparently no data on this factor.

As shown in Table 3.2 some data are available for certain factors (food matrix, effector of absorption), but there are limited data for others (interactions, genetic factors, etc.).

III. MEMBRANE TRANSPORTERS INVOLVED IN INTESTINAL ABSORPTION OF FSM

It has long been assumed that FSMs are absorbed by passive diffusion. However, in 2000 it was established that most newly absorbed phytosterols are effluxed, i.e. pumped back to the intestinal lumen, by two apical membrane transporters: the (ATP)-binding cassette G5 (ABCG5) and ABCG8 (Berge et al., 2000). Recent studies have also shown that membrane transporters are involved in carotenoids and vitamin E absorption. More precisely, our team was the first to establish that the Scavenger Receptor class B type I (SR–BI) is involved in apical uptake, i.e. absorption, of lutein (Reboul et al., 2003, 2005) and vitamin E (Reboul et al., 2006). SR–BI has also been involved in the absorption of other carotenoids: β–carotene (During et al., 2005; van Bennekum et al., 2005) and lycopene (Moussa et al., 2008). Very recently a membrane transporter called STRA6 (Kawaguchi et al., 2007) has been shown to be responsible for retinol (vitamin A) uptake by retinal pigment epithelium, and it is possible that it is the intestinal vitamin A transporter not yet identified (Hollander, 1981). After being absorbed by enterocyte, FSMs must be secreted into the body at the basolateral side of the cell. It has recently been found that the ATP-binding cassette A1 (ABCA1) is implicated in vitamin E secretion at the basolateral side of the enterocyte (Reboul et al., 2009). In conclusion several proteins have recently been involved in FSM trafficking across the enterocyte, and other membrane transporters (NPC1-L1, CD36, etc.) and intracellular proteins (FABP, CRBP, etc.) are suspected to be involved in the uptake, intracellular traffic, and secretion of FSM by the enterocyte. Table 3.3 and Figure 3.1 show a summary of current knowledge on absorption efficiency and absorption pathways of FSM.

IV. POTENTIAL PHYSIOLOGICAL AND PATHOPHYSIOLOGICAL CONSEQUENCES OF PROTEIN-MEDIATED ABSORPTION OF FSMs

The involvement of transporter(s) in FSM absorption has potentially important physiological and pathophysiological consequences. First, it is reasonable to hypothesize that mutations or genetic polymorphisms (most

Table 3.3 Current knowledge on mechanisms involved in intestinal absorption of fat-soluble food microconstituents

Fat-soluble microconstituents species	Absorption efficiency	Transport characteristics across enterocytes (at nutritional doses)
Retinyl-palmitate	75–99%	Apical uptake of retinol is a facilitated diffusion process involving a not yet identified membrane transporter
β-carotene α-carotene β-cryptoxanthin	3.5–90%	Apical uptake of provitamin A carotenoids is a facilitated diffusion process involving, at least, the membrane transporter SR-BI
d-α-tocopherol d-γ-tocopherol	10–95%	Apical uptake of tocopherols is a facilitated diffusion process involving, at least, the membrane transporter SR-BI Basolateral efflux of tocopherols is an energy-dependent process involving ABCA1
Cholecalciferol	55–99%	Apical uptake assumed to be passive
Phylloquinone Menaquinones	13–80%	Phylloquinone apical uptake mediated by an energy-requiring transporter Menaquinone apical uptake assumed to be passive
Lycopene Lutein Zeaxanthin		Apical uptake of nonprovitamin A carotenoids is a facilitated diffusion process involving, at least in part, the membrane transporter SR-BI
Phytosterols	0.04–1.9%	Apical uptake involves NCP1-L1 and SR-BI. Apical membrane transporters ABCG5 and ABCG8 assumed to pump most of phytosterols back into the lumen

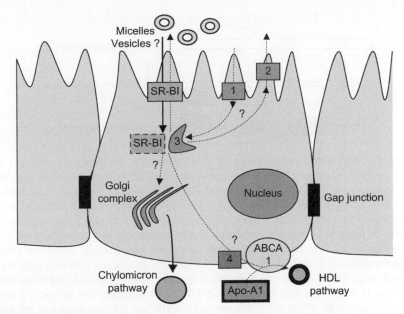

Figure 3.1 Possible pathways for FSM transport across the human enterocyte. FSMs are delivered by mixed micelles, and possibly by vesicles (liposomes), to apical membrane transporters (SR-BI has been involved in uptake of vitamin E, carotenoids and phytosterols; NPC1-L1 in that of phytosterols). FSM are either (i) effluxed back to the intestinal lumen via apical membrane transporters (ABCG5/G8 has been involved in the efflux of phytosterols; SR-BI in that of vitamin E), or (ii) secreted in the body into chylomicrons (called the apo-B-dependent pathway) or into HDL via ABCA1 (called the apo-AI-dependent pathway; has been involved in the efflux of a fraction of newly absorbed vitamin E), or other basolateral membrane transporters. 1: putative apical membrane protein mediating FSM uptake/efflux, 2: putative apical membrane protein mediating efflux, 3: putative intracellular binding protein, 4: putative basolateral membrane protein mediating efflux.

common genetic polymorphisms being SNPs – single nucleotide poly-morphisms) in several transporter(s) modulate the absorption efficiency of these microconstituents. This hypothesis is explained in Figure 3.2, with three theoretical transporters as examples; one is involved in apical uptake (T1), one in intracellular transport (T2), and one in basolateral secretion (T3). This hypothesis can explain the high interindividual variability in FSM absorption efficiency as well as the gaussian distribution of absorption efficiency observed in the population (Figure 3.3) (Borel *et al.*, 1998). Planned experi-ments by our team in the European project called 'Lycocard' (on the role of lycopene in preventing cardiovascular diseases; IP n°016213 from the sixth Framework Program, FP6) will assess the involvement of transporters in absorption efficiency of lycopene (a carotenoid responsible for the red color

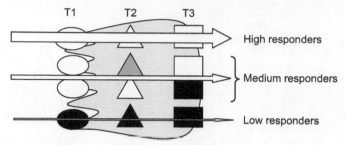

Figure 3.2 Interindividual variability of FSM absorption explained by variation in efficiency of transporters. T1: putative apical membrane transporter of FSM (e.g. SR-BI). T2: putative intracellular transporter of FSM (e.g. FABP). T3: putative basolateral membrane transporter of FSM (e.g. ABCA1). A white transporter indicates that it works with maximal efficiency. A black transporter indicates that a genetic polymorphism or a mutation leads to an impaired efficiency. A gray transporter indicates that a genetic polymorphism leads to an intermediate efficiency. The hypothesis is as follows: when all the transporters of an FSM are efficient the subjects absorb the FSM very well; they are 'high responders.' When one or several key transporters of an FSM are ineffective the subjects absorb the FSM very poorly (a diffusion mechanism can still occur); they are 'low responders.' When some transporters are efficient and some are less efficient the subjects absorb FSMs with an intermediate efficiency; they are 'medium responders.' This hypothesis can explain the gaussian response to β-carotene observed in Figure 3.3.

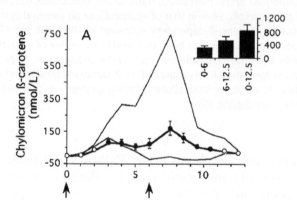

Figure 3.3 Interindividual variability in absorption efficiency of FSM. Example of β-carotene. Interindividual variability in chylomicron β-carotene response obtained after the intake of two successive meals by 16 healthy subjects (data from Borel *et al.*, 1998). The first meal provided 120 mg β-carotene and 40 g triacylglycerols, while the second meal provided 40 g triacylglycerols without β-carotene. The arrows show the times at which the meals were ingested. Inserts: area under the curve (AUC) of the response obtained after the first meal (0–6 h), after the second meal (6–12.5 h), or after both meals (0–12.5 h). The thick curves show the mean responses ± SEM of the 16 subjects, while the thin curves show the mean responses of the subjects who had the lowest and the highest responses (as estimated by their 0–12.5 h AUC).

of tomatoes). To attain our objective we will measure absorption efficiency of lycopene in a postprandial study similar to that described in Figure 3.3. Then we will compare the absorption efficiency of lycopene between groups of subjects bearing different SNPs in transporters of this carotenoid.

The effect of genetic polymorphisms in FSM transporters on FSM absorption efficiency may in turn lead to modification of the long-term plasma and tissue status of these microconstituents. This hypothesis is supported by recent results of our team, which showed in a cohort of 128 volunteers that SNPs in SCARB1 (gene coding for SR-BI) involved in vitamin E and carotenoid absorption (Reboul *et al.*, 2005, 2006; van Bennekum *et al.*, 2005) are related to the plasma status of α-tocopherol and γ-tocopherol and carotenoids (α-carotene, β-carotene, and β-cryptoxanthin) (Borel *et al.*, 2007).

Since most FSMs are involved in the prevention of age-related diseases (Christen, 1999; Nicoletti *et al.*, 2001; Fairfield and Fletcher, 2002; Mares-Perlman *et al.*, 2002; Sesso *et al.*, 2003; Knekt *et al.*, 2004), it can be suggested that SNPs in FSM transporters directly modulate the plasma and tissue status of FSMs, and indirectly the preventive effect of these micro-constituents on the development of diseases. This hypothesis is currently being tested in a French ANR-PNRA project called 'Compalimage,' in which our team is also involved.

V. POTENTIAL ECONOMIC CONSEQUENCES: PERSONALIZED NUTRITION

The fact that genetic variants may modulate the individual response to FSM has potential important consequences with regard to RDA. Indeed, these allowances have been elaborated to meet the requirement of 95% of the population. This means that, whereas most people who respect RDA have adequate intake, a minor fraction (5%) of the population has an intake lower than required for good health. On the opposite end, an important fraction of the population has a higher intake than required and is, perhaps, at risk for adverse effects (Omenn *et al.*, 1994; Rautalahti *et al.*, 1997). Because it is likely that most of the interindividual variability in absorption is due to genetic differences among individuals, personalized nutrition, i.e. RDA appropriate to subgroups of the population bearing common genetic variants, is assumed to be the future of human nutrition. Some companies propose adjusting the intake of drugs with regard to genotypes of xenobiotic metabolism enzymes (phase I and phase II enzymes: CYP450, GSTM, GSTP). For example, subjects carrying the CYP2D6*3/*4 variant are poor metabolizers of drugs such as metoprolol, a beta blocker used to treat high blood pressure, or fluoxetine (Prozac), an antidepressant (Terra *et al.*,

2005). It can be anticipated that the same approach will be used in nutrition, particularly for FSMs, which can be beneficial at low doses and hazardous at higher doses. It is therefore reasonable to anticipate that RDA or doses of FSM in supplements will be adapted to subgroups of subjects carrying key SNPs that govern FSM bioavailability. Knowledge of the nutrigenetic factors that affect FSM bioavailability can be of marketing value to food companies. Indeed, it will be possible to state that the dose of micronutrients incorporated in foods is not harmful, even for higher responders to the micronutrient incorporated. It will also be possible to elaborate different doses of FSM supplements for low responders and high responders, keeping in mind that the high doses of FSM that might be recommended for low responders should be checked for possible harmful effects on gastrointestinal mucosa.

In conclusion, the discovery that transporters are involved in the intestinal absorption of FSMs is a key finding in the field of FSM bioavailability. Interindividual variations of absorption efficiency can be partly attributed to mutations or genetic polymorphisms of gene coding for these transporters. This may have physiological and pathophysiological consequences with regard to the beneficial role of some FSM in diminishing the incidence of certain age-related diseases. One can speculate that ongoing results in the field of nutrigenetic absorption efficiency will lead to more personalized RDA, which will take into account the genetic characteristics of subgroups in the population.

ACKNOWLEDGMENTS

Special thanks to Dr. Claude-Louis Léger (UMR 476 INSERM/ 1260 INRA) for helpful criticism.

REFERENCES

Berge, K. E., Tian, H., Graf, G. A., et al. (2000). Accumulation of dietary cholesterol in sitosterolemia caused by mutations in adjacent ABC transporters. *Science, 290*, 1771–1775.

Booth, S. L., Broe, K. E., Gagnon, D. R., et al. (2003). Vitamin K intake and bone mineral density in women and men. *Am. J. Clin. Nutr, 77*, 512–516.

Borel, P. (2003). Factors affecting intestinal absorption of highly lipophilic food microconstituents (fat-soluble vitamins, carotenoids and phytosterols). *Clin. Chem. Lab. Med., 41*, 979–994.

Borel, P., Grolier, P., Mekki, N., et al. (1998). Low and high responders to pharmacological doses of beta-carotene: proportion in the population, mechanisms involved and consequences on beta-carotene metabolism. *J. Lipid Res., 39*, 2250–2260.

Borel, P., Moussa, M., Reboul, E., et al. (2007). Human plasma levels of vitamin E and carotenoids are associated with genetic polymorphisms in genes involved in lipid metabolism. *J. Nutr., 137*, 2653–2659.

Christen, W. G. (1999). Antioxidant vitamins and age-related eye disease. *Proc. Assoc. Am. Phys., 111*, 16–21.

During, A., Dawson, H. D., & Harrison, E. H. (2005). Carotenoid transport is decreased and expression of the lipid transporters SR-BI, NPC1L1, and ABCA1 is downregulated in Caco-2 cells treated with ezetimibe. *J. Nutr., 135*, 2305–2312.

Fairfield, K. M., & Fletcher, R. H. (2002). Vitamins for chronic disease prevention in adults: scientific review. *JAMA, 287*, 3116–3126.

Gale, C. R., Hall, N. F., Phillips, D. I., & Martyn, C. N. (2003). Lutein and zeaxanthin status and risk of age-related macular degeneration. *Invest. Ophthalmol. Vis. Sci., 44*, 2461–2465.

Holick, M. F. (2005). Vitamin D: important for prevention of osteoporosis, cardiovascular heart disease, type 1 diabetes, autoimmune diseases, and some cancers. *South. Med. J., 98*, 1024–1027.

Hollander, D. (1981). Intestinal absorption of vitamin A, E, D, and K. *J. Lab. Clin. Med., 97*, 449–462.

Jiménez-Jiménez, F. J., Molina, J. A., Fernandezcalle, P., et al. (1993). Serum levels of beta-carotene and other carotenoids in Parkinson's disease. *Neurosci. Lett., 157*, 103–106.

Kawaguchi, R., Yu, J., Honda, J., et al. (2007). A membrane receptor for retinol binding protein mediates cellular uptake of vitamin A. *Science, 315*, 820–825.

Knekt, P., Ritz, J., Pereira, M. A., et al. (2004). Antioxidant vitamins and coronary heart disease risk: a pooled analysis of 9 cohorts. *Am. J. Clin. Nutr., 80*, 1508–1520.

Mares-Perlman, J. A., Millen, A. E., Ficek, T. L., & Hankinson, S. E. (2002). The body of evidence to support a protective role for lutein and zeaxanthin in delaying chronic disease. Overview. *J. Nutr., 132*, 518S–524S.

Moussa, M., Landrier, J. F., Reboul, E., et al. (2008). Lycopene absorption in human intestinal cells and in mice involves scavenger receptor class B type I but not niemann pick C1-like. *J. Nutr., 138*, 1–5.

Nicoletti, G., Crescibene, L., Scornaienchi, M., et al. (2001). Plasma levels of vitamin E in Parkinson's disease. *Arch. Gerontol. Geriatr., 33*, 7–12.

Omenn, G. S., Goodman, G., Thornquist, M. R., et al. (1994). The beta-carotene and retinol efficacy trial (CARET) for chemoprevention of lung cancer in high risk populations, smokers and asbestos-exposed workers. *Cancer Res., 54*, S2038–S2043.

Osganian, S. K., Stampfer, M. J., Rimm, E., Spiegelman, D., Manson, J. E., & Willett, W. C. (2003). Dietary carotenoids and risk of coronary artery disease in women. *Am. J. Clin. Nutr., 77*, 1390–1399.

Quilez, J., Garcia-Lorda, P., & Salas-Salvado, J. (2003). Potential uses and benefits of phytosterols in diet: present situation and future directions. *Clin. Nutr., 22*, 343–351.

Rao, A. V., & Agarwal, S. (1999). Role of lycopene as antioxidant carotenoid in the prevention of chronic diseases: A review. *Nutr. Res., 19*, 305–323.

Rautalahti, M., Albanes, D., Virtamo, J., Taylor, P. R., Huttunen, J. K., & Heinonen, O. P. (1997). Beta-carotene did not work: Aftermath of the ATBC study. *Cancer Lett., 114*, 235–236.

Reboul, E., Abou, L., Mikail, C., et al. (2003). Lutein is apparently absorbed by a carrier-mediated transport process in Caco-2 cells. *Clin. Nutr., 22*, S103.

Reboul, E., Abou, L., Mikail, C., et al. (2005). Lutein transport by Caco-2 TC-7 cells occurs partly by a facilitated process involving the scavenger receptor class B type I (SR-BI). *Biochem. J., 387*, 455–461.

Reboul, E., Klein, A., Bietrix, F., et al. (2006). Scavenger receptor class B type I (SR-BI) is involved in vitamin E transport across the enterocyte. *J. Biol. Chem., 281*, 4739–4745.

Reboul, E., et al. (2009). ATP-binding cassette transporter A1 is significantly involved in the intestinal absorption of alpha- and gamma-tocopherol but not in that of retinyl palmitate in mice. *Am. J. Clin. Nutr., 89*, 177–184.

Sesso, H. D., Liu, S., Gaziano, J. M., & Buring, J. E. (2003). Dietary lycopene, tomato-based food products and cardiovascular disease in women. *J. Nutr., 133*, 2336–2341.

Stahl, W. (2005). Macular carotenoids: lutein and zeaxanthin. *Dev. Ophthalmol., 38*, 70–88.

Terra, S. G., Pauly, D. F., Lee, C. R., et al. (2005). Beta-adrenergic receptor polymorphisms and responses during titration of metoprolol controlled release/extended release in heart failure. *Clin. Pharmacol. Ther., 77*, 127–137.

Traber, M. G. (2007). Heart disease and single-vitamin supplementation. *Am. J. Clin. Nutr., 85*, 293S–299S.

van Bennekum, A., Werder, M., Thuahnai, S. T., et al. (2005). Class B scavenger receptor-mediated intestinal absorption of dietary beta-carotene and cholesterol. *Biochemistry, 44*, 4517–4525.

West, C. E., & Castenmiller, J. J. J. M. (1998). Quantification of the "SLAMENGHI" factors for carotenoid bioavailability and bioconversion. *Internat. J. Vit. Nutr. Res., 68*, 371–377.

CHAPTER 4

Food Security: Local and Individual Solutions to a Global Problem

Albert McGill

Contents

Abstract

The concept of food security is a global issue, and as defined by the Food and Agriculture Organization (FAO, 1996), includes elements of access, affordability, safety, and choice for all people. At present the world situation is far removed from a state of security, with more than 850 million people starving (FAO, 2006). Tackling a problem of this size and complexity seems far beyond the scope of most individuals and even organizations. This chapter examines how the issue may be approached successfully from three perspectives: (1) The Global Perspective, by a professional organization, the International Union of Food Science and Technology (IUFoST), whose main areas of concern are that 'all people have access to sufficient, safe food to meet their dietary needs'; (2) The Urban Perspective, as provided by motivated groups at local and urban levels in Japan and London, whose concerns are to obtain 'safe and nutritious food to meet their... food preferences for a... healthy life'; and (3) The Individual Perspective, which considers the demands of the modern individual for personalized and tailored diets 'to meet their dietary needs and food preferences for an active and healthy life'. The background of

Global Issues in Food Science and Technology
© 2009 Elsevier Inc.

each perspective is described, as are the processes and procedures adopted so far, for these on-going activities, and the results achieved at this stage.

Any professional organization embarking on such a project must be aware of the range and diversity of others already active in the field. The demand for understanding and sensitivity in establishing collaborative networking of resources is paramount. IUFoST has the scope for such a project but neither the communications network nor the management protocols to engage actively. Both the London Food Strategy and Daichi-o Mamoru Kai have demonstrated how local urban projects can be established effectively, as continuing projects. Nutrigenomics is a technology in the early stages of development, but promises a better understanding of nutrition at a molecular and individual level. Although the problem of global food security is immense, the increasing importance of the cult of the individual may offer a key to a solution based on the efforts of individual scientists being integrated through their respective professional bodies and building successes at a local level into international solutions.

I. INTRODUCTION

The International Bill of Human Rights (ECOSOC, 1999) enshrines the principle that everyone has the right to adequate food, to be free from hunger, and to enjoy general human dignity. The Food and Agriculture Organization (FAO), defines Food Security as existing '… when all people, at all times, have physical and economic access to sufficient, safe and nutritious food to meet their dietary needs and food preference for an active and healthy life' (FAO, 1996). In its most recent report on The State of Food Insecurity in the World (FAO, 2006), the failure to reduce the *number* of people starving remains an indictment on all efforts to alleviate the problem. There is some indication that the *proportion* of starving people is reducing, but not quickly enough to meet the goals of either the World Food Summit, 1996 (FAO, 2004, 2006) or the Millennium Development Goals (MDGs) (FAO, 2005). Many organizations engaged in strategies to tackle these issues have made some progress. A summary of key factors in meeting the MDGs was made by Sanchez and Swaminathan (2005), who noted that although the hunger target of the MDGs will require unprecedented levels of effort, the cost of additional assistance amounts to 60 US cents per month for every person living in a developed country in the period 2010–2015. Perhaps an opportunity exists to address the problem from a different perspective and with a more specific task force.

II. THE GLOBAL PERSPECTIVE

The International Union of Food Science and Technology (IUFoST) is the sole existing global professional food science and technology organization. It is a voluntary, non-profit association of national food science organizations that links the world's best food scientists and technologists. It was inaugurated formally during the Third International Congress of Food Science and Technology convened in Washington, DC, USA, in 1970. At the first International Congress of Food Science and Technology in London in 1962, the Congress President, Lord Rank, a flour miller by profession, delivered this message in his closing address, 'If the potentialities of... food science and technology are to... culminate in the peoples of the world receiving a sufficiency of food that is... appealing and nutritionally adequate, then there must be international collaboration.' The present membership of IUFoST comprises 67 countries, each of which is represented through its Adhering Body (AB). The food professions of each member country contribute to their national AB as best fits their individual circumstances. This allows for wide differences between groups of countries, depending on their types of professional organizations and stage of economic development. IUFoST is best known for its Congresses, Conferences, and Symposia and is a full Scientific Member of the International Council of Scientific Unions (ICSU).

At the IUFoST Congress in Budapest in 1995, a declaration was approved that extended the intentions of the organization put forward by Lord Rank in 1962, and drew upon the outcomes of the International Conference on Nutrition (ICN, 1992). The key points of the declaration emphasized that '... the determination to work for the elimination of hunger and the reduction of all forms of malnutrition throughout the world.... ensure sustained nutritional well-being for all people in a peaceful, just and environmentally safe world... recognizing that access to nutritionally adequate and safe food is the right of each individual' (IUFoST, 1995). However, this was more a statement of principle by IUFoST, as it has no mechanisms for the development of strategic plans or for their implementation. Such sentiments as expressed in Budapest were soon followed by the World Food Summit in Rome, Italy, that contained the definition of Food Security and included the World Food Summit Plan of Action (FAO, 1996), and subsequently by the International Bill of Human Rights (ECOSOC, 1999).

During a plenary lecture in Seoul, Korea, at the 2001 IUFoST Congress, Owen Fennema surprised many delegates by criticizing the abject failure of IUFoST to make any progress towards meeting the obligations of the Budapest Declaration (IUFoST, 2001). Two years later at the Chicago

Congress, the continuing failure to make progress was highlighted by Ismael Serageldin (IUFoST, 2003). Serageldin drew an analogy between the challenges of tackling hunger and those of slavery, with both being immense problems in the world historically but as the necessary evil about which society as a whole could and can do little. Only the total commitment of people like William Wilberforce, the British abolitionist, and his supporters in the 18th and 19th centuries, to politicize the problem of slavery and its solutions, eventually made a difference. Serageldin proposed the recruitment of a new group of 'abolitionists,' with world hunger as its target.

Discussions among some delegates in Chicago supported Serageldin's view and emphasized that the lack of effort in tackling world hunger was an international problem on a grand scale, and often seen as too overwhelming. However, the problem of world hunger could be broken down into local components, which might engage the voluntary work of individual scientists, complemented by their professions and the work of their national Adhering Body of IUFoST, the existing and progressive work of major agencies (e.g. WHO/FAO/WFP), and other voluntary groups such as OXFAM. The move to develop a local and collaborative approach was first proposed by Lord Rank in 1962, but only by using the organization and contacts provided by IUFoST. To improve the alignment of targets the term 'hunger' has been replaced in IUFoST's vocabulary by the wider concept of 'food security' as defined by FAO.

To translate the Budapest Declaration from a statement of principle into a plan of action with identified outcomes, it must be re-defined as a project, and considered in three parts as: collaboration, communication, and activation.

II.A. Collaboration

Both the size and the focus of organizations, such as IUFoST and the professions accommodated within its Adhering Bodies (ABs), limit the inputs and impact that can be directed onto the issue of Global Food Security. Much of what may be achieved will depend on a considerable knowledge of the activities and programs of the major participants, namely the UN agencies of WHO and FAO. In addition, and at a more local level, the impact of the work of voluntary agencies such as OXFAM, World Vision and the Save the Children Fund, must be considered in determining what can be offered by the professions of the ABs of particular countries. Consideration for, and sensitivity towards, those already in the field are necessary to avoid territorial disputes and wastage of effort and resources. Identifying the active agencies, understanding their programs and negotiating supporting activities are the key factors for any project of this kind.

An excellent starting point and continuing reference for any country's AB wishing to participate in such a project is the FAO annual report of 'The State of Food Insecurity in the World', which is published in November each year. More than 80 countries have data listed and a first task would be to validate and possibly enhance those data. Details of the impact and use of this approach is given in more detail elsewhere (McGill, 2006a). In examining the reports of the FAO Special Programme for Food Security (SPFS) (FAO, 2005a) and their comprehensive work at regional and national level, there is little evidence of significant work directed at postharvest and supply-chain food processing for either improved local usage and nutrition or for the value adding necessary to bring surplus export income to reduce poverty among primary producers. A similar criticism could be made of the reports of the International Food Policy Research Institute (IFPRI) for the Hunger Task Force and the Millennium Project. Their projections for 5–10 years give little consideration for food beyond crop production (IFPRI, 2005). A lack of collaboration, so far, has prevented any input from IUFoST to the ICSU-sponsored project on Global Environmental Change and Food Systems (GECAFS, 2005), in which the emphasis on crops rather than food is again dominant. That IUFoST is a member of ICSU makes this more surprising, but not unique (Rosegrant and Cline, 2003), and indications are that closer ties will be fostered in future.

II.B. Communication

Although communication between organizations may prove difficult, it would be expected to be better within organizations, such as IUFoST. However, it must be remembered that the structure and administrative protocols of this organization were not designed for interactive project management. The only related activity by IUFoST was that of Joseph Hulse as Chair of an Integrated Food Systems Task Force, which has produced only a draft outline report and no further action. Provided at least some of the constituent ABs were interested in participation in a Food Security project, the mechanisms for recording objectives, progress, and achievements could be established. The ABs have not been made aware of any need to collect, record, and report such progress as there might be in activities of their constituent professions that could be perceived as contributing to the aims of the Budapest Declaration. The failure of the administration of IUFoST to capture, record, and publish such progress, as may exist, derives from the absence of any operational project plan rather than a decision to abandon the Declaration. To implement a project plan would place demands on the professions represented within each AB, would be more demanding and diverse than exists at present, and would rely ultimately on the commitment of *individuals* at a *local* level. Their ability and resource-fulness in co-ordination of efforts with other local agencies and their own programs will determine what progress can be made.

II.C. Activation

A project as outlined here was approved by the Governing Council of IUFoST in Kuala Lumpur in 2005. Since then, much information has been gathered for collaboration and many of the areas of ignorance of the food professions in Food Security exposed. Partners have been identified and strategy established, but the IUFoST has yet to decide to move from a more reactive stance to an active stance on this issue and may still lack the operational protocols to deliver such a change.

III. THE URBAN PERSPECTIVE

The definition of Food Security according to FAO (1996) not only involves '... physical and economic access to sufficient, safe, and nutritious food to meet dietary needs... for an active and healthy life...' but also to '... food preference...' or indeed, *food choice*. This has great importance to all people, whether they are suffering extreme food shortage or experiencing a more sustainable lifestyle. Although hunger is generally concentrated in rural areas where the majority of poor and food-insecure people live, the seemingly inexorable movement of people from rural to urban environments has been ongoing with as many people now living in the cities as in the countryside. Even in Africa, an urban majority is likely to be reached by 2020 (FAO, 2005). Consequently the battle for food security will be fought as much in cities as in the countryside and efforts are being made to ensure that, even in the cities, farming and food production are developed and expanded (RUAF, 2006). Interests in self-sufficiency and the use of local produce have added to the growing demands for *organic* produce on a worldwide basis. For example, in Australia, *organic* produce generated A$28 million in sales in 1990 and A$300 million by 2004 (The Weekend Australian, 2006). This development causes great anxiety for committed environmentalists, gen-erating demand for the *right* food to such an extent that such demand can be met only by importing *organic* produce from many developing countries. Those countries are now being encouraged by market forces to grow food for export, putting a strain on food supplies for their own citizens. In addition, the cost of global transport, particularly by air, is adding greatly to greenhouse gas accumulation. The pressure to resolve these issues has been relieved through two very different urban projects: the London Food Strategy and Daichi-o-Mamoru-Kai.

III.A. The London food strategy

This was published as 'Better food for London: the Mayor's Draft Food Strategy' (LDA, 2005). Following some web-based comments, the outcomes of a number of public forums and inputs from theme-specialized committees, a revised document was published in operational format as 'Healthy and Sustainable Food for London' in the following year (LDA, 2006). The Mayor's initial observation was 'that the food system in the capital is not functioning in a way that is consistent with the ambition that London should be a world-class, sustainable city.'

Many features of the London food system are positive:

- Every day, millions of Londoners are able to access the food they want and can afford.
- World-class retailers contribute to the commercial prosperity of the city, and provide employment for many thousands of people.
- Hundreds of businesses throughout London and the surrounding regions prosper by processing and manufacturing the food the city needs.
- The city's extraordinary social and cultural diversity is reflected in a restaurant culture that has seen it crowned the 'gastronomic capital of the world.' Half of all the restaurants in the UK are in London.

Many features are much less positive:

- A rising number of Londoners, particularly children are becoming obese.
- In some parts of London, people struggle to access affordable, nutritious food (cf. Food Security definition).
- The safe preparation of food, both in the home and in London's plethora of restaurants and cafes, remains a key issue.
- Many small or independent enterprises, such as retailers, farmers, food manufacturers, and marketers, struggle to survive, and their business fragility may affect both the diversity and resilience of London's food system.
- The environmental consequences of the way London's food is grown, processed, transported, and disposed of are profound and extensive (LDA, 2005a).

As with all other cities in the world London operates within a globalized setting, and is equally affected by the negotiations of the World Trade Organization (WTO) and the corporate strategies of global enterprises such as Wal-Mart, Tesco, or Nestlé. The effect of the power of general market forces will also impact consumer information and choice. Both market forces and consumer preferences are dynamic and subject to change. Policy instruments, regulatory frameworks, information campaigns, targeted investment, and political leadership can actively shape and encourage the direction of change. London has the means to do this and has accepted the responsibility to act. However, the capital's food system is inextricably linked to bodies, individuals, and places outside London, and often in other countries. This is why the need for networking and collaboration are critical to the progress of the strategy.

Few attempts have been made to consider food as part of an integrated and interdependent system. Diet, health, and well-being are inextricably linked and food lies at the heart of the Mayor's cross-cutting themes of health, equalities, and sustainable development. Food waste, as a byproduct, requires new composting facilities but also offers the potential for emerging renewable energy initiatives.

The importance of food within London is summarized best through the five principal themes of the Strategy framework:

III.A.1. Health

The relationship between diet/food and health is now regularly acknowledged (Rayner and Scarborough, 2005). The improvement of Londoners' diets could deliver significant benefits ranging from a reduction in the incidence of cancer and coronary heart disease to the onset of type-2 diabetes. While many in cities around the world suffer from under-nutrition or malnutrition, in London, diseases related to over-consumption are increasingly common. Air pollution in London, associated with road freight, has a number of major health implications, replicated in many other major cities. The incidence of food-borne disease remains a key concern.

III.A.2. Environment

The food system has significant environmental impacts. In the case of London, it is responsible for 41% of the ecological footprint, while food preparation, storage, and consumption account for 10–20% of the environmental impact on the average household. More than one third of domestic waste in London is food related. In total, it has been estimated that close to half of the human impact on the environment is directly related to the operation of the food system. Informed consumer choice and more efficient food transportation could reduce this impact.

III.A.3. Economic

In London the agri-food sector is significant, employing nearly 500 000 people; the food and drink sector is the capital's second-largest and fastest-growing manufacturing sector; food retail businesses in the city employ tens of thousands of people; and London's thriving restaurant culture accounts for around one quarter of total UK activity in the sector and is a major attraction for tourists to the city. Londoners spend GBP 8.8 billion each year in retail food outlets.

III.A.4. Social and cultural

Although food provides the essentials for survival, this is surpassed by its central role in socializing, providing pleasure and sustaining health and lifestyles. Many Londoners now prefer to eat outside the home using the 12 000 restaurants (half the nation's total), 6000 cafes, and 5000 pubs and bars.

III.A.5. Food security

Unlike the FAO definition of food security (FAO, 1996), this strategy reserves the term for issues associated with the ability of the food system to withstand an emergency or crisis event, such as flooding, disruption of energy supplies, or terrorist attack.

The five principal themes of the framework are presented as columns in Table 4.1. Matched with these principal themes are eight stages of the food chain: (1) Primary Production, (2) Processing and Manufacturing, (3) Transport, Storage and Distribution, (4) Food Retail, (5) Purchasing Food, (6) Food Preparation, Storage and Cooking, (7) Eating and Consumption, and (8) Disposal.

These eight stages of the framework are presented as rows in Table 4.1. The framework is used throughout the strategy to structure the detailed plans and allocate timelines and responsibilities.

III.B. Daichi-o-Mamoru-Kai (Association to Preserve the Earth) (iNSnet, 2005)

In Japan, most foods are derived from imported crops and the calorie-based self-sufficiency ratio is only about 40%, the lowest of all OECD countries. Land available for agriculture has been declining annually, dropping to 12.8% of the total land area in 2002. The primary industry workforce (agriculture, forestry, and fisheries) also declined to 4.7% of the total. Under these circumstances, various groups, large and small, have been formed to protect and develop local food traditions and primary industries, by establishing closer relationships among producers who farm organically, processors who maintain traditional manufacturing methods and their supporting consumers. One of these groups, Daichi-o-Mamoru-Kai, was established in 1975 as a result of an encounter among farmers disillusioned with agrichemicals, which they believed were damaging both the ecosystem and consumers of their produce, to seek ways of growing agrichemical-free vegetables and rice and delivering them to consumers in urban areas. With some success the number of participating producers and consumers increased and the association established a distribution company,

Table 4.1 The food strategy framework for the mayor's draft food strategy (London Development Agency, 2005)

Stages of the food chain	Key organizations	Health	Environment	Economic	Social & cultural	Food security
Stage 1: Primary production	Abattoirs, allotments, city farmers, community growers, cooperatives, fisheries, market gardens	Public access to nature (health benefits), labor standards & pesticides exposure, health & safety, farmer welfare, public health & antibiotics use, nutrient content, animal feed (quality & sourcing of)	Biodiversity, energy/water use, climate change, agri-environmental schemes, pesticides, GM crops, flooding, soil erosion, environmental management, soil fertility, quality assurance standards, pollution (air, water & soil), fishing by-catch, fish stocks, bush meat trade	Income, employment, labour skills, access to markets, farming methods, diversification, non-food crops, crime (inc. vandalism, fly-tipping, theft), subsidies, economies of scale & farming intensity, quality assurance standards	Public access to nature (educational benefits), labor standards, animal welfare, migrant labour & gang masters, quality assurance standard, ethnic food production, skills	Biodiversity & genetic crop diversity, energy & water scarcity, GM crops, climate change, flooding, soil erosion, skills, potential for self sufficiency, animal & human disease, food scares
Stage 2: Processing & manufacturing	BME processors, farmers, large processors, packaging companies, SME processors	Health & safety, public health (nutrition, additives & flavoring), labor standards	Energy use (including Heating & cooling), climate change, air quality, water use, packaging, waste & recycling	Employment, skills, income, access to markets	Labor standards	Disruption to fuel supplies, human & animal diseases, food scares

Stage 3: Transport, storage & distribution	Distribution companies, farmers, logistics companies, retailers & supermarkets	Mode of transport impacts, vehicle design, schedule, pollution (noise & air), congestion, infrastructure maintenance, nutrition	Mode of transport, vehicle design, load profile, driver training, fuel type, air quality, food miles & CO_2 emissions/climate change, energy use, packaging	Mode of transport & costs, employment, vehicle design, load profile, information & communication technology, refrigeration, storage & warehousing	Labor standards, skills & training	Emergency/ disruption, oil dependency, 'Just-in-Time' delivery, mode of transport, infrastructure maintenance, international relations, climate change
Stage 4: Food retail	BME retailers, catering companies, convenience retailers, direct selling (box schemes, internet, markets), importers/ exporters, target groups, markets (street & farm), off licences, public sectors, restaurant, SME retailers, cooperatives, social enterprises	Transport impacts, food safety & hygiene	Transport impacts, congestion, climate change, production methods	Price, employment, pricing system (e.g. farm gate price) & contract criteria, quality, reliability, WTO rules, import/ export duty, quality assurance standards	Eating out & 'on the go'	Emergency disruption to supplies, price, quantity, reliability, food access, nutritional value, transport infrastructure, climate change, biodiversity, energy supply, diversity of supply

(continued)

Table 4.1 The food strategy framework for the mayor's draft food strategy (London Development Agency, 2005)—Cont'd

Stages of the food chain	Key organizations	Health	Environment	Economic	Social & cultural	Food security
Stage 5: Purchasing food	Consumer, public procurement	Nutrition, consumer preference, labeling	Transport mode, vehicle efficiency, journey profile, air pollution, congestion, energy, consumer demand for organic food	Household incomes, food price, consumer demand & preferences, emerging markets (e.g. ethical goods, internet shopping)	Lifestyles/habits, income, convenience & physical access, work patterns, cooking skills, nutrition/food knowledge, education, consumer preference, labeling	Skills, facilities, disruption to utilities
Stage 6: Food preparation, storage & cooking	Catering companies, community groups, individuals, public sector, restaurants, take-away outlets	Lifestyle/habits, nutrition/vitamins, skills, ethnic food & ethnic food skills, health & safety, food safety & hygiene, target groups (age, ethnicity, pregnant mothers)	Energy & water use, climate change, air quality, cooking skills & shopping preferences	Skills, cooking equipment, employment	Lifestyles/habits skills, cooking clubs, ethnic food & ethnic food skills, work patterns, target groups, cultural/special events	Skills, facilities, distribution to utilities

Stage 7: Eating & consumption	Business, care homes, community groups, individuals, public sector, restaurants	Lifestyle habits, nutrition/vitamins, health/well-being, breast-feeding, dieting, nutrition standards	Climate change, food related litter & disposable packaging	Eating out, tourists, ethnic food, corporate food procurement, public procurement, taste/quality, take-away, employment	Lifestyle habits, family groups, breast-feeding, recipes, work patterns, cultural/special events, Maslow, dieting, books & magazines, recipes, work patterns, take-away, cultural/special events	Contamination of food & water supplies
Stage 8: Disposal	Community groups, households, individuals, local authorities, markets, manufacturers, public sector, retailers, restaurants, waste companies	Possible health impacts from landfill as well as visual pollution & smell: possible health impacts of incineration	Loss of land to accommodation landfill, leachates from landfill, methane & CO_2 emissions from incineration, congestion & air quality issues from the transport of waste	Transport costs of collection/infrastructure, increasing cost of waste management, need for investment in new facilities, job creation through recycling	Waste recycling & composting collections, home composting, lifestyle/habits (e.g. convenience food & eating out), propensity to compost influenced by lifestyle/habits	Threat of distribution to the collection & disposal of waste

Source: London Development Agency. (2005). (from Table 3.1, www.lda.gov.uk and 'Ethics and the Politics of Food', 2006; ISBN-10: 90-8686-008-7, pp. 371–373). [Reproduced by kind permission of Wageningen Academic Publishers, The Netherlands]

moving from a group-purchasing system to a door-to-door delivery service for individual households. It now handles up to 3500 items, including meat, fish, and processed foods.

Having encouraged its primary producers to collaborate and share knowledge of their particular production methods, the association ran into difficulties with its expanding customer network. Customers, who had taken for granted attractive, uniform, unblemished vegetables, complained immediately about misshapen, varied produce, which was frequently marked or superficially worm-damaged and sometimes in short supply or seasonally unavailable. The association turned these complaints into a marketing strategy through the education of its customers in balancing appearance against the advantages of chemical-free foods from known producers. It now holds about 100 events each year during which consumers and customers visit farms, where farmers hear complaints and explain the processes necessary to supply their vegetables. This approach re-unites consumers with the point of production of their food and helps to minimize ignorance of food quality.

The next stage of the association's plans has been to raise the awareness of consumers to the energy consumption of providing their food, particularly through transport. This is the 'food miles' concept of multiplying the mass of the food by the distance it travels. According to the Ministry of Agriculture, Forestry and Fisheries, Japan's total food mileage for 2001 was about 900 billion ton-kilometers. This was 8.6 times that of France, three times that of the USA and 2.8 times that of South Korea. This is because of the large volume of Japan's food imports and the great distances covered, and the amounts of energy consumed and carbon dioxide (CO_2) produced are huge.

Another critical problem, arising from the raising of consumer demand for domestically produced foods that are considered fresher and healthier, is that they are more expensive than imported foods. This is a result of the relative quantities concerned and the economics of supply contracts obtained. With a national interest in trying to reduce global warming since the ratification of the Kyoto Protocol, the association decided to enhance their image by showing that purchasing their produce also helped to reduce CO_2 emissions. They estimated the CO_2 amounts produced from the transport of various types of food, coming from different production areas, to compare domestic with imported foods. Tokyo was used as the destination for all produce and they developed the concept of a unit called the *poco*, where 1 *poco* = 100 grams CO_2. So, choosing to buy asparagus from Hokkaido means a reduction of 4 *pocos* of CO_2 compared with buying the same produce from Australia. Daichi-o-Mamoru-Kai disseminates this kind of information through its public relations magazines and website as a way of advocating domestically grown and processed foods to reduce CO_2

emissions. It has estimated that by eating only domestically grown and processed food they could reduce per capita CO_2 emissions by 20 000 tons during a one year campaign involving its 70 000 members.

Both of these cases show attempts to resolve global problems at a local or urban level. Daichi-o-Mamoru-Kai has been established since 1975, and it has developed a range of more than 3500 food items for home delivery to an active, consulted 70 000 membership. It has extended its activities from giving its consumers a choice on purchasing locally sourced foods virtually free from chemical and pesticide contamination, to making an informed ethical choice that connects production, processing and their environmental impacts on CO_2 production through its concept of *pocos*.

The Mayor of London (1999–2008), Ken Livingstone, has developed a Food Strategy for London that considers food as an integrated and interdependent system, one that sits alongside, and is expected to network with, other similar strategies on Spatial Development, Transport, Municipal Waste Management, and Economic Development. It is supported and enhanced by the detailed study on providing a Food Hub for the storage and distribution of food throughout London (LDA, 2005b). Both are complex and detailed proposals with extensive consultation through both public forums and specialized thematic committees. It is not expected there will be an easy consensus on the implementation of the Food Strategy, but its creators are committed to a process of consultation and, where necessary, compromise. This Strategy, when implemented, should give Londoners access to a diversity of local produce and the capacity for informed choice that can allow them to make eating both healthily and ethically, a norm. It will also go some way towards answering a major criticism found in the Food Ethics Council's report (2005) on UK government policy on health and diet (DOH, 2004, 2005), namely, that good food choices cannot be made in ignorance or when suitable produce is unavailable.

Both studies show that, by developing the commitment of consumers, whether on the basis of location or belief, choices of food purchase and consumption can be made on ethical as well as health, safety, or economic grounds, which can be harnessed to support, and indeed drive, changes in policy more effectively than might have been expected.

IV. THE INDIVIDUAL PERSPECTIVE

There can be little doubt that in the developed world, this can be seen as the era of the individual. The erosion of the family unit, increasing affluence, rejection of organized religion in favor of the cult of *self,* the spread of democratic government and society's acceptance of the one person unit, have made the importance of individual choice paramount.

This development of narcissism has had a major effect, not only on the way people live, but also on how they choose their food and how such food experiences are marketed to them. Many new housing developments are being constructed as single dwellings with only microwave cooking facilities and no recognized kitchen. The concept of eating-out or buying take-away meals is so common as to be unremarkable. Speculation on how such eating habits and food practices may develop are explored, at least in part, by Greenfield (2004). The pursuit of longevity coupled with good health looms large on the agendas of today's individuals and this has been encouraged by the potential for individual prescription through the outcomes of such events as the sequencing of the human genome.

The Human Genome Project has provided the background information and new tools that enable researchers to take a more global view of gene-based interactions, including those with dietary components. Nutritional genomics is the study of how food and genes interact, with food being a complex mixture of thousands of different compounds and the human body being run by around 40 000 genes. Associations between diet and chronic disease have long been recognized through epidemiological studies. New genomic technologies are now allowing more to be discovered about the basis of these associations through *Nutrigenomics,* the functional interactions of food with the genome at the molecular, cellular, and systemic levels, and *Nutrigenetics,* the ways in which individuals respond differently to different diets, according to their individual genetic make-up. In his overview of the subject Gibney (2005) regrets the new terminology, pointing out that *Molecular Nutrition* instead of *Nutrigenomics* would have been more in keeping with the established disciplines such as *Molecular Biology* or *Molecular Medicine.* Unfortunately, popularity or notoriety may derive faster from a more dramatic name in the eyes of both public and practitioners (compare the attraction of similar university courses, *Home Economics* and *Ekotrophology*). The two new areas are of significant importance to both the health-care system and to the food industry, generating potential for future conflict and controversy. They are also very important to those seeking to understand the biology underlying normal homeostasis and disease. The term *Nutrigenomics* is being used to cover almost all studies in this field, where many of the technologies are new, still being refined and a rethink and standardization of approaches is required (Astley, 2006). This may well come from the establishment of the European Nutrigenomics Organization (NuGO), which involves 22 partner institutions from 10 EU member states, and with funding of 17.3 million Euros over six years. Its work will complement the work of the biomedical and pharmacological research communities in genome-based development of curative therapies.

However, at the same time, services are being advertised on the internet for genotyping as a basis for the recommendation and delivery of personalized dietary advice and even as a service for the delivery of meals, similar to those sold as part of weight control regimes. While Gibney's overview of the field (2005) is interesting, the most authoritative evaluation has been given in a report by the Public Health Genetics Unit (PHGU) (2005) of Cambridge University, UK, which has cast doubt on the value of any individualized dietary advice based on present genomic information.

For the individual living in a developed country, there are a number of questions that need answers:

- Are current advertised services and diets based on *nutrigenomic* information merely fanciful or grossly misleading?
- Is the food industry rushing into this new market too early, in search of quick profits, as happened with dietary supplements (SNE, 2004)?
- What are the ethical implications of a general exchange of individuals' genetic information?
- Will this information and its related dietary advice be used to judge compliance with the more individual responsibility for diet-related health, as outlined in successive UK Department of Health White Papers (DOH, 2004, 2005)?

The pros and cons of applying current *nutrigenomic* information for dietary advice seem to derive from the food industry and medical geneticists, with many other scientists and a large part of the general public found somewhere between the two extremes.

The food industry, its spin-off companies, and its entrepreneurs have a responsibility to their shareholders and financial backers to make a profit on their operations. New raw materials, processes, products, and services provide fertile ground for these developments. One of the earlier developments in this field was *functional foods,* closely followed by *nutraceuticals.* Karla (2003) defines both concepts as well as clarifying their differences from dietary supplements. He refines the original definition of *nutraceutical* by its originator, DeFelice (Brower, 1998), as '... a food (or part of a food) that provides medical or health benefits, including the prevention and/or treatment of a disease,' but would run into opposition on functional foods as defined by Just-Food (2006) as '... those foods or food components (whole, fortified or enriched with functional food components) that are scientifically recognized as having physiological benefits beyond those of basic nutrition.' The current market for such foods is estimated by Just-Food to be US$7–63 billion and is expected to grow to US$167 billion by 2010. Its advantage is in being seen as a 'natural' medicine and that it requires no testing as a pharmaceutical product. If each potential customer could be tested and charged heavily for the test (US$150–1500), and then 'prescribed' a diet, which, as it is tailored, would be done at a special price, then considerable business could be done and individual customers would be assured of taking care of their own health.

Many health professionals see *nutrigenomics* as a supplementary tool and not an individualized replacement for general dietary guidelines (SNE, 2004). Indeed there are strong possibilities of sowing confusion and frustration among consumers by providing advice, when their confidence in existing nutritional guidelines is not high. However, the European Nutrigenomics Organization (NuGO) and its outreach arm, The Nutrigenomics Society, are dedicated '... to develop, integrate and facilitate genomic technologies and research for nutritional science, to train a new generation of nutrigenomics scientists, in order to improve the impact of nutrition in health promotion and disease prevention' (NuGO, 2006). Their motivation and funding will enable rapid developments in this field to be pursued vigorously.

Not all scientists are convinced of the readiness of *Nutrigenomics* to play a significant role in the prevention and/or treatment of the major chronic diseases (McGill, 2006b). An investigation by the Public Health Genetics Unit (PHGU) of Cambridge University, UK, was sponsored by The Nuffield Trust to explore '... this gene environment interaction in the important area of diet and susceptibility to chronic conditions such as heart disease, cancer, obesity and diabetes.' Eminent national researchers were brought together with representatives of key policymaking bodies to examine the evidence for nutrient–gene interaction, to consider clinical applications and to make recommendations for policy implications. These recommendations (PHGU, 2005) '... urged caution in promising too much from this very new science too soon and emphasized the need not to confuse the population public health messages.' The consensus of their meeting was that it was much too soon to develop any new 'healthy eating' messages derived from *nutrigenomics* research. At the individual level there is a danger of scaring people about increased risks. The group noted that genotyping as a basis for the delivery of personalized dietary advice is available over the internet. It was not thought that the level of such activity in the UK was sufficient at present to justify the issuing of public advice.

However, in 2004 and 2005 the Department of Health in the UK published two sequential White Papers developing the idea of a greater individual responsibility of all citizens for dietary choice and resulting consumer health (DOH, 2004, 2005). The Food Ethics Council (2005) criticized the papers on the basis that good food choices cannot be made in ignorance or when suitable produce is unavailable for economic, seasonal, or geographic reasons. A subsequent launch symposium discussed their own report and debated how responsibilities for diet and health should be shared among the state, companies, and citizens (Food Ethics Council, 2006). One of its three conclusions asked was how relevant personalized nutrition was with regard to science and the food industry and how such an approach would impact public health policy. They expressed a general skepticism that *nutrigenomics* was likely to deliver widespread health benefits, even in the long run.

Thus the individual is left with the desire and the means to make personal choices on food consumption, directed by government to take more responsibility for their own health-related food-choice behavior, encouraged to embrace new technology that may enable such choices to be almost perfect, and yet warned about the efficacy of such technology.

ENDNOTE

Few individuals can fail to be moved by images of children dying from starvation (at the rate of one every five seconds) or be appalled by the concept of more then 850 million people around the world suffering hunger while food is plentiful and developed countries worry about obesity. The images become too common, their impact dilutes, and it is much too difficult an issue for the individual to resolve. The annual publication of The State of Food Insecurity in the World (FAO 2004, 2005, 2006) gives much information, country by country on the progress of programs towards the targets of the World Food Summit (1996) and the Millennium Development Goals (2000). There have been some successes in the Caribbean, Peru, Ghana, and Mozambique with failures in Guatemala, the Yemen, Jordan, and sub-Saharan Africa (FAO, 2006). It is clear that although targets are set for the world as a whole, the greatest changes occur, for both good and ill, at a more local level. Actions of comparatively small groups of active individuals make comprehensive impact at country and regional level.

The continuing flow of people from rural to urban areas has severe implications for both food production and demands for environmental resources of water and energy. Progress made through the London Food Strategy points to the approach the people in cities can take in obtaining the food system they desire, integrated with a sustainable system for the city. Those who wish to reject the produce distributed from mass farming and retail systems can follow the route of Daichi-o-Mamoru-Kai, and indeed the growth of farmers' markets across Europe is quite significant. Again, the demands and commitment of individuals and small groups make impacts on large and complex systems.

The cult of the individual, with its demands for instant and personal gratification, may deserve criticism but can become a driver in the development of the applications of genomic research into diets and health. While the technology is considered to be premature at present, the growing demand for personalized dietary advice is unlikely to disappear, fuelled as it may be by government encouragement for individuals to take responsibility for their own diet-related health (DOH, 2004, 2005). Information derived from this type of work will improve overall understanding of nutritional needs.

Food security as a global issue is enormous, complex, and too difficult for most people to comprehend when taken as a whole. There is evidence that great progress can be made at a local level and through the ideas, planning, and implementation of small groups of *individuals*. Those in the food-related professions, through their special skills and knowledge, have even greater potential to make a contribution to the larger task. If the Adhering Bodies (ABs) of IUFoST could be encouraged to participate, who knows what might be achieved?

The growing strength of the cult of the individual may act as a spur to drive much greater personal involvement in addressing local issues relating to food availability, quality, safety, and nutritional efficacy, while maintaining the essential element of choice. It may be that by integrating the effects of small local successes, through professional bodies, larger problems will be solved and the issue of global food security addressed. It is worth remembering the old African saying: 'A person can eat an elephant, if they take it one bite at a time!'

REFERENCES

Astley, S. (2006). The European Nutrigenomics Organisation – NuGO. Website: http://www.ifr.bbsrc.ac.uk/Science/ScienceBriefs/nugo.html [accessed: 01/10/06].

Brower, V. (1998). Nutraceuticals: poised for a healthy slice of the healthcare market? *National Biotechnology, 16,* 728–731

DOH (Department of Health, UK). (2004). *'Choosing health: making healthier choices easier.'* London: Department of Health.

DOH (2005). *'Choosing a Better Diet: a food and health action plan.'* London: Department of Health.

ECOSOC (United Nations Economic and Social Council). (1999). 'International Bill of Human Rights.' New York.

FAO (Food and Agriculture Organization of the United Nations). (1996). 'Rome Declaration on World Food Security and World Food Summit Plan of Action.' Adopted at the World Food Summit, November 13–17, Rome.

FAO (2004). *'The State of Food Insecurity in the World 2004. Monitoring progress towards the World Food Summit and Millennium Development Goals.'* Rome: FAO.

FAO (2005). *'The State of Food Insecurity in the World 2005. Eradicating world hunger – key to achieving the Millennium Development Goals.'* Rome: FAO.

FAO (2005a). 'The Special Programme for Food Security. Support to national programmes.' Website: http://www.fao.org/tc/spfs/support [accessed: 12/01/05].

FAO (2006). *'The State of Food Insecurity in the World 2006. Eradicating world hunger – taking stock ten years after the World Food Summit.'* Rome: FAO.

Food Ethics Council. (2005). 'GETTING PERSONAL: shifting responsibilities for dietary health.' Website: http://www.foodethicscouncil.org [accessed: 12/20/05].

Food Ethics Council (2006). 'GETTING PERSONAL: shifting responsibilities for dietary health. Launch symposium report.' Website: http://www.foodethicscouncil.org [accessed: 04/21/06].

GECAFS (Global Environmental Change and Food Systems) (2005). 'Science Plan and Implementation Strategy.' Website: http://www.gecafs.org [accessed: 02/20/06].

Gibney, M. J. (2005). Nutrigenomics in human nutrition – an overview. S.A.J. *Clin. Nutr.,* *18*, 115–118.

Greenfield, S. (2004). *Tomorrow's People.* London: Penguin Books Ltd.

ICN (International Conference on Nutrition) (1992). #Final Report of the Conference on Nutrition.' ICN/92/3 and PREPCOM2.ICN/92/FINAL REPORT. FAO and WHO, Rome.

IFPRI (International Food Policy Research Institute) (2005). 'IFPRI's Strategy: Toward food and nutrition security: Food Policy research, capacity strengthening, and policy communications.' IFPRI, USA. Document may be downloaded at http://www.ifpri.org/about/ifpristrategy.pdf

iNSnet (2005). 'Unique NGOs in Japan. Daichi-o-Mamoru-Kai.' Website: http://www.insnet.org/ins_spoton [accessed: 11/25/05].

IUFoST (International Union of Food Science and Technology) (1995). 'World Food Congress, Budapest, Hungary.' Website: http://www.iufost.org. [accessed: 01/05/05].

IUFoST (2001). 'World Food Congress, Seoul, South Korea.' Website: http://www.iufost.org. [accessed: 01/05/05].

IUFoST (2003). 'World Food Congress, Chicago, USA.' Information available on http://www.iufost.org. [accessed: 01/05/2005].

Just-Food (2006). 'Global market review of functional food – forecasts to 2010.' Website: http://www.just-food.com/store/product. [accessed: 04/24/06].

Kalra, E. K. (2003). Nutraceutical – Definition and Introduction. *AAPS PharmSci 5(3), Article 25,* DOI:10.1208/ps050225.

LDA (London Development Agency) (2005). 'Better food for London: the Mayor's Draft Food Strategy.' London Development agency, London. Website: http://www.lda.gov.uk/londonfood. [accessed: 10/25/05].

LDA (2005a). 'London Sustainable Food Hub. Opportunities for a sustainable food logistics centre in London,' p. 1. Website: http://www.lda.gov.uk. [accessed: 10/30/05]

LDA (2005b). 'London Sustainable Food Hub. Opportunities for a sustainable food logistics centre in London.' Website: http://www.lda.gov.uk. [accessed: 10/30/05].

LDA. (2006). *'Healthy and Sustainable Food for London. The Mayor's Food Strategy, May 2006.'* London: LDA.

McGill, A.E.J. (2006a). Technological Interventions for Global Food Security. In: Kaiser, M., and Lien, M. (eds) Ethics and the Politics of Food, pp. 363–367. Wageningen Academic Publishers, The Netherlands.

McGill, A.E.J. (2006b). Nutrigenomics: A bridge too far, for now? In: Kaiser, M., and Lien, M. (eds) Ethics and the Politics of Food, pp. 330–333. Wageningen Academic Publishers, The Netherlands.

NuGO (2006). 'The nutrigenomics Society, the Outreach of the European Nutrigenomics Organisation – NuGO.' Website: http://www.nugo.org/everyone. [accessed: 01/10/06].

PHGU (Public Health Genetics Unit, Cambridge University, UK). (2005). *'Report of a workshop hosted by The Nuffield Trust and PHGU on 5 February 2004.'* London: The Nuffield Trust.

Rayner, M., & Scarborough, P. (2005). The burden of food related ill health in the UK.J. Epidemiol. *Community Health, 59,* 1054–1057.

Rosegrant, M. W., & Cline, S. A. (2003). Global food Security: Challenges and Policies. *Science, 302,* 1917–1919.

RUAF (Resource Centres on Urban Agriculture & Food Security). (2006). 'What is Urban Agriculture?' Website: http://www.ruaf.org/node/512. [accessed: 04/22/06].

Sanchez, P. A., & Swaminathan, M. S. (2005). Cutting world hunger in half. *Science, 307*, 357–359.

SNE (Society for Nutrition Education) (2004). 'Nutrition and Genes Report. Comments – Society for nutrition Education.' Website: http://www.sne.org. [accessed: 12/20/2005].

The Weekend Australian (2006). 'Jury out on chemicals v. organic battle.' 21–22 January, p. 22. Website: http://www.theaustralian.com.au.

Consumer Trends

CHAPTER 5

Sensory Science and Consumer Behavior

Herbert Stone *and* **Joel Sidel**

Contents

Abstract

Measurement of consumer attitudes and perceptions is a critical part of a product's success in the marketplace. It is generally recognized that sensory information, when combined with attitudinal and imagery measures, has the potential to significantly enhance product appeal and to achieve a greater market presence. Historically, these information sources have been separate and without appreciation for the important link between perceptions, attitudes, and product purchase behavior. This is further complicated by the lack of appreciation for the complexities of the human perceptual system when capturing different kinds of responses. This presentation explores the current practices for selecting consumers for a sensory task vs. those selected for attitudinal tasks; and how their differences in judgments can be integrated to provide a more detailed picture of the marketplace. While most consumers are able to express a preference, a smaller percentage, ~ 65%, can describe their perceptions with sensitivity greater than chance. With this knowledge one can better understand how product characteristics influence, and are influenced by, product imagery, and ultimately consumer purchase and development of brand loyalty. This enables a company to be more effective in developing new products and competing in today's global market.

I. INTRODUCTION

Of the many challenges that food companies face, one of the most important, besides safety, is to develop products that best satisfy consumer expectations. In recent years globalization and industry consolidation have intensified competition for the consumer's purchase choices. While competition is not new, consolidation has created companies with greater resources, making it

Global Issues in Food Science and Technology
ISBN 9780123741240

more difficult for any one company to realize a competitive advantage, at least not by following traditional practices. Globalization also has meant that innovation through technology alone is not enough to provide an advantage. One result of these changes has been the continued high failure rate of new products, leading companies to initiate a wide range of research activities in order to be more competitive, and to re-focus on the consumer to better understand the factors that influence product choice. It is generally agreed that there are many inter-related factors that influence consumer choice. Some can be measured directly, while others require an indirect approach and some are much more difficult to define, let alone, measure. Factors such as habit, impulse, brand loyalty, imagery, preferences, nutritional benefits, price/value ratio, perceived quality, and the actual eating experience all are believed to impact purchase behavior. Some of these, such as brand loyalty and preferences are relatively easy to measure; perceived quality and impulse are more difficult. Consumers, however, present their own set of challenges; for example, using a product in different ways than envisioned by technologists and marketing professionals. This results in the obtained information being subject to misinterpretation. To better anticipate such situations, some researchers are accompanying consumers while shopping as well as videotaping meal preparation in an effort to uncover behaviors not captured through use of traditional marketing research techniques. These activities have received considerable media attention and will certainly add to our understanding of consumer behavior but such activities also have their own set of challenges; for example, overcoming the researchers' biases when interpreting the information. Nonetheless, such techniques expand our knowledge about consumer behavior.

Consumers can readily express their preferences and respond to statements about a product's image, package, and so on; however, when asked to explain the basis for their choice, consumers often repeat advertising information and/or make statements they think the researcher wants to hear. For example, when asked about the vanilla in vanilla ice cream, invariably the responses indicate more vanilla flavor is needed. But when more is added, the same response occurs as if there were no change. With continued additions of as much as 40–50%, the acceptance decreases dramatically. This is a very typical result and should lead the researcher to conclude that, in general, consumers either do not understand the question or are unclear as to how to respond; i.e. is it the quality of the flavor, or is it the amount, or is it a combination of the two? A further degree of complexity is that consumer preferences are not homogeneous and their perceptual skills, about which we will have more to say later, are not the same. This means that a researcher must exercise caution before deciding on which consumers to recruit, what kinds of questions will be asked and how the responses will be captured. In effect, a test is always in the context of a specified population (or sub-populations) of consumers, the products, the information/orientation presented to the consumers at the start of the test, the specific questions and the sequence on the scorecard and not least, the means by which the responses are captured (e.g. scaled). As previously noted, researchers often make assumptions about consumers that are without basis. Assuming consumers will treat the test situation in the same way as the researcher is a common problem. Each test situation has its own set of complexities. Consumers arrive for a test but their minds are often engaged in matters unrelated to the test. An initial orientation is intended to start everyone with the same basic information. The scorecard also has its own impact. As the consumer evaluates a product and responds to various questions, a pattern of behavior develops (in most instances), reflecting what was 'learned' from the previous questions, a kind of hypothesis forming as to the purpose for the test. If the questions are not understood or the consumer is not appropriate for the task, results will be imprecise and the decision-making process seriously compromised. One must always keep in mind that any consumer (qualified or not) will respond to a question (or series of questions), making the process appear deceptively simple, when it is not. This discussion focuses on how these interactive components of behavior can be better managed such that the consumers' responses can be measured with greater confidence leading to more informed product market decisions.

II. CONSUMER BEHAVIOR

Volumes have been written about measuring and describing consumer behavior and it is not our intent to repeat or even attempt to summarize this information (see e.g., Aaker, 1996; Kotler, 1999; Stone and Sidel, 2004). However, it is useful to highlight certain aspects of behavior impacted by the product testing process. In this instance we refer to foods, beverages, and other disposable consumer goods; however, the techniques are applicable to any product with which the consumer has contact (sporting goods, clothing, etc.).

The following list is not intended to be complete; its purpose is to highlight consumer behavior most associated with the product testing process:

1. There is a wide range of sensory sensitivities in any population. For example, sensitivity to low concentrations of acids in foods can vary by as much as 100% or more: this is observed for males, females, young and old, and is independent of geography, education, etc.
2. About 30% of any given population cannot discriminate differences among products they regularly consume.
3. Females are generally more sensitive to differences than males.

4. High-frequency users of a product are more sensitive to differences vs. infrequent users.
5. Brand loyal users are better able to recognize differences among their brands vs. the price- or sale-based users.
6. Demographic criteria such as age, household income, and family size exhibit low correlation with blind test preference differences.
7. Consumers have to learn to use their senses; when asked to respond to specific product characteristics such as sour taste, fruit flavor, etc. it usually requires 4 or more hours for a consumer to learn how to verbalize perceptions. Some consumers are far superior to others with little or no training/education; however, these are a very small segment of the population and not at all representative of the general population.
8. Consumers differ in their food preferences to the extent that unique preference segments can be identified in almost all competitive categories. Preference segments are meant to refer to groups of consumers who satisfy all typical qualifying criteria, yet their responses to coded products indicate clear preferences for a style of product not defined by their stated brand purchasing habits.
9. As the number of scale categories increases from 3 to 5 to 7 to 9, measurement sensitivity increases. When the number of categories increases beyond about 9 or 10, sensitivity decreases.
10. The 3-, 5-, 7- point quality/quantity scales (e.g. poor quality, good quality, etc.) are confusing to consumers and relatively insensitive compared to intensity scales.
11. Just–about–right (JAR) and semantic differential scales are not appropriate for measuring sensory attributes or preferences. They are less sensitive than category and interval scales.
12. The 9-point hedonic scale is the most useful scale for measuring liking/ preference and approximates an equal interval scale.
13. Face scales (Figure 5.1) are culture specific and not useful for cross-cultural studies. Face scales are assumed to be understood by children; however, there is no evidence that this is true, especially among younger children.
14. Graphic rating or line scales are best suited for measuring attributed strength by trained panels. A description of the scale and its use can be found in the literature (Sidel and Stone, 2006).

As noted earlier, consumers are not homogeneous in their preferences when testing products without knowledge of the brand. As shown in Figure 5.2, examination of the aggregate results does not reveal much product differentiation, nor is there any better differentiation in results reported by brand use (not shown here). However, when the same data are

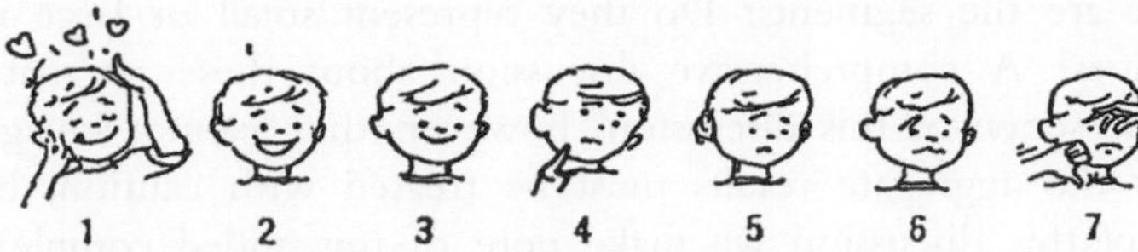

Figure 5.1 An example of a face scale. The task for the younger test participant is to mark the face that best represents how each one feels about the product just sampled. The numerical values usually are not included but are shown here for illustrative purposes.

re-analyzed using cluster analysis, results are much more informative, as shown on the right in Figure 5.1. The preferences are not explained by typical demographic or brand usage information but rather are more likely associated with imagery and lifestyle. These results provide opportunities for new products and a better understanding of the competition; however, they also represent a business challenge. In the latter situation there are the questions whether additional products should be developed to exploit the segments, and if so, will they impact sales of the existing products? And

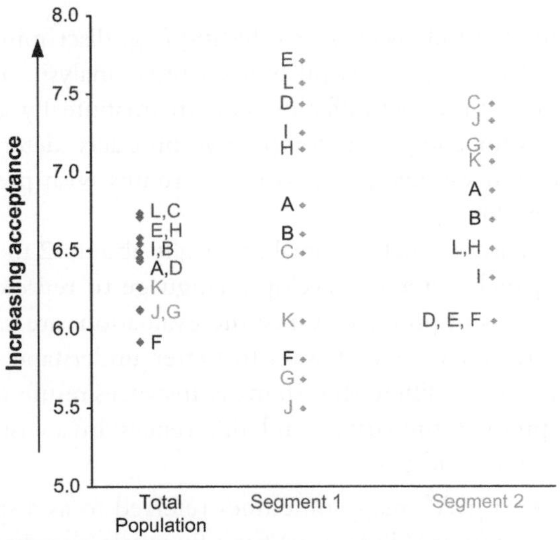

Figure 5.2 Example of preference segmentation. The y-axis is the truncated portion of the 9-point hedonic scale (1 – dislike extremely, 9 – like extremely). The letters are the designations for an array of competitive wines evaluated by 215 qualified consumers who qualified for the test in response to questions about wine consumption, frequency of usage, brand choices, and other criteria. The aggregate results on the left, shown as total population, show limited wine preference differentiation, while the results on the right reflect the results from a cluster analysis of these same responses. See text for additional comments.

how large are the segments? Do they represent small or large market opportunities? A comprehensive discussion about these alternatives is beyond the scope of this discussion; however, they cannot be ignored. Essentially the aggregate results must be treated with caution, but for purposes of this discussion we make note of this added complexity of consumer behavior.

In addition to measuring preferences and related attitudinal information, we also are interested in measuring the perceptions that people experience when sampling a product. We refer to this as sensory evaluation; defined here as 'a science that measures, analyzes and interprets the responses of people to products as perceived by the senses of sight, sound, smell, taste and touch,' and characterized for use as follows:

- small panel procedures, relying on not more than about 25 people in most instances
- analytical tests; participants must be qualified based on sensory skill
- products usually evaluated without brand identification
- repeated trials for analytical tests (difference and descriptive)
- products can be evaluated in appropriate environments (laboratory, at home, at work, etc.).

Of the many methods available for testing (e.g. discrimination and descriptive methods), the most useful is descriptive analysis, a method that provides measures of product differences on an attribute-by-attribute basis, provides a means for comparing competitive products, determines the effects of formulation variables, and correlates results with preferences, etc. (Stone and Sidel, 2004).

Descriptive analysis panels are small, typically about 12 members. Once organized, the panel members develop a language to represent their perceptions for an array of products. Once the evaluations are completed, results can be used in a variety of ways to better understand the extent of product similarities and differences. In many instances results are mapped to communicate product similarities and differences in a visual display, as shown in Figures 5.3 and 5.4.

Figure 5.3 is a QDA® map, sometimes referred to as a spider or radar plot. Each attribute (word/descriptor) has a line scale associated with it and each subject scores each product on multiple occasions, yielding the means (and related statistical measures) that are directly placed on the plot. The attributes are placed equidistant around a center point (none detected) outward to where the product line crosses, representing the mean value for that product. This provides an easily understood display of product similarities and differences, and when combined with knowledge as to the

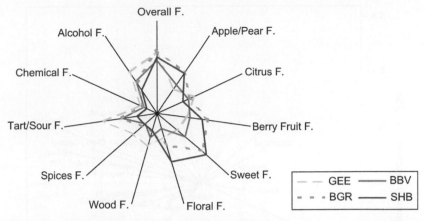

Figure 5.3 A descriptive analysis plot from a QDA® (Quantitative Descriptive Analysis method) test of competitive white wines. The spokes from the center represent the attribute (fruity flavor, floral flavor, etc.); the intensities were obtained by measuring the distance from the center to the place at which the product passes. So, wine labeled SHB was perceived as being significantly sweeter and more floral than wines GEE and BBV, etc. Significance was based on the analysis of variance (not shown here).

formulation and technology of the product, it provides the viewer with insight to that product. An equally if not more important benefit is realized when the sensory results are combined with preferences and attitudinal information obtained from targeted consumers. Figure 5.4 is a plot of all attributes using a Principle Components Analysis (PCA). This map displays the attribute groupings that best define the products that were evaluated. Adding the products provides further useful information. For more discussion about this and related information the reader is referred to Stone and Sidel (2004).

III. A WAY FORWARD

As mentioned at the outset of this discussion, there is a growing appreciation for the complexities of consumer behavior; the importance of connecting a product with consumer expectations based on the imagery as can be conveyed through advertising in addition to other more typical kinds of product information. The challenge has been to create a test situation/protocol in which all these product-related information sources can be incorporated into a testing environment. In a protocol that we developed, an array of products are subjected to sensory, chemical, and physical analyses along with measures of preference, purchase intent, and imagery. The sensory measures are obtained from a trained descriptive analysis panel and the preference and other

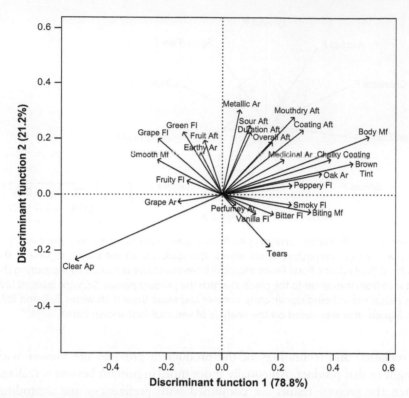

Figure 5.4 A discriminant function map displaying all attributes and their relationships to each other based on the product similarities and differences. When products are superimposed, the map provides another perspective as to which attribute grouping best defines which product. Attributes defining the products are those extending the farthest from the center. For example, the attribute 'clear appearance' was extended in the lower left quadrant, indicating it was a 'defining' attribute across the wines. Attributes closer to one another (e.g. with almost parallel lines) or superimposed on one another are highly related to each other. See text, and Sidel and Stone (2006) for more details about this map.

attitudinal information is obtained from qualified consumers. This approach provides a more in-depth understanding of how attitudes and perceptions interact and this, in turn, leads to a better knowledge of which product formulations are most likely to succeed. In situations where product changes are warranted, it provides a road map for those changes.

Figures 5.5, 5.6 and 5.7 are a series of maps that display some of this type of information. The first, Figure 5.5, depicts ratings of product descriptions vs. a series of concepts.

The concepts were derived from surveys, focus groups, and related sources. For purposes of this discussion, most of the lower-rated concepts/

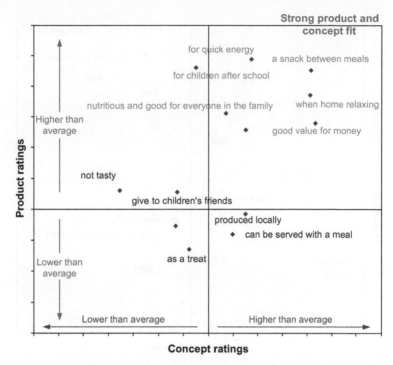

Figure 5.5 A map displaying the responses of consumers to a series of statements prior to product evaluations. Consumers were given a list of product descriptions and a list of concepts, and asked to rate these. Statements located to the upper right quadrant were scored by the consumers as having a higher than average rating on a scale of agreement (agree–disagree). A similar conclusion can be drawn from the product ratings on the y-axis on a scale of liking (1 – dislike extremely and 9 – like extremely). The 'Strong product and concept fit' is in the upper right quadrant and the responses close to this location represent a best fit. See text for more explanation.

benefits have been omitted. The second map, Figure 5.6, includes preference ratings (the y-axis) obtained from qualified consumers served unbranded products.

As shown in the map, the various benefits/concepts were not equally important, and some of them were more important in relation to specific products. In other words, the location of a product in relation to specific benefits is of major importance. If advertising a product promotes certain benefits, it is of importance that the actual eating experience reinforces the benefit. Otherwise the consumer's expectations will not be realized and at some point sales will decline. Figure 5.6 shows the responses to the benefits after sampling the product. This determines the impact of the product tasting on the benefits. Finally, one can now integrate the sensory analysis data within the map. As shown in Figure 5.7, the important sensory

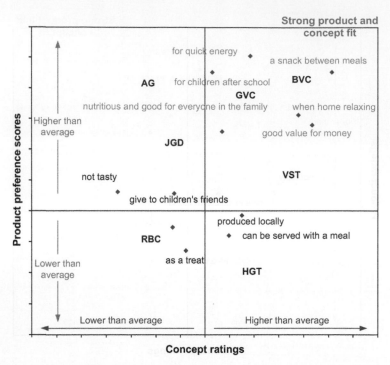

Figure 5.6 The same map as in Figure 5.4, only now the blind preference scores (y-axis, using 9-point hedonic scale: 1 – dislike extremely and 9 – like extremely) are included for some of the products (identified with 3-letter abbreviations). After evaluating a product each consumer responds to a series of statements/concepts using an agree/disagree scale. This provides a basis for deciding which products best match specific statements; equally important is the location of each product relative to others and the statements, and whether these statements have changed after actual product evaluation.

attributes are identified through regression. For details about procedures used see Sidel and Stone (2006).

Not only does this help to define the attributes that are most important, but it also provides a road map for the development effort. In this way, one can better understand what formulations best match a specific market strategy and population segment. The immediate impact of this knowledge is an increase in products that better match consumer expectations. This approach to reaching more informed product decisions is possible through advances in measuring the sensory characteristics of products and through the integration of imagery and related attitudinal measures with more traditional preference testing models. While these efforts represent a major step forward in understanding how and why products succeed, they cannot

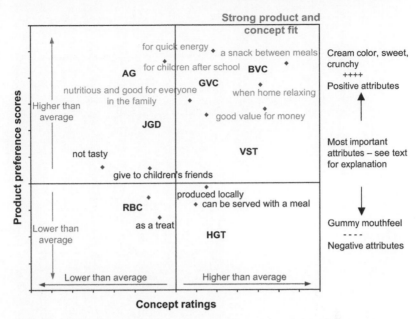

Figure 5.7 This map repeats the information reported in Figure 5.5, but now includes the important positive and negative sensory attributes (placed outside the map for ease of reading). The important attributes were determined from regression analysis. For details on the methodology, see Sidel and Stone (2006). The two maps integrate concepts before and after product evaluation, and preferences. Other information that could be mapped includes purchase intent and advertising, among other. See text for more details.

guarantee success, since that would require decision-makers who understand the value of these approaches and are able to act on the information given.

REFERENCES

Aaker, D. A. (1996). *Building Strong Brands*. New York: The Free Press.

Kotler, P. (1999). *Kotler on Marketing*. New York: The Free Press.

Sidel, J. L., & Stone, H. (2006). Sensory Science: Methodology. In Y. H. Hui (Ed.), *Handbook of Food Science, Technology, and Engineering* (pp. 57, 1–24). Boca Raton, FL: CRC Press, Taylor & Francis Group.

Smith, P. G., & Reinertsen, D. G. (1997). Developing Products in Half the Time: New Rules. *New Tools* (2nd Edn).. New York: Van Nostrand Reinhold.

Stone, H., & Sidel, J. L. (2004). *Sensory Evaluation Practices* (3rd Edn).. San Diego, CA: Elsevier Academic Press.

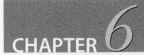

CHAPTER 6

Designing Foods for Sensory Pleasure

Mark Kerslake

Contents

Abstract

The arrival of industrialized food brought an increase in scientific understanding, new inventions, and great product development. In the 1970s and early 1980s, when techniques measuring and quantifying the hedonic or pleasurable aspects of food were first used on a wide scale, it was easy to find products with large disparities in consumer appreciation, but nowadays it is much less likely. Companies today have scientists working hard to formulate good products for their consumers, and whether large or small, these companies are doing consumer research of one kind or another. As a result many companies can now identify (at the same time) an idea that will satisfy consumer expectations. The challenge is no longer to identify an idea that will deliver business, but rather to be the first to develop this idea, and to deliver the best product fit that meets concept expectations. Food scientists often approach formulation of products in the same way as those in other sciences, with a good dose of creativity but mainly with lots of intuition, i.e. with 'the art' gained from years of experience in their profession. It is often overlooked that today's science has the capability to help better design this art. Through discussion of product examples designed, the key approach now is to exploit product opportunities by identifying the levers for formulation that could potentially turn a consumer *on* or *off*. It is no longer enough to be satisfied with the current way products are developed. It is necessary to re-think preference mapping approaches and to deploy more

Global Issues in Food Science and Technology
© 2009 Elsevier Inc.

ISBN 9780123741240

experimental designs, as well as to discover how these evolving techniques can be used to design foods that deliver optimized consumer preference.

I. INTRODUCTION

In antiquity, survival was the primal need and man was busy chasing protein to provide sustenance to fight off hunger, the cold, aggressors, etc. As basic needs were met more time was spent in the pursuit of pleasurable experiences often associated with food. Under Louis XIV, *Le Potager du roi at Versailles*, the king's vegetable garden, was established as the leading innovative food-producing institution of its time. It was capable of producing pineapples and figs, while *les petits pois* (garden peas) became the new food of the 17th century eagerly awaited each summer. However, the arrival of industrialized food in the late 1800s and early 1900s led to the establishment of completely different values.

First, consistency of product became important with natural foods, like cheese for example, being selected and sold against a certain quality standard using the retailer's name for authority (e.g. J.L. Kraft in Chicago and Fred Walker in Australia, both in 1903); as product trials generated repeat purchases from satisfied customers, 'retailer names' became valuable 'brand names'. Secondly, advances in science and technology led to development of new ways of processing existing foods (e.g. flaked cereal in 1906 by co-inventor W.H. Kellogg) or completely new raw ingredients and foods that had not been seen before (e.g. margarine in 1878, first commercially produced by Unilever). The era also saw the birth of many of the great food companies known today and the great brands they sell. This industrialization was first led by increased scientific understanding, inventions, and great product development, which brought with it the development of our profession, Food Science and Technology. Excellent examples can be seen in the scientific approaches of these great food companies across the world.

As time passed it was realized that development of ways to measure and quantify the hedonic or pleasurable aspects of food was needed, and the sensorial sciences developed slowly to meet this need. Several important milestones in this domain laid down the fundamental tools used today and some should be mentioned, such as the nine-point hedonic scale for consumer assessment (Jones *et al.*, 1955; Peyram *et al.*, 1957), the graphic line scale (Anderson, 1970), and the development of quantitative descriptive sensory analysis (Stone and Sidel, 1974). Initially this work often involved calculations done by hand, or as later done, with mainframe computers and programming in FORTRAN on

Table 6.1 Range and standard deviation of multiproduct consumer test of confectionery in the early 1990s

	Products	Base N	Min	Max	Mean	Range	Std dev
Test 1	20	200	3.11	6.65	6.32	3.54	2.04

cards or tickertape. In the late 1970s and early 1980s when consumer evaluation techniques were first used on a wider scale, due to the impact of wider computer availability, it was easy to find products with large disparities in consumer appreciation, as in the above example (Table 6.1) where the range in mean scores on a nine-point scale is 3.54.

Today researchers are much less likely to find products that are actually rejected (i.e. intensely disliked) by consumers, and much more likely to find products that are very well accepted (Table 6.2). In this second example one observes that the range in mean scores is smaller at 1.51, as is the observed standard deviation of responses compared to the first test. This evolution is not surprising because every serious food company has food scientists that are working hard to formulate good products for consumers; every company, large and small, is doing consumer research of one kind or another wherein many companies can identify at the same time (in competition) an idea that will satisfy consumer expectations.

The result of this activity means that the rate of new product launches is increasing and the speed at which products reach market is increasing. Therefore it is becoming much less likely to easily find a product that pleases a wide audience in an environment where product development experience and consumer understanding are rapidly spreading across all companies. As the competition has become fierce, the challenge is no longer to identify an idea that will generate profit, but rather to be the first in its development and to deliver the best product fit to meet concept expectations, thus securing a lasting competitive advantage.

Table 6.2 Range and standard deviation of multiproduct consumer test of confectionery in 2006

	Products	Base N	Min	Max	Mean	Range	Std dev
Test 2	14	200	5.77	7.28	6.43	1.51	1.89

II. THE PERSONAL TOUCH

Designing a food does not mean just making a product that is healthy and good to eat, but making one that actually looks healthy and good to eat, and therefore more appetizing. In parts of today's Western society food is often elevated to an 'art form' and cooking is considered an artistic activity, rather than as only a necessity to satisfy hunger and survival for another day. Our observation over the years is that many food scientists often approach formulation in the same way, with some science and a good dose of creativity, but mainly with lots of intuition, 'the art' knowledge gained from years of experience in their profession. This knowledge is based on trial and error, but often cannot be explained by the user in a formal scientific way, as the key underlying principles are not understood well enough to know why these 'tricks of the trade' work in practice.

Sometimes we observe an attachment to 'the art' in product formulation that would do more justice to a French impressionist painter, than to constructing product formulation on a solid scientific foundation. It should not be overlooked that today's food scientists have the capability of designing better food by harnessing in a positive manner, using scientific principles, the creativity or intuition ('the art') built up over the years from practical experience without denigrating or rejecting a personal approach in artistically designing food. It is hoped this chapter will convince readers of that possibility.

III. THE EXPERIMENTAL APPROACH

In the 21st century we need to look for openings in the market more thoroughly and opportunistically than ever before. The foundation of this approach is that consumer satisfaction begins with product experience. Advertising or image-building can drive a consumer's first purchase; however there is no compensation for an unsatisfactory sensory or functional experience with a product. Consumer satisfaction is the key factor, and a pleasurable sensory experience is critical for repurchase of food products, without which there is no chance to build a business before delisting occurs.

Several product examples will be discussed later. The key now to exploiting product opportunities is to identify the levers for formulation that turn a consumer on and off. Some food scientists may think, 'Yes we know, we do that,' which may be the case, but this response attracts some skepticism as that is not what we observe in general with clients of Numsight (a marketing consulting service) large and small. Even with large multinational clients, we observe less often than one would expect the application of scientific principles to product design in the field due to time and cost constraints. Time to market often means cutting corners and unfortunately in product design this is precisely the corner that should not be cut, but simply used more intelligently. It is therefore no longer enough to be satisfied by the current way products are often developed.

IV. WHAT IS THE POTENTIAL POSITIVE IMPACT?

Looking at a preference mapping approach, we can see how it is necessary to re-think the philosophy of that approach and deploy more experimental designs, and along the way, discover how these evolving techniques can be used to design foods that deliver optimized consumer preference. The classical preference mapping approach is to take products from the market, put them into a consumer and sensory test, and relate the data sets statistically to derive what the drivers are for consumer preference and describe the ideal product. The product set is chosen from leading products in the market and a competitive set; after selection it looks like the product set in Figure 6.1.

After looking at the distribution of products across the sensory space represented, one might ask, would you go fishing with that many holes in your fishing net – if you were serious about catching fish? This initial preference map of products is actually quite unbalanced in the way the nature of the sensory space is represented (Figure 6.2), as there are a large number of unoccupied spaces, making it difficult to map the consumer response correctly to the sensory areas in the empty spaces. It certainly looks like a lot of fish get away with that approach.

What if we tried something different, such as the proposition presented in Figure 6.3? It is not perfect but we have reduced our chances of missing the fish by 100% simply by tasting beforehand, and then selecting and

Product a	Product d		
Product b		Product f	Product h
	Product e	Product g	Product i
			Product j
Product c			Product k

Figure 6.1 Initial schematic of a preference map product set, chosen from leading products in the market.

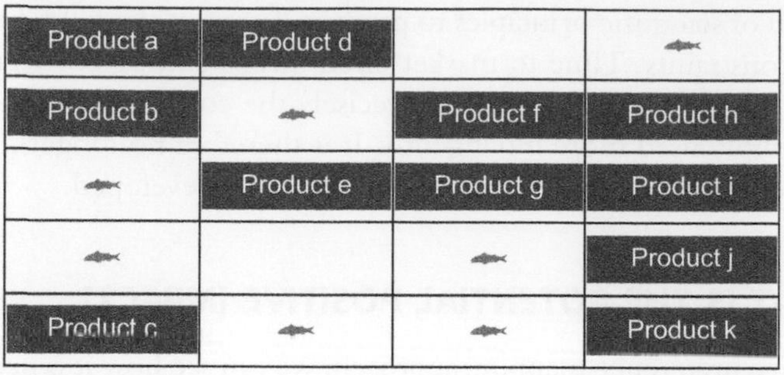

Product a	Product d		🐟
Product b	🐟	Product f	Product h
🐟	Product e	Product g	Product i
🐟		🐟	Product j
Product c	🐟	🐟	Product k

Figure 6.2 Initial schematic of a preference map product set, chosen from leading products in the market, illustrating the unbalanced nature of the sensory space represented.

placing the same number of products in an ordered manner across the observed sensory space. Only the approach has changed, simply by carefully spacing the products chosen across the sensory space.

Here is a real-life example from the beverage industry of a sensory map with and without the tasting effort. The effort needed to construct such a data set was substantial, as it required a small team of five very experienced people spending three sessions over 3 days tasting more than 40 beverages and being asked to choose the most appropriate ones spanning all the sensory directions identified in this category of products. However the quality of the outcome is certainly worth the extra effort. The original sensory space (Figure 6.4), without pre-selection of the products, is quite compressed with a few extreme outlying products that distort the space.

The revised sensory space is illustrated in Figure 6.5, after selection and removal of four products (#s 9, 22, 24, and 25). The tasting effort produced a set of samples that are much better balanced in sensory terms. In this case the products removed were those showing only specific characteristics already known to be disliked by consumers or represented by other products

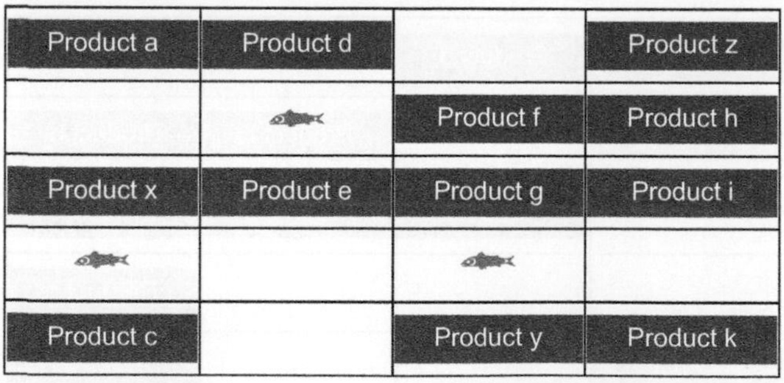

Product a	Product d		Product z
	🐟	Product f	Product h
Product x	Product e	Product g	Product i
🐟		🐟	
Product c		Product y	Product k

Figure 6.3 Revised schematic of a preference map product set, where products chosen from leading products in the market were supplemented in a more organized manner.

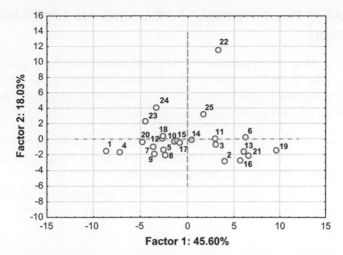

Figure 6.4 Initial preference map product set, where products were chosen from leading products in the market.

at lower levels. The consequent alignment of sensory attributes along the factors was much more coherent and certainly made the data analysis easier to complete.

If this relatively simple approach can have such a positive impact on a complex preference mapping study then it seems sensible to think about applying an organized scientific approach to more product development applications.

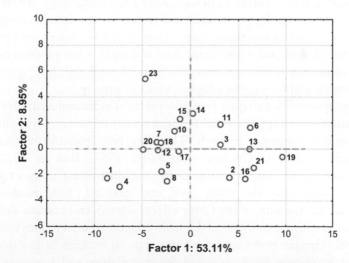

Figure 6.5 Revised preference map product set, where products chosen from leading products in the market were supplemented or removed in a more organized manner.

V. WHAT IS THE RATIONALE FOR USING AN EXPERIMENTAL DESIGN?

Before discussing this approach in detail, we should step back and ask the question: why should we use an experimental design in the first place? Scientists at Numsight, for example, believe that an experimental design is invaluable in answering efficiently three questions:

- Which product parameters are most influential and over what range?
- How can we find the optimum settings taking into account conflicting demands of different responses?
- Once the optimum is found, how can we guarantee it is robust, or do we need to change the specifications to achieve robustness?

VI. THE KEY PRINCIPLES OF EXPERIMENTAL DESIGN

The first question to ask is: what are the project objectives and how does that translate into a strategy for product development? The experimental design technique is best used very early in the sub-projects, when the product development team has to deal with multiple product parameters and has limited time to do so. The approach is to carefully evaluate any existing data (e.g. technical, sensory, or consumer data) that may exist and to extract winning assumptions for development of an enriched product brief. At this point one can evaluate what type of experiment design, if any, is appropriate. There is usually a choice, either a screening stage, where there is a large set of variables to evaluate and screen to sort out those with the most impact, or an optimization stage, where variables chosen are varied in more detail. A designed experiment does not replace the product developer's expertise, but builds on it to organize their knowledge, leveraging it for faster progress and to achieve a higher value in time spent.

The purpose of this chapter is to give practical industrial examples of the experimental design approach and to more widely encourage its application in sensory and consumer research. The basic principles can be found in many textbooks (e.g. Cochran and Cox, 1957; Box et al., 1978; Montgomery, 2000). Implementing those principles means using a rigorous logic to ensure that the plans used are balanced, fully randomized, orthogonal (i.e. the sum of any factor effects is zero), and contain sufficient degrees of freedom to estimate all the main effects and interactions built into the model. This last point is of great importance, because in practice, one often has to choose between using fewer variables and having a more complete design that will estimate most interactions, or using more variables but having a less complete design that will estimate fewer interactions. This is where the product developer's expertise and careful upfront analysis to determine winning assumptions for an enriched product brief are critical.

Take for example the planning choices in the following design; five main variables were desired:

- Raw material base (3 qualitative levels – categorical variable)
- Flavor A (3 concentrations of one flavor style A – continuous variable)
- Flavor B (3 concentrations of one flavor style B – continuous variable)
- Sweetener level (2 levels – continuous variable)
- Ratio of sweetener types (2 levels – continuous variable).

In this example there are 108 possible experiments in all, but if only 15 experiments could be done, there would still be several choices of design. The design takes into account all variable main effects and allows examination of some three-level interactions, which seems quite good. Note though, one would not be able to look at the interactions for all three variables with three levels, as there are not enough degrees of freedom with the 15 experimental products to be made. For estimating the main effects of the five variables we have already used up 8 degrees of freedom available, 2 for the three-level categorical variable, 1 for each three-level continuous variable, 1 for each two-level variable, 1 for global mean, and 1 for error estimation. This leaves only 7 degrees of freedom available. The linear interaction of the sweetener level and ratio takes 1 more, leaving 6 degrees of freedom. So to examine the other interactions, one has to choose carefully, as each three-level variable x three-level variable interaction requires 4 degrees of freedom and each two-level variable x three-level variable interaction requires 2 degrees of freedom. Clearly there are a number of options to choose, the choice being made according to the product developer's expertise and careful upfront analysis to determine winning assumptions.

VII. A SNACK FOOD EXAMPLE

Another real-life snack example illustrates the effectiveness of an experimental design to more product development applications. One client of Numsight came to us after working on a big business opportunity in snack products; they had been trying to win this account (worth some millions of Euros annually) for more than 12 months, during which time years of product development resources had been exhausted with limited progress towards the goal. When we entered the project a recipe had been identified that had some promise, however the product was not crunchy enough and had several mouthfeel defects. The client was having particular difficulty because the type of snack product involved was on the periphery of the type of product they usually developed and required different technology and know-how to construct.

Table 6.3 Probability of a significant effect for each variable on the measured sensory response in a snack example

Variable	Hardness	Crunchiness	Gritty
Flour type	5.1%	24.0%	57.9%
Raising agent	0.6%	0.2%	17.5%
Starch source	4.1%	4.7%	60.5%
Sugar type	51.1%	24.0%	57.9%
Raising agent (starch source)	22.7%	5.9%	78.7%

Such a case is ideally suited to an experimental design approach. The problem is clear and the potential recipe variables are known even if the effect of those variables on sensory attributes is not. We first set out to harness in a positive manner the existing knowledge or intuition built up over years of practical experience, and tried to set a priority importance on variables that might or might not have a recipe effect together with the technology team. In the first round of experimentation it was decided to examine three components of the recipe (flour type, raising agent, starch source, and sugar type) and their interactions. This gave a design with nine experimental products to examine the main effects of the three design components and 16 products in total, and in addition, the first order interactions. The results were surprising. Of the four recipe components considered 'important,' only two were very important, as was the interaction between them, which had not been considered the case by the development team so far. This was good news as it increased the pool of known knowledge. The probability of each effect having a significant effect on the measured sensory response results is detailed in Table 6.3. It can be clearly seen that the raising agent, starch source, and the interaction of these two components have a strong effect on crunchiness.

VIII. A CONFECTIONERY EXAMPLE

Another case design is based on a real-life confectionery example. Again the problem is very clear; the potential recipe variables were known even if the effect of those variables on sensory attributes was not. We again first set out to list the existing variables that extensive knowledge built up over years of practical experience suggested would have a recipe effect with the technology team. This led to a list similar to that in Table 6.4 (list has been

Table 6.4 Potential recipe variables in a confectionery example

Process variables	Number of levels
Variable 1 – Milk	3
Variable 2 – Cocoa	3
Variable 3 – Lactose	3
Variable 4 – Flavor added	2
Variable 5 – Mixing time	3
Variable 6 – Cocoa type	2
Variable 7 – Milk powder type	2
Total possible combinations	648

modified for publication due to confidentiality reasons) and the identification of the levels of each variable that could be sensibly altered. A quick calculation led to a total of 648 possible combinations! Clearly no one could make that many experiments nor could one imagine making maybe more than 20 experimental products for a consumer evaluation.

So what should one do in this difficult situation? The answer is to simply mix some scientific rigor with a good dose of common sense to see how it might be possible to cut the problem into more manageable pieces. From the variables under study it was clear that the first three have a clear impact on flavor and it would be very unwise not to consider all the possible interactions of these three ingredients with each other. While it would be ideal not to separate the other variables, it was clearly necessary to make some decisions, so it was decided to treat them apart in two additional add-on designs that would provide supplementary information on linear effects. If the contribution of those factors seemed important, then the strategy would be to reintegrate them in a secondary step later on with the mix design variables. An alternative design strategy could have been to use response surface designs and to combine all variables at once; however the biggest practical drawback to this strategy is that it often makes the control of what actually takes place in the pilot plant more difficult, as more variables have to be controlled at once, which is not always straightforward and can lead to a level of complexity that induces errors in fabrication of the experimental samples.

In our case we chose to first construct a mix design to measure the impact of changing milk, cocoa, and lactose content in the recipe. It is then possible to estimate what happens for any mixture of these three ingredients in the recipe area tested. Secondly, two small supplementary add-on designs were then constructed. The purpose of the first supplementary design was to measure the linear impact of flavor added and mixing time with the previous mix design center point; whereas the second add-on design measures the interaction of cocoa type and milk powder type on the recipe and their interactions with the previous mix design center point.

The mix design is a simple three-parameter design type in which the sum total is 100%, so it is very useful for recipe systems where one fixes part of the recipe and works on varying the other proportion. In this case we fixed the $1 - Y\%$ of the recipe and worked on the other $Y\%$, which was varied in ranges as follows:

- Milk level up to 5% more than standard
- Cocoa level 3% more or less than standard
- Lactose up to 6% more than standard (Figure 6.6).

The design space constructed looks like that in Figure 6.6. Secondly, two small supplementary add-on designs were then constructed around the mix design center point as indicated in Table 6.5. The purpose of Design 1 is to measure the linear impact and interactions of cocoa type and milk powder type with the previous mix design center point. The purpose of Design 2 is to measure the linear impact of flavor added and mixing time with the previous mix design center point.

These designs are a good compromise between what is practical and what is theoretically possible. The information is gained in a systematic way and developed in the context of the surrounding designs, and the outcomes can be used for the following stages of the project. The results of the mix design are illustrated in Table 6.6; cocoa solids and milk have an impact on the market sample, and on each cluster, with interaction between the two in some clusters. Lactose has virtually no impact. Cocoa solids have a strong influence on Cluster 1, with a limiting level apparent.

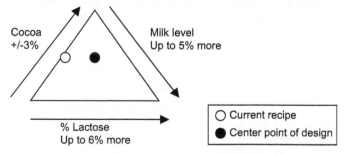

Figure 6.6 Mix design to measure the impact of changing milk, cocoa, and lactose content in the recipe.

Table 6.5 Two small supplementary add-on designs constructed around the mix design center point

Design 1				
Cocoa type – Std or New	Center point recipe	Standard	New	
Milk powder type – Whole or Skim	Center point recipe	Whole	Skim	
Design 2				
Flavor added – Std or New	Center point recipe	Standard	New	
Mixing time – Std, + 10% or 20%	Center point recipe	Standard	+ 10%	+ 20%

Table 6.6 Impact of changing milk, cocoa, and lactose content in recipe from mix design across total sample and different consumer segments

Overall impression	Market sample	Cluster 1	Cluster 2	Cluster 3
Milk	+	+/−		+
Lactose				+
Cocoa milk	+	+		+
Cocoa solids	+	+	+	+
Cocoa solids quadratic	−	−		
Purchase intent after trial	**Market sample**	**Cluster 1**	**Cluster 2**	**Cluster 3**
Milk	+	+/−	+	+
Lactose	+	+		+
Cocoa milk				+
Cocoa solids	+	+	+	+
Cocoa solids quadratic	−	−		

Table 6.7 Impact of changing cocoa type and milk powder type from the first add-on design, as shown across total sample and different consumer segments

Overall impression	Market sample	Cluster 1	Cluster 2	Cluster 3
Cocoa type		+		
Milk powder type				
Purchase intent after trial	**Market sample**	**Cluster 1**	**Cluster 2**	**Cluster 3**
Cocoa type		+		
Milk powder type				

The first add-on design with cocoa type and milk powder type showed the following results (Table 6.7): cocoa type has an influence on Cluster 1, with no impact apparent on milk powder type.

From these results precise recipes have been calculated and are ready for further development; if they pass implementation benchmark hurdles they will be launched.

To close, the key benefits of using an experimental design are:

• To determine with precision which product parameters are most influential and over what range.
• To find the optimum settings taking into account conflicting demands of different response.
• To guarantee once the optimum is found that the design is robust, or if the specifications need to be changed to achieve robustness.

REFERENCES

Anderson, N. H. (1970). Functional measurement and psychological judgment. *Psychol. Rev.*, 77, 153–170.

Box, G. E. P., Hunter, W. G., & Hunter, S. J. (1978). *Statistics for Experimenters*. New York: John Wiley and Sons, Inc.

Cochran, W. G., & Cox, G. M. (1957). *Experimental Designs*. New York: John Wiley and Sons, Inc.

Jones, L. V., Peyram, D. R., & Thurstone, L. L. (1955). Development of a scale for measuring soldier's food preferences. *Food Res.*, 20, 512–520.

Montgomery, D. C. (2000). *Design and Analysis of Experiments*. New York: John Wiley and Sons, Inc.

Peyram, D. R., & Pilgrim, F. J. (1957). Hedonic scale method of measuring food preferences. *Food Technol.*, 11(9), 9–14.

Stone, H., Sidel, J. L., Oliver, S., Woolsey, A., & Singleton, R. C. (1974). Sensory evaluation by quantitative descriptive analysis. *Food Technol.*, 28(1), 24–34.

CHAPTER 7

The Influence of Eating Habits on Preferences Towards Innovative Food Products

Kai Sparke *and* **Klaus Menrad**

Contents

Abstract

The food industry in European countries is a sector with comparably low research and development efforts. Many companies belong to small and medium-sized enterprises and have limited resources. Subsequently, many new food products fail in the market. Integrating consumer research into product development enhances the success rate of products. Consumer segmentation is a suitable approach for effective target group identification. The aim of this project was to develop and test a survey tool that determines the consumer's food consumption style in a simple and efficient way. Using lifestyle and imagery research, visual stimuli for different food products and meals were elaborated. In a consumer survey, 330 people assessed the stimuli and also rated different product cards for two innovative food products. Cluster analysis resulted in ten different consumer segments, which could be well sourced with socio-demographic characteristics. Analysis of the segmentation tool bore good discriminatory power. Additionally, the consumers' preferences towards food product features were examined with conjoint analysis. Differences in preferences could be observed for diverse food consumption style clusters and were the basis for identifying the target group's specific food product design.

Global Issues in Food Science and Technology
© 2009 Elsevier Inc.

ISBN 9780123741240
All rights reserved.

I. INTRODUCTION

The food industry in Germany and elsewhere in Europe is a sector with comparably low research and development (R&D) efforts. On the input side of the German food industry, expenditures for R&D were 0.2% of the total turnover in 2003. Taken as a whole, the processing industry spends more than 4% of the total turnover for R&D, which is more than ten times higher. In the entire processing industry around 4.5% of its employees worked in R&D in 2003, while the analog figure reached only 0.5% in the food industry (Stifterverband fuer die Deutsche Wissenschaft, 2005; Federal Statistical Office, 2006).

Concerning the output side of food innovations there is also significant room for improvement. Since the beginning of this decade three out of four new food products fail in the German food and retail market within one year after launching. Either these products are listed as out of supply or they generate insufficient sales (Menrad, 2004; Rosada, 2005). One reason for this situation can be found in the food sector's structure. In 2003 about 76% of all German food processing companies had less than 100 employees. Only 2.5% of the companies have more than 500 employees (Federal Ministry of Food, Agriculture, and Consumer Protection, 2004). Thus the majority of enterprises in the food sector belong to the small and medium-sized enterprises (SMEs). Long-lasting research and substantiated new food product development seem to be too expensive and labor-intensive for these companies.

According to Grunert et al. (1996) a strong market orientation is crucial for successful new product development. Schmalen (2005) identified target group market research as one of the few key factors for successful innovation policy in the food industry. Prior active consumer research, including analysis of customer desire, trends, and niches in the market, boosts the likelihood of a new food product's success. Earle et al. (2001) mention the different stages of product development process in which consumer research should be integrated. In the beginning small consumer focus groups evaluate the prototypes, whereas sizeable consumer surveys that include product acceptance tests are conducted immediately before a commercial product launch. Linnemann et al. (1999) propose an integral model of food product innovation, which includes steps like analysis of market development, categorization of consumers regarding preferences and perceptions, and development of adequate product assortments for several consumer segments. Thus consumer segmentation seems to be a suitable approach for successful target group identification during new product development processes.

Socio-demographic attributes of consumers are often used for segmentation because they fulfill many segmentation criterion requirements, such as simple ascertainability and measurability. However a major disadvantage is their limited relevance in analyzing consumer behavior (Meffert, 2000). Loudon and Della Bitta (1993) point out that the socio-demographic characteristics are less and less suitable as determinants for consumer behavior, whereas lifestyle and psychologically oriented approaches offer promising opportunities regarding segmentation. The segmentation procedure of the SINUS market research institute (Flaig *et al.*, 1993) has gained importance and wide implementation in marketing in Germany and other European countries. SINUS divides a population into lifestyle milieus based on people's statements about products or different aspects of their lives, e.g. preferred products or leisure time activities. Brunsoe *et al.* (1996) conducted consumer segmentation studies in the four European countries Denmark, France, Germany, and the United Kingdom concerning a food-related lifestyle. Furthermore, Stiess and Hayn (2005) worked out a representative classification of the nutrition styles of the German population.

The aim of a research project recently conducted at the University of Applied Sciences Weihenstephan (Sparke, 2008) was to develop a survey tool that classifies consumer segments according to their eating habits or food consumption style, respectively, and thus enables identification of consumer target groups that bear high acceptance towards new food products. The tool should be easy to handle with regard to possible application to SMEs in the food industry.

II. METHODOLOGY

II.A. Consumer segmentation with regard to eating habits

Development of the main features of the segmentation tool was geared towards the results of Brunsoe *et al.* (1996), and Stiess and Hayn (2005). Brunsoe *et al.* (1996) concentrated on the cognitive components of human behavior by combining them with several dimensions of nutrition when developing their instrument for a food-related lifestyle. Ways of shopping, cooking methods, quality aspects, consumption situations, and buying motives were converted into statements in a standardized questionnaire for oral interviews. Five clusters of food consumers were obtained in Germany: 'uninvolved' (21%), 'careless' (11%), 'rational' (26%), 'conservative' (18%), and 'adventurous' (24%). Stiess and Hayn (2005) used people's purchasing and quality orientation, cooking orientation, overall nutrition orientation, and socio-demographic information for consumer segmentation. Seven nutrition style clusters were obtained: 'uninterested fast-fooder' (12%), 'cheap- and meat-eaters' (13%), 'joyless habitual-cooks' (17%), 'ambitious fitness-oriented'

(9%), 'stressed-out daily life-managers' (16%), 'sophisticated nutrition-conscious' (13%), and 'conventional health-oriented' (20%).

Keeping in mind the tool's usability for consumer research of SMEs in our research project, attention was turned to the consumer's affinity towards food itself, thereby excluding the issues of positioning foods in the context of human nutrition. Dimensions such as how to prepare a meal or consumer purchasing behavior lose their importance, compared to the consumers' direct attitudes towards certain food products. For the actors on the supply side (e.g. the food industry and catering services) all dimensions mentioned above finally concretize in consumer or user acceptance or their rejection of offered products at the point of sale or point of consumption, respectively.

The segments investigated by Brunsoe et al. (1996) and Stiess and Hayn (2005) contain the demand side in its entirety, and also the SINUS-milieus (Flaig et al., 1993) with their attitude indications in the field of food. These people are the demanders of all food products and meals offered on the market. Thus, the spectrum of food-related lifestyles can be regarded as congruent with the spectrum of possible consumption situations and consumed products.

The developed approach to segment consumers may be called a 'food consumption style.' Operationalization of this food consumption style was not carried out by abstract statements about eating habits but by concrete food products in order to diminish the survey efforts. Choice of these food products was guided by characteristics of the above-mentioned consumer segments of food-related lifestyle. By means of literature review (e.g. cookbooks and food magazines), group discussion, and creative techniques typical food products and meals were compiled for all of those segments. They should not just contain a 'basic product,' but also symbolize potential consumption situations and represent trends in nutrition like ethno-food, convenience, organic production, functional food, or regional food specialties. Finally, 13 food products and meals were chosen to establish the instrument needed to investigate the consumer's food consumption style (see Table 7.3).

Methods of imagery research were used to design the survey tool. Imagery is regarded as the quasi-sensory experiences a person is consciously aware of and that can exist even in the absence of stimulus conditions that produce genuine sensory experience. Images are regarded as some kind of tool of thought that provides a temporary representation of memories, and thus can be used in a functional way (Childers et al., 1983). Imagery may be presented and processed verbally or non-verbally. The latter can be divided into visual, auditory, gustatory, olfactory, or tactile imagery of which visual imagery is most important in marketing applications.

To determine the consumer's food consumption style in a simple, effective, and valid way, a well-founded communication approach is essential at the same time, as shown by Holbrook (1983) in an imagery communication model. A sender transmits a message through a channel to a receiver and activates an effect. Reduced to the aspects most interesting for developing a food consumption lifestyle tool, the approach can be described as follows: the researcher presents an object to the consumer using imagery techniques and annotates the caused reaction. The object, which can be a food product or meal, stimulates both a cognitive and an affective pre-occupation with the food and a subsequent judgment of the involved interviewees. Food is a product that contains utilitarian as well as emotional components. Nutritional value or requirements for preparation or consumption may be considered in a cognitive way, whereas taste, pleasure, and certain food products (as a status symbol) pertain to the affective aspects.

Presenting an object in a visual manner offers diverse advantages. Childers et al. (1983) cite the greater variability in the appearance of pictures compared to the appearance of words. Holbrook (1982) states that one might investigate the symbolic, hedonic, emotional, and esthetic components of an object better with non-verbal imagery methods. Finally, Childers et al. (1983) emphasize the possibility of improving the effectiveness of marketing communications using imagery. One advantage with respect to practical procedure in marketing research is that images can be understood easier by respondents who demonstrate an inferior ability upon exposure to verbal presentations, e.g. by children or foreigners. Additionally, differences exist among domestic adult consumers instructed to imagine an object based on a verbal description (Rossiter, 1982).

Operationalization of food products and meals into visual imagery stimuli was conducted according to Rossiter (1982), who gives suggestions for application of imagery techniques in marketing. Pictures should contain objects as concrete as possible and with high imagery content. Color enhances the suitability to effect emotional reaction. Pictures were taken from databases to illustrate all 13 different food products and meals exactly, attractively, and in an equal way. Pictures were presented in a mixed order to the respondents, who were asked to assess the attractiveness of the illustrated meals and food products on a rating scale.

II.B. Preference testing of innovative food products

When dealing with food products still in the developmental stage and not yet launched in the market, the consumer's reaction cannot be measured based on consecutive purchase behavior. For consumer goods not yet available in the marketplace, preferences are regarded as an important factor in the decision-making process of potential consumers (Kotler et al., 2003). To make consumers prefer one product alternative against another, the alternative has to offer some benefits to the consumer that cannot be copied by the competing product. In this sense, consumer preferences reveal benefits or utilities. Utility is a crucial criterion for every rational decision-making process of consumers (Krelle, 1968). Taking into account the financial restrictions of consumers and their individual necessity structure, consumers try to maximize the benefit of purchasing products. Thus, investigating and analyzing the consumers' preferences, and the extent to which products contribute to their preference building/shaping, is a suitable way to explain consumer behavior towards products in the development phase.

Two different products were studied to analyze consumer preferences. Both were combined with the consumer's food consumption style; one product was a dried fruit snack and the other an assortment of chocolates. Tests of both products were in co-operation with SMEs located in southern Germany. The novelty character of the dried fruit snack in this case refers to the drying process that the fruit underwent via an innovative microwave technique. This technology resulted in fruit pieces with a crispy consistency and thus enabled a new snack experience by consumers, while there were no deviations in the original fruit taste (Heindl, 2003). Furthermore, this kind of snack served the market's mega trend in health and wellbeing (Heimig, 2005). In recent years, sales of candy, in particular new chocolate products, increased in Germany especially among children – an important consumer group. Within this development there was also a trend towards high-quality chocolate bars and assortments (Duerr, 2004).

Conjoint analysis was used to obtain consumer preference towards these products. Such analysis was introduced to marketing in the early 1970s (Green and Srinavasan, 1978) and is considered a suitable method for assessment of product concepts regarding the needs of a consumer target segment (Backhaus *et al.*, 2003). In conjoint analysis it is assumed that the product being assessed can be defined in terms of a few important characteristics. Furthermore, it is assumed that the consumer decision related to such a product is based on tradeoffs among these product characteristics. The purpose of conjoint analysis is to estimate utility scores (i.e. part-worths) for these characteristics. Utility scores measure the importance of each single characteristic to the interviewee's overall preference for a product. The characteristics of a product are explained in terms of its factors and factor levels. The factors are the general attribute categories of a product, whereas the factor levels are the specific values of the factors (SPSS, 1997).

Table 7.1 gives a setup overview of the conjoint design for the two products: dried fruit snack and chocolate assortment. The first product consists of the following factors (i.e. general attribute categories):

Table 7.1 Set-up of conjoint study on dried fruit snack and chocolate assortment

Dried fruit snack		Chocolate assortment	
Factor	**Factor levels**	**Factor**	**Factor levels**
Basic product	Naturally dried fruit Microwave dried fruit	Chocolate type	Dark chocolate Whole milk chocolate White chocolate
Reference to drying processing	No Yes	Calorie content	400 kcal ('light') 600 kcal (normal)
Production type of Fruit growth	Conventional Organic	Filling	Yogurt Nougat Fruit Alcohol
Consumption suggestion	Sports snack Healthy alternative snack Exotic treat	Packaging	Single packaging Blister packaging
Final product	Pure Chocolate coated With nut mix	Packaging design	Precious Simple Trendy
Price (Euro)	0.79 € (low) 1.99 € (medium) 3.19 € (high)	Price (Euro)	1.19 € (low) 2.99 € (medium) 4.79 € (high)

Source: Sparke, K. (2008). Verbrauchersegmentierung bei der Neuproduktbeurteilung von Lebensmitteln. PhD thesis at the Faculty of Economy of the Technical University of Munich.

- basic product – using conventional or microwave drying technology
- reference to drying processing
- fruit growing: differentiated in conventional and organic production
- consumption suggestion
- final product
- price.

As the term 'microwave dried' might sound negative to consumers, the reference to drying process was established as another factor to evaluate potential impacts. Next to the core factor, namely the basic product (factor levels: naturally dried and microwave–dried fruit), the factors 'consumption suggestion' and 'final product' should bear further indices for product design. The second product, i.e. chocolate assortment, consists of the following attributes:

- chocolate type – dark, white, or whole milk chocolate
- calorie content
- filling
- packaging type – either single or blister packaging
- packaging design
- price.

Factors and factor levels of the chocolate assortment partially tried to reflect current market trends in Germany. In particular, dark chocolates with a high cocoa content have significantly gained market shares since 2000 (Duerr, 2004; Rosbach, 2005). The idea of 'light' products with lower calorie content has also reached the confectionary market in Germany and thus it was tested in our project. Fillings like fruit and yogurt are relatively new in this market, while modern and convenient kinds of packaging have also found their way into the confectionery chocolate segment.

With six factors and about two to four factor levels each, the product design is composed of 216 different possible product options for the dried fruit snack and 432 different options for the chocolate assortment. By means of the statistical program SPSS (1997), these numbers were reduced to 18 products each, which were illustrated as product cards and presented to the consumers. A rating of the product cards was made up according to their personal preferences.

II.C. Overall framework and empirical procedure

The combination of visual imagery stimuli and the consumers' affective and cognitive reactions and preferences towards newly developed food products results in the framework shown in Figure 7.1. A consumer judges the different pictures of food products and meals and thus reveals one's individual food consumption style. Consumer segments can be drawn from this investigation and combined with socio-demographic and other characteristics.

A further path of analysis is a set of product alternatives for the dried fruit snack and chocolate assortment, respectively. Their evaluation results in individual preferences, which can be aggregated to preferences of a certain consumer segment and give insight to a target group for specific new product development and marketing.

In spring 2005 a consumer survey was carried out in several hyper-markets and supermarkets in southern Germany. The tasks of the survey were, as mentioned above, to investigate consumer food consumption style,

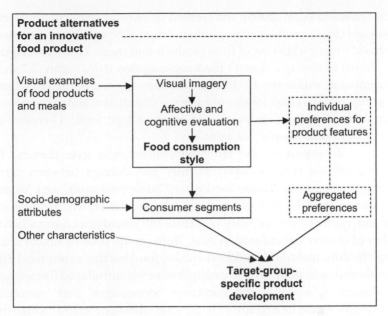

Figure 7.1   Framework of innovative food-related consumer segmentation for new product development (Sparke, 2008).

their preferences with respect to the presented innovative food products, dried fruit snack and chocolate assortment, their previous purchasing pattern referring to these product groups, and their information behavior with respect to new food products placed on the market. Additionally, some socio–demographic information was collected from the interviewees. Consumers evaluated 13 pictures of food products and meals, rated 18 product cards for an innovative food product, and answered orally further standardized questions. Altogether, 327 persons were asked about their food consumption style. Of these, 170 persons rated the chocolate assortment and 155 respondents assessed the dried fruit snack. The latter additionally tasted some naturally and microwave–dried apple bits to get a gustatory impression of the diverse basic products.

III. RESULTS AND DISCUSSION

III.A. Food consumption style

Assessments of the 13 food products and meals of all 327 respondents were basis for the execution of a cluster analysis. This procedure resulted in a ten–cluster solution. For this number of segments the dendrogram of the clustering process still showed a low index of heterogeneity. Additionally, this quantity was in the dimension of cluster solutions of Brunsoe et al. (1996), as well as Stiess and Hayn (2005), and seemed to be manageable for preference analysis and target-group-specific product design. The naming of clusters was only based upon evaluations of food products and meals. Cobweb-diagrams were chosen to display a cluster's food consumption style. Figure 7.2 shows the orientation within the 13 different modules for one of the ten clusters. This group exhibits a high affinity towards traditional food and to some extent to simple food and refuses exotic and modern type food. Therefore this segment was called 'simple fare eater.'

Table 7.2 summarizes the ten food consumption style clusters. The clustering process records display vicinity and distance between certain clusters. The segments 'fusion food eater,' 'junk food eater,' and 'canteen eater' were more stable than others during the procedure, while 'home-style eater' and 'exotic food eater' were very close and joined one cluster when the number of clusters was reduced to nine. Both groups showed similar likings towards healthy, traditional, and high-quality food but the 'exotic food eater' also evaluated sushi positively, an example of newly introduced foreign food. The clusters 'wholesome and conscious connoisseurs' and 'convenient connoisseurs' would be aggregated in the case where one might change from ten to eight segments. Both appreciate the spectrum, from the traditional over fresh and healthy food towards delicatessen food, but the 'convenient connoisseurs' also favored food that was easy to prepare and could be consumed quickly.

However, next to the results of the cluster analysis and the establishment of ten consumer segments, the suitability of the developed instrument has to be reviewed and discussed. An evaluation of the survey tool for determining

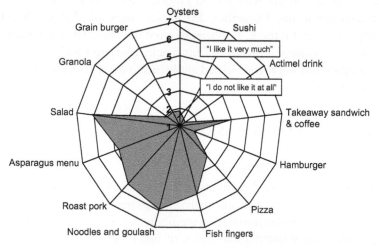

Figure 7.2 Food consumption style of 'Simple fare eater' (Sparke, 2008).

Table 7.2 Summary of the ten food consumption styles

Name (and relevance) of the cluster	Orientation within the food consumption style cobweb	Socio-demographic characteristics (in tendencies)
Home-style eater (15.5%)		Persons older than 50 years Increased professional training and university certificates Medium and higher incomes Two-person households
Fusion food eater (13.1%)	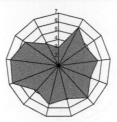	Persons younger than 30 years increased University entrance certifications and completed studies Single- and two-person households, also families with children
Wannabe wholesome eater (16.6%)	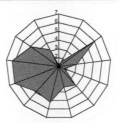	Elderly persons Increased professional training and university certificates, few graduated Medium incomes Above average share of families with children
Wholesome and conscious connoisseurs (14.3%)	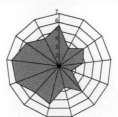	Persons aged 30 to 60 years Increased proportion of graduated persons Above average share of single-person households and families with two children

(continued)

Table 7.2   Summary of the ten food consumption styles—Cont'd

Name (and relevance) of the cluster	Orientation within the food consumption style cobweb	Socio-demographic characteristics (in tendencies)
Simple fare eater (6.1%)		Persons older than 70 years Above average share of common school certificates Low and medium incomes Single- and two-person households
Wholesome and convenient eater (9.6%)		Young and middle-aged persons Increased share of school certificates yet without further education Enlarged share of low incomes Households with one or two children
Exotic food eater (3.5%)		Ages 30 to 50 years, few elderly persons Increased share of university entrance certifications and completed studies Households with one or two children
Convenient connoisseurs (5.8%)		Ages 30 to 70 years Increased share of university entrance certifications and completed studies Higher incomes Single persons and families with one child

Table 7.2   Summary of the ten food consumption styles—Cont'd

Name (and relevance) of the cluster	Orientation within the food consumption style cobweb	Socio-demographic characteristics (in tendencies)
Junk food eater (7.9%)		Persons younger than 30 years Above average share of common school and professional training certificates Lower incomes Households with up to four persons
Canteen eater (7.6%)		Persons younger than 30 years University entrance certifications and completed studies Low and very high incomes Households with several persons

The arrangement of food variables in cobwebs is identical to that in Figure 7.2.

Source: Sparke, K. (2008). Verbrauchersegmentierung bei der Neuproduktbeurteilung von Lebensmitteln. PhD thesis at the Faculty of Economy of the Technical University of Munich.

a consumer's food consumption style was conducted through discriminant analysis. This statistical procedure is used to determine which variables discriminate between two or more groups and thus enables one to choose the best suitable variables. Food consumption style clusters were formed using the 13 food product stimuli and consumer likings towards them. These stimuli can be taken as variables of a discriminant function, calculated to reflect the structure between groups as good as possible, i.e. to achieve an optimal separation between the segments (Backhaus et al., 2003).

Nine discriminant functions were incorporated to describe the inter-segment structure of the 13 variables of which the first eight functions are significant ($p < 0.001$). When assigning the individual respondent to a food consumption style cluster by means of the nine discriminant functions, 87.5% of all respondents are assigned in the same way they were grouped by cluster analysis. A randomized procedure would result in 11.8% correct assignments. The mean discriminant coefficient of a variable describes how this variable discriminates between the groups considering all discriminant functions. Table 7.3 lists the values of the mean discriminant coefficient of the 13 food products and meal variables sorted by its discriminant value.

'Oysters' represent the variable with the greatest importance for discrimination, followed by 'Actimel drink,' 'Sushi,' and 'Asparagus menu.' The variable 'Salad' separates the worst between the consumer segments. Also the variable 'Granola' has comparably little importance in this respect. The cobwebs shown in Table 7.2 confirm these results. Many food consumption style clusters bear likings towards salad and granola, thus those variables do not make the real difference between the groups. Further, the variable 'Roast pork' has widespread acceptance. This might be due to the fact that the survey was conducted in southern Germany where this type of food is very traditional and typical. Even young people and persons without an affinity for cooking, who would never prepare roast pork as it is a time-consuming

Table 7.3 Mean discriminant coefficients of food variables

Food product and meal variable	Mean discriminant coefficient
Oysters	0.388
Actimel drink	0.375
Sushi	0.311
Asparagus menu	0.309
Takeaway sandwich & coffee	0.218
Hamburger	0.214
Grain burger	0.178
Fish fingers	0.165
Pizza	0.162
Noodles and goulash	0.155
Roast pork	0.146
Granola	0.122
Salad	0.084

Source: Sparke, K. (2008). Verbrauchersegmentierung bei der Neuproduktbeurteilung von Lebensmitteln. PhD thesis at the Faculty of Economy of the Technical University of Munich.

dish, like it and consume it at their parents' home or in restaurants. However, in this case regional food is not really suitable to discriminate between consumer groups.

When improving the survey tool one should think about exchanging or revising some variables. A variable that symbolizes food consumption in a person's work context might be useful. Many people have lunch in company or university canteens. Furthermore, the integration of snacks (e.g. sweets, ice cream, or fruit) might be considered. Nevertheless it can be concluded that in general the survey tool is a suitable way to examine people's food consumption style in a rather simple and efficient way.

When looking at the socio-demographic background of the diverse consumption style clusters, differences become obvious. A χ^2-test regarding age, gender, education, income, and whether a person has children or not, reveals significant differences between single food consumption style clusters. Significant differences can be observed concerning gender and the existence of children ($p < 0.05$). Differences regarding the respondent's education and age ($p < 0.001$) and their income ($p < 0.1$) are also significant, but the share of cross table cells with an expected frequency below 5 exceeds the defined limit of 20%. Descriptions of all clusters regarding their socio-demographic characteristics can be found in Table 7.2.

As listed in Table 7.2, cluster size varies clearly. The smallest segment, 'exotic food eater,' shares only 3.5% of the sample whereas the largest one, 'wannabe wholesome eater,' accounts for 16.6%. Small clusters may be regarded as unsuitable because product development and marketing strategy for a small consumer group might be only slightly efficient. But on the other hand such 'niche clusters' do exist among the consumers. Furthermore, the two smallest groups, 'exotic food eaters' and 'convenient connoisseurs,' are close to other segments like 'home-style eater' or 'wholesome and conscious connoisseurs,' respectively, and thus could be commonly targeted if required. Further, the small clusters in particular seem to be interesting for target-group-specific product development, as their food consumption styles show likings towards high-quality food and their socio-demographic background indicates high purchasing power.

Due to time and budgetary restrictions, the sample was not drawn representatively concerning several socio-demographic criterions. The main aim was to question consumers, but a balanced composition of the sample regarding the characteristics of age and gender was attended during the survey. Ex-post examination of the samples generated an overall good fit with publications of official regional censuses. Sample size was kept small since the food consumption style tool should be applicable to SMEs. Calculations of optimal sample sizes (Bortz, 2005), which were conducted after the survey and considered findings presented in part B. Preferences for innovative food products yielded the results that single food consumption style clusters may even consist of just less than ten individuals without loosing statistical relevance. However, the sample size of 327 interviewees was too small to generate significant results of cross table analysis for socio-demographic characteristics. Therefore a kind of sample of adequate size could be surveyed to obtain basic findings about food consumption style clusters. Testing of innovative food products could be conducted with smaller samples, in which case discriminant analysis is suitable to classify respondents of these surveys into the basic food consumption styles.

III.B. Preferences for innovative food products

The respondents' individual rankings of 18 cards evaluating product alternatives for the dried fruit snack and chocolate assortment respectively were the data basis for conjoint analysis. SPSS statistical software was used to calculate the importance of each factor and the part-worth utility values for the factor levels (Backhaus *et al.*, 2003). The sum of importance of all factors was 100%. For the factor price, negative linearity was implied because a product with a higher price level leads to a lower benefit for the consumer. Table 7.4 lists the importance of the used factors and part-worth utilities of the factor levels for the total sample.

For the dried fruit snack, the character of the final product has the highest importance, followed by price, basic product, and consumption suggestion. References to drying processing technology and fruit growing are less important. Part-worth utilities were also aggregated for the ten food consumption style consumer segments. The results of the analysis of dried fruit are shown in Figure 7.3. The factor price is not included as there are no differences between the different clusters.

Standard error bars show that food consumption style clusters differ significantly from each other with regard to preferences towards product attributes. The 'wholesome and conscious connoisseurs' prefer organically grown fruit and the consumption suggestion of a 'healthy alternative snack' more than other groups do. Design of the final product is less important to them, while this factor is the most important taken together among all respondents. The 'exotic food eaters' favor organically grown fruit as well and like it as either natural or chocolate-coated. On the other hand, both groups do not have a previous high purchase frequency of dried fruit snacks. According to answers on questions related to purchasing behavior within the survey, both groups bought dried fruit snacks less than once a month. However, they may be attracted to a snack concept based on organic fruit. Further, both groups show comparably little price sensibility. If regarded as one consumer group, they would be about 18% of all consumers, which might be sufficient for target-group-specific product development and marketing.

Table 7.4 Preferences and part-worth utilities of the tested product concepts

	Dried fruit snack				Chocolate assortment			
Factor	Importance (%)	Factor levels	Part-worth utility	Factor	Importance (%)	Factor levels	Part-worth utility	
Basic product	15.26	Naturally dried fruit	0.4806	Chocolate type	21.95	Dark chocolate	−0.0880	
		Microwave dried fruit	−0.4806			Whole milk chocolate	0.5609	
						White chocolate	−0.4728	
Reference to drying processing	11.14	No	0.3121	Calorie content	6.61	400 kcal ("light")	0.0744	
		Yes	−0.3121			600 kcal (normal)	0.1489	
Fruit growing	9.19	Conventional	−0.2873	Filling	36.41	Yogurt	0.6110	
		Organic	0.2873			Nougat	1.3848	
						Fruit	−0.8357	
						Alcohol	−1.1601	

(continued)

Table 7.4 Preferences and part-worth utilities of the tested product concepts—Cont'd

| Dried fruit snack | | | | Chocolate assortment | | | |
Factor	Importance (%)	Factor levels	Part-worth utility	Factor	Importance (%)	Factor levels	Part-worth utility
Consumption suggestion	13.45	Sports snack	0.0197	Packaging	7.89	Single packaging	0.1320
		Healthy alternative snack	−0.1752			Blister packaging	−0.1320
		Exotic treat	0.1555				
Final product	28.73	Pure	−0.1832	Packaging design	16.85	Precious	0.2107
		Chocolate coated	−0.3315			Simple	−0.0990
		With nut mix	0.5148			Trendy	−0.1117
Price (Euro)	22.23	0.79 € (low)	−1.4568	Price (Euro)	10.29	1.19 € (low)	−0.0373
		1.99 € (medium)	−2.9136			2.99 € (medium)	−0.0746
		3.19 € (high)	−4.3704			4.79 € (high)	−0.1118

Source: Sparke, K. (2008). Verbrauchersegmentierung bei der Neuproduktbeurteilung von Lebensmitteln. PhD thesis at the Faculty of Economy of the Technical University of Munich.

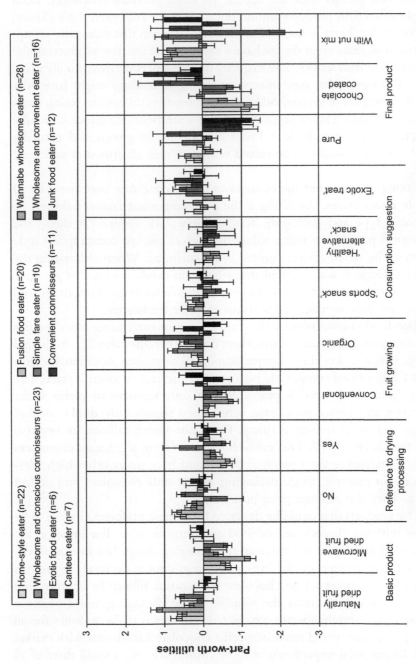

Figure 7.3 Dried fruit snack part-worth utilities for ten food consumption styles. Error bars indicate standard errors (Sparke, 2008).

The 'wannabe wholesome eater' and 'wholesome and convenient eater' are the two groups with the highest previous purchase frequency. Both bought dried fruit products around once a month. However, these clusters give only a few hints about their preferences. It seems that these core groups of dried fruit consumers do not have a strong desire for new products in the field of dried fruit snacks. The majority of consumers prefer naturally dried fruit. This means that new microwave drying technology might have only a limited prospect of success. Additionally, consumers do not like seeing a reference to the drying processing as it appears to scare them. At a glance, a broadly accepted dried fruit snack is based on organically grown and naturally dried fruit. It should also be mixed with nuts and advertised as an exotic treat.

Taking into account the evaluations of the chocolate assortment by all sample interviewees, the filling is the most important factor, followed by chocolate type and packaging design (Table 7.4). Figure 7.4 shows the aggregated part-worth-utility values for each of the ten consumption style clusters. The factor calorie content is not included. When elaborating the product design it was assumed that a less rich candy type might generate a certain acceptance, but it turned out that this factor was not important and did not produce significant deviations or other tendencies.

The food consumption style cluster 'convenient connoisseurs' prefers dark chocolate much more than others do. Further, they also like an alcohol filling, which makes them unique among all segments. Additionally, they show a distinct and comparably high positive preference towards price. The group 'junk food eater' is practically the only segment to prefer white chocolate and blister packaging, while they significantly dislike alcohol fillings and prefer nougat as filling. Price part-worth utilities are negative with this cluster as well. The 'exotic food eaters' show significant differences in their preference for a precious design and have many other high preferences, for example, single packaging, whole milk chocolate, and nougat filling. They also evaluate price positively.

The products that could be designed from these preferences seem to be in line with the cluster's general food consumption style. For instance, the 'junk food eater' pays no attention to packaging design but instead to the product's cheap price; costly single packaging is not as necessary to them as would be the sweet white chocolate and nougat filling. In contrast, the 'exotic food eaters' prefer the valuable single packaging and a precious design. Target-group-specific product design seems to be possible for all three groups and would result in specific chocolate assortments with market or niche potential respectively, as those clusters are only a small share of all consumers. Observing the consumer segment's previous chocolate-related purchase pattern, the 'home-style eater' group purchases chocolate

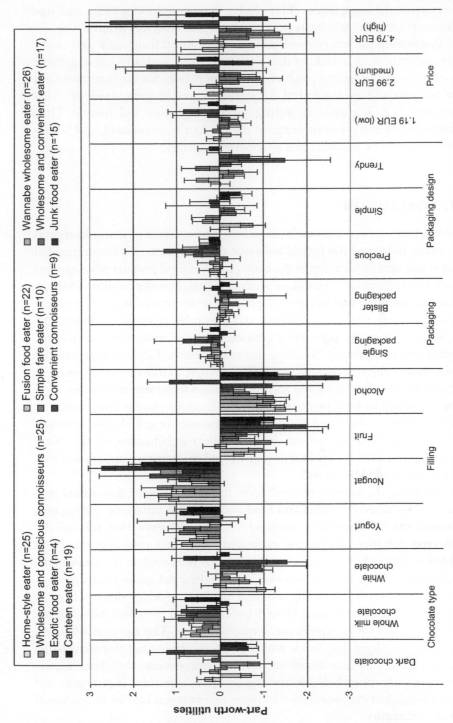

Figure 7.4 Chocolate assortment part-worth utilities for ten food consumption styles. Error bars indicate standard errors (Sparke, 2008).

assortments the most often. This cluster does not show specific and significant preferences but bears positive preferences towards price.

Comparing the interviewee evaluations of dried fruit snack and chocolate assortment, it is striking that the factor price is clearly less negatively assessed for the chocolate product. The main reasons for buying chocolates are for purchase as a gift and for consumption at some kind of social gathering, e.g. at parties or during visits with family and friends. Thus a high-quality and precious image of the product is appreciated, and higher prices seem to be regarded as a parameter of quality.

IV. CONCLUSION

Taking into account the nearly stagnated food markets in many European countries today and the hybrid behavior of consumers, it is recommended that consumers be included in the early phases of food product development due to cost reasons and the risk of product failure. In order to allow SMEs to evaluate consumer requirements in product development at a higher extent than in the past, new easy-to-handle and reasonable tools are necessary. The approach used in the pilot study conducted at our institution has demonstrated one possibility for fulfilling this requirement. Taken all together, the reduction of different dimensions of food-related lifestyles to the consumer's affinity towards food itself and the use of images to present the various meal options seems to be a practicable way to segment people into consumer clusters, which then can be used in consumer-oriented food product development. The instrument used to collect information on the 'food consumption style' can be easily implemented in consumer surveys and thus can be used by SMEs as well.

However, the instrument needs to be further developed, fine-tuned, and checked for different market and food consumption situations. Taking into account the food products analyzed, one can observe that part of the observed deviations cannot be explained with food consumption styles, which means additional factors must exist that help explain the consumers' preferences. Conjoint analysis is a widely accepted and applied instrument in consumer research, providing valuable data for product design. Further surveys and product testing should be conducted in which the product concepts tested have a 'significant' innovation level. This aspect was fulfilled in testing the dried fruit snack with microwave drying but this technology did not have a positive consumer reaction. In contrast, the chocolate assortment concept included interesting and trendy features (e.g. yogurt and fruit filling), but the overall level of innovativeness was limited in this tested product category.

A food company's strategy for consumer integration in product development using the food consumption style approach could be structured as follows: (1) potential target groups are selected from all food consumption style clusters taking into account their previous purchase frequency and general attitude towards the product group of interest; and (2) target group preferences towards specific product features of the innovative product concept are analyzed, and significant results or clear tendencies are then taken as evidence for final product design. Thus, the food consumption style tool is applied to obtain consumer segment-related preferences towards product features and represents a good starting point for product development. Finally, the success of product development activities and marketing of newly developed products are also dependent on appropriate marketing strategies, including consumer communication.

REFERENCES

Backhaus, K., Erichson, B., Plinke, W., & Weiber, R. (2003). *Multivariate Analysemethoden*. Berlin: Springer.

Bortz, J. (2005). *Statistik fuer Human- und Sozialwissenschaftler*. Heidelberg: Springer.

Brunsoe, K., Grunert, K. G., & Brehdahl, L. (1996). An analysis of national and cross-national consumer segments using the food-related-lifestyle instrument in Denmark, France, Germany and Great Britain. *MAPP Working Paper no. 35, Aarhus*.

Childers, T. L., & Houston, M. J. (1983). Imagery paradigms for consumer research: alternative perspectives from cognitive psychology. *Advances in Consumer Research, 10*(1), 59–64.

Duerr, H. (2004). Gute Zeiten fuer die Besten. *Lebensmittelzeitung, 56*(42), 39–42.

Earle, M., Earle, A., & Anderson, A. (2001). *Food product development*. Boca Raton: Woodhead Publishing Limited.

Federal Statistical Office.. (2006). *Statistical Yearbook for the Federal Republic of Germany*. Wiesbaden: Federal Statistical Office of Germany.

Flaig, B. B., Meyer, T., & Ueltzhoeffer, J. (1993). *Alltagsaesthetik und politische Kultur*. Bonn: Dietz Verlag.

Green, P. E., & Srinavasan, V. (1978). Conjoint Analysis in Consumer Research: Issues and Outlook. *Journal of Consumer Research, 5*, 103–123.

Grunert, K. G., Larsen, H. H., Madsen, T. K., & Baadsgard, A. (1996). *Market Orientation in Food and Agriculture*. Boston: Kluwer.

Heimig, D. (2005). Gesund gesnackt. *Lebensmittelzeitung, 57*(30), 34–36.

Heindl, A. (2003). Mikrowelle kann mehr. *Lebensmitteltechnik, 1–2*, 60–63.

Holbrook, M. B. (1982). Some further dimensions of psycholinguistics, imagery, and consumer response. *Advances in Consumer Research, 9*(1), 112–117.

Holbrook, M. B. (1983). Product imagery and the illusion of reality: some insights from consumer esthetics. *Advances in Consumer Research, 10*(1), 65–71.

Kotler, P., Armstrong, G., Saunders, J., & Wong, V. (2003). *Grundlagen des*. Muenchen: Marketing. Pearson Education.

Krelle, W. (1968). *Praeferenz- und Entscheidungstheorie*. Tuebingen: J.C.B. Mohr.

Linnemann, M., Meerdink, C. H., Meulenberg, M. T. G., & Jongen, W. M. F. (1999). Consumer-oriented technology development. *Trends in Food Science and Technology, 9*, 409–414.

Loudon, D. L., & Della Bitta, A. J. (1993). *Consumer Behavior: Concepts and Applications*. New York: McGraw-Hill.

Meffert, H. (2000). *Marketing*. Wiesbaden: Gabler.

Menrad, K. (2004). Innovations in the food industry in Germany. *Research Policy, 33*, 845–878.

Rosada, M. (2005). Neueinfuehrungen zwischen Top und Flop. Talk on the LP Innovations Congress, Bonn.

Rosbach, B. (2005). Suesswaren. *Lebensmittelzeitung Spezial, 3*, 54–57.

Rossiter, J. R. (1982). Visual imagery: applications to advertising. *Advances in Consumer Research, 9*(1), 101–106.

Schmalen, C. (2005). Einflussfaktoren der Markteinfuehrung von Produktinnovationen klein- und mittelstaendischer Unternehmen der Ernaehrungsindustrie. Utz, München.

Sparke, K. (2008). Verbrauchersegmentierung bei der Neuproduktbeurteilung von Lebensmitteln. PhD thesis at the Faculty of Economy of the Technical University of Munich.

SPSS (Statistical Package for the Social Sciences).. (1997). *SPSS Conjoint 8.0*. Chicago: SPSS Marketing Department.

Stiess, I., & Hayn, D. (2005). *Ernaehrungsstile im Alltag: Ergebnisse einer repraesentativen Untersuchung. Working Paper*. Frankfurt: ISOE Institute.

Stifterverband fuer die Deutsche Wissenschaft. (2005). FuE Info 2. Essen.

CHAPTER 8

Consumer-Targeted Sensory Quality

Margaret Everitt

Contents

Abstract

Defining a meaningful sensory component within the overall product specification continues to pose challenges for the food and drink industry. Product quality can easily deviate away from a product set for launch onto market if the key sensory criteria are not clearly specified and communicated to the production team. This chapter advocates maximum use of the product sensory learning gained during the development process to help produce sensory specifications more relevant to the consumers' needs. Two areas in particular are discussed: first, how to establish the key criteria for a target product, and more specifically, how to define the quality range of these criteria; and second, transfer of this information into realistic, practical sensory specifications.

I. INTRODUCTION

Specifications relating to the implied or expected quality of a product have been required since commercial trading first began. Yet defining

Global Issues in Food Science and Technology
© 2009 Elsevier Inc.

a meaningful sensory component within the overall product specification continues to pose challenges for the food and drink industry, largely because of the difficulty in establishing standard criteria that are objectively understood and applied for features of a product notoriously prone to individual subjective interpretation.

Whether a food product is purchased on the basis of being healthy, to provide an indulgence or simply to stave off hunger, it is vital that the sensory quality fits with the expectations of the target market if acceptance is to be gained, and re-purchase assured. Taking a consumer-focused approach right from the start of the product development process not only benefits product guidance, but also provides a wealth of valuable information from which more valid sensory specifications can be built.

The benefits of utilizing sensory research to aid product development are now widely acknowledged across the industry, and many businesses possess a sophisticated 'tool box' of sensory and consumer techniques, which enable far greater insight and understanding of their target consumers' behavior, attitudes, needs, and tastes. In addition, much has progressed at the production level to improve sensory assessment procedures and implement sensory quality control systems (Beckley and Kroll, 1996). A special issue of *Food Quality and Preference*, 'Advances in Sensory Evaluation for Quality Control' (Munñoz, 2002), is also a recommended 'read' as this publication contains ten excellent papers that address advances and applications related to the whole subject area.

In order to ensure that the sensory quality of a routinely produced product is maintained as originally established, there needs to be clear communication during its development and production about the key sensory features. This information link however creates a hot spot of challenges, in particular, how to effectively gain and standardize production's understanding of the 'target' product's sensory quality criteria, irrespective of internal views and preferences. This chapter looks at how to best bridge this communication gap and therefore is focused primarily on:

- Identification and definition of a 'target' product's key sensory criteria based on the consumers' requirements
- Transference of the information into realistic, workable sensory specifications.

The approach therefore advocates making maximum use of information gained from consumers and sensory research during the product development process.

II. SENSORY QUALITY: THE CONSUMERS' PERSPECTIVE

II.A. Target product quality

Two aspects in particular need to be addressed when establishing target product quality: (a) the key sensory attributes affecting liking and (b) the sensory criteria that need to be modified to satisfy different groups of consumers within a market sector (i.e. determine if more than one 'target product' specification will be defined).

II.A.1. Cluster analysis and preference mapping

Cluster analysis along with preference mapping techniques are now well recognized, and they are the established statistical methods readily employed by research and development teams to benchmark consumer liking; both enable the key drivers of preference to be identified and defined. It is this information that provides the essence of sensory specification. A wealth of literature is available that explains preference mapping and clustering methods and their application in detail. A good appreciation of these methods is required to employ them effectively and also to interpret the output confidently. Essential aspects to appreciate are the various preference mapping models (Greenhoff and MacFie, 1994) that can be used along with the most popular types of clustering measures. There are many variants of cluster analysis and different grouping criteria, hence choosing the most appropriate method can be difficult. Meilgaard *et al.* (1991) provides a good, concise overview of the topic with references to more detailed reading. The advice and assistance of a statistician experienced in this area is not to be belittled.

II.A.2. Product set

Careful consideration is also required when selecting the set of products to be researched. Importantly, the range should comprehensively (but realistically) illustrate the current commercial sensory quality that is available for a specific product sector. If the sensory range is too restrictive or too wide it can result in a distortion of both the resulting hedonic and sensory measures for each product, with subsequent misrepresentation and misinterpretation of the market's 'ideal' product. Although not essential for all preference mapping analyses, trained sensory panel data are usually obtained in addition to consumer liking, to provide detailed measurement of the sensory attributes. The examples discussed in the following relate to data generated by extended internal preference mapping.

II.B. Extended internal preference mapping output

Extended internal preference mapping is most commonly used when the focus of the research is to understand the most important sensory drivers of consumer liking and thereby improve an existing product. The data are usually pretreated using a cluster analysis technique, typically a hierarchical method such as average linkage or Ward's method, to formally test for the presence of segments (clusters) of consumers with different preference patterns.

Figure 8.1 illustrates an extended internal preference map to which cluster analysis has been applied. The analysis provides a correlation co-efficient for each sensory attribute in relation to consumer liking, and is done for each cluster. This enables the key attributes (both negative and positive) to be confidently defined in relation to the most-liked products. The analysis also provides an indication of the level of importance, i.e. the weighting that each of these attributes should be given in the sensory specification.

Looking at the example of the ANOVA table for consumer 'preference' Cluster 1 in Figure 8.2, the product mean scores listed to the right suggest that this consumer group is quite decisive about which products they like and dislike. Product 301 is distinguished from all others as the most liked statistically, with products 741 and 893 clearly the least liked.

It is also clear that flavor is far more important (Figure 8.3) in this group of consumers than appearance or texture, as the most significant liking correlations mainly relate to flavor attributes. In this example, cinnamon,

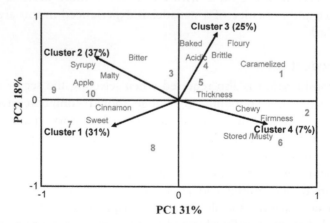

Figure 8.1 Illustration of extended internal preference map to which cluster analysis has been applied. The preference direction of four clusters is indicated by the black arrows. The PCA (principal component analysis) map is created based on consumer liking data; the sensory attributes are then regressed into the liking space in relation to each preference axis.

Overall liking means for segment 1				Mean	N	Sample
LSD						
A				7.62	64	301
B				6.76	64	268
C		B		6.2	64	579
C		B	D	6.09	64	803
C			D	6.08	64	327
C		E	D	5.98	64	154
C	F	E	D	5.56	64	296
	F	E	D	5.43	64	532
	F	E		5.35	64	965
	F			5.24	64	619
G				4.56	64	741
H				3.81	64	893

Figure 8.2 Example of ANOVA table for a consumer 'preference' cluster; e.g. Cluster 1 shows the mean liking scores for each of the 12 products, plus the statistical differences between products, as defined using least significant difference (LSD) test.

apple, and sweet flavors are the most important positive attributes, with apple (correlation 0.69) and cinnamon (correlation 0.72) being the most similar, while 'musty' flavor (correlation −0.81) is the most important of the five key negative attributes.

Each consumer cluster can be analyzed in this way to determine whether development of one or more products would satisfy the market sector. Figure 8.4a displays an extended internal preference map that shows the positioning of the 12 products referred to in Cluster 1, for a total of four 'preference' clusters. The dots represent the positioning of individual

Segment 1
Correlations with sensory attributes

Positive (attributes liked a lot)
Apple flavor	**0.69**
Cinnamon flavor	**0.72**
Sweet	0.57

Negative (attributes disliked)
Baked odor	−0.62
Acidic taste	−0.60
Stored/musty flavor	**−0.81**
Caramelized flavor	−0.60
Floury texture	−0.56

Figure 8.3 Consumer liking and sensory attribute correlations showing the key positive and negative sensory attributes, i.e. preference drivers for Cluster 1.

consumers in relation to the product liking space. It can be seen that the direction of overall liking for the total consumer sample moves slightly upwards from the bottom left to top right of the plot towards product 301.

Figures 8.4b and 8.4c show the ANOVA tables with LSD for the other two main clusters, Cluster 2 (31%) and Cluster 3 (25%). Cluster 4 (illustrated on the left in Figure 8.4a) accounted for a very low percentage of the total

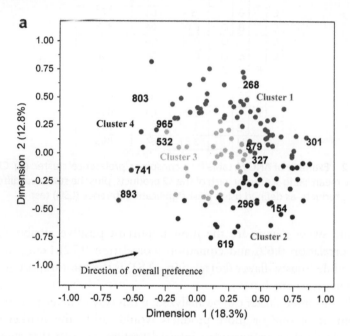

b

Overall liking means for segment 2				
	LSD	Mean	N	Sample
	A	6.83	45	154
B	A	6.70	45	301
B	A	6.70	45	296
B	A	6.53	45	619
B		6.05	45	327
B		6.03	45	579
	C	5.23	45	893
D	C	4.79	45	965
D	C E	4.55	45	532
D	E	4.40	45	741
	E	3.95	45	268
	E	3.83	45	803

c

Overall liking means for segment 3				
	LSD	Mean	N	Sample
	A	7.16	31	619
	A	6.94	31	301
	A	6.80	31	327
B	A	6.39	31	532
B	C	5.74	31	268
B	C	5.71	31	154
B	C D	5.68	31	296
B	C D	5.52	31	803
B	C D	5.42	31	741
	C D	5.10	31	893
	C D	4.74	31	965
	D	4.63	31	579

Figure 8.4 a. Extended internal preference map showing positioning of the 12 products referred to in Cluster 1, for a total of four 'preference' clusters. **b.** ANOVA table with LSD showing overall liking means for Cluster 2. **c.** ANOVA table with LSD showing overall liking means for Cluster 3.

consumer sample, i.e. 7%; thus a decision was made not to develop the product for this particular group.

Figure 8.4a also shows that product 301 is well liked by three consumer clusters (1–3), Cluster 1 notably liking it the most distinctly (Figure 8.2). This was selected as a common 'target' product suitable for the three main clusters (1–3). If Clusters 2 and 3 had disliked or been indifferent towards product 301, there would have been a strong case to develop at least two launch products and hence define two target specifications, e.g. product 301 for Cluster 1 and possibly product 619 for Clusters 2 and 3.

II.C. The sensory quality range and limits of consumer tolerance

Once the most liked product(s) has been defined, a range of acceptable deviation around this 'target' needs to be specified. In this approach the degree of sensory quality variation tolerated by consumers before their acceptance of the product and subsequent loyalty and loss of, is the guide by which this range is set.

Historically this deviation or 'tolerance' range is possibly the most challenging and controversial aspect of a sensory specification. Without sufficient appreciation and 'calibration' within a production workforce to a common standard, the sensory specification is wide open to individual interpretation, which in turn can result in fluctuating degrees of quality control. Again consumer and sensory research information gained during the development process can help eliminate these issues.

II.C.1. Hedonic scales

Hedonic scales are well tried and tested in consumer research for capturing liking data (Stone and Siddel, 1985). Figure 8.5 shows a typical example of a nine-point hedonic scale, a version regularly used with consumers in preference mapping studies to capture liking scores.

A mean liking score of 7 or higher on a nine-point scale is usually indicative of highly acceptable sensory quality; hence, a product achieving this score could be used confidently as a good illustration of 'target' quality. On this basis, products from a research set can then be selected to provide physical references to illustrate the sensory quality that realistically represents the consumers' acceptance limits.

Dislike Extremely	Dislike Very Much	Dislike Moderately	Dislike Slightly	Neither Like nor Dislike	Like Slightly	Like Moderately	Like Very Much	Like Extremely

Figure 8.5 Nine-point hedonic scale with verbal anchors.

II.C.2. Reference products

Referring again to Figure 8.2, product 301 indicates a mean liking score of 7.6 and would provide a good target quality reference. Product 268 has a mean score of 6.8 and although it records a statistically significant decrease in liking it still represents a high level of liking. From a practical stance, if this product was selected to illustrate the acceptance limits it may set the quality range too tight. However, product 579 has a mean score of 6.2 and shows a decrease in liking from product 301 of two statistical levels. This product could therefore, from a business perspective, represent a more realistic example of the limits of sensory variation. Further insight regarding the degree of variation that consumers will tolerate, especially with reference to the key drivers of preference, can be gained by comparing the sensory profiles of those products selected to show the target quality range.

Figure 8.6 shows the profile of each of the three products selected for Cluster 1 (i.e. Segment 1). The sensory panel mean score is plotted on to each attribute axis for each product (low scores are central; high scores are peripheral). Where the profile lines overlap there are no significant differences between the products. Where they diverge the differences are significant, and

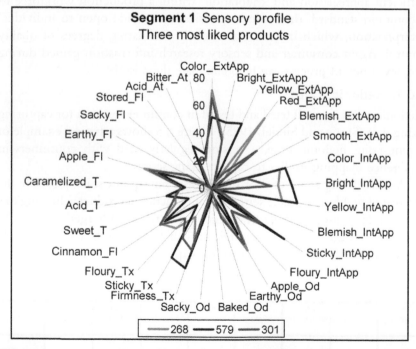

Figure 8.6 Sensory profile as generated by the trained sensory panel, for the three products selected to represent the sensory acceptance range in Cluster 1, i.e. products 301, 268, and 579.

the greater the gap between each score the greater the difference in the intensity between each product for that specific attribute. It can be seen that the variation in intensity of the key flavor attributes (left, upper left of Figure 8.6) is tighter than that seen for the majority of the appearance and texture attributes. It further verifies the higher weighting (see section II.A) these should be given in the sensory specification. Armed with this in-depth knowledge a business is better equipped to define realistic sensory specification targets that are sufficiently tight enough to ensure maintenance of consumer satisfaction, while being flexible enough to avoid unnecessary product wastage. A valuable benefit of this approach is that, either way, whether the decision is made to set the quality target range fairly tight or fairly loose, it can be done from a more informed perspective and appreciation of the business risk involved.

II.D. Visual defects and off-flavors

In addition to the key sensory drivers identified by the preference mapping analysis, a vital component of a sensory specification is that which relates to the risk associated with likely occurring product defects and off-flavors. Some of these characteristics may be identified under the negative key drivers, for example as 'stored/musty' in Cluster 1 (Figure 8.3), but ordinarily these features are not likely to be represented in the preference mapping product set. Again it is important not only to specify the type of defects and off-flavors but to also allocate to each one an importance weighting, i.e. major versus minor importance. For example, is an uneven pie crust more or less important than a cracked one; is a burnt flavor from caramelized sugars more or less important than fumy or acrid notes from hydrolyzed fats. Further consumer research can be conducted to verify the importance of these factors but for many companies neither budget nor time allow this, so the responsibility for these particular decisions tends to fall internally. The pooling of technical knowledge and processing expertise along with applicable sales and consumer complaint information will aid the process.

III. DEFINING THE SENSORY SPECIFICATION

III.A. Sensory grading

Using the concept of a sensory acceptance 'range' during product development, helps set the scene for the transference of information into product specification. For routine assessment of sensory quality, be it daily QC (Quality Control) checks or longer-term monitoring, grading is a highly practical method. Providing appropriate training is given to the assessors, grading provides a relatively quick and simple technique to apply that is not overly sensitive to small differences, i.e. changes in quality.

Grading uses the principle of ranges to represent different levels of quality. Typically, in most grading systems there are three to five levels, ranging from optimum through typical/target quality, down to a level considered to be sub-standard. These ranges are usually given verbal labels, some popular examples being *Gold Standard*, *A-Star*, *Superior* at the top end, *Borderline*, *Just Acceptable*, and *C-Grade* at the lower end. Each grade is converted into a numeric value for recording and analysis purposes. As discussed in section II.A, defining the boundary of each range usually proves the most challenging and is prone to vagueness and poor differentiation. This is where the consumer-targeted approach proves especially effective by identifying products that can be used as physical training aids.

III.B. Defining the quality grades
III.B.1. Assessment scale

Figure 8.7 illustrates how the product development information can be transferred into the sensory specification. Using a ten-point category scale where $0 =$ none, and $9 =$ high intensity, first the target range (represented by the white zone) is defined on the scale taken from the intensities depicted by the products selected, as discussed in section II.C.2, in this case products 301, 268, and 579. The 'borderline' range (light gray area) is then defined and is best set quite tight, i.e. as one category on the quality scale, for this should be treated as the 'warning' zone indicating that quality is in danger of deviating too far away from the target. So for Cluster 1, good examples to illustrate this change in quality would be products 803 and 327, which both recorded

Flavor										
Apple	0	1	2	3	4	5	6	7	8	9
Sweet	0	1	2	3	4	5	6	7	8	9
Cinnamon	0	1	2	3	4	5	6	7	8	9
Acid	0	1	2	3	4	5	6	7	8	9
Baked	0	1	2	3	4	5	6	7	8	9
Caramelized	0	1	2	3	4	5	6	7	8	9
Off-Flavors, specify	0	1	2	3	4	5	6	7	8	9
Texture										
Crisp	0	1	2	3	4	5	6	7	8	9
Floury	0	1	2	3	4	5	6	7	8	9
Grainy	0	1	2	3	4	5	6	7	8	9

Figure 8.7 Sensory specification with defined target (white), borderline (light gray), and unacceptable (dark gray) quality ranges (zones) for specific sensory criteria.

a similar drop in consumer liking from the target range. The remainder of each attribute scale, highlighted in dark gray, indicates quality deviations that would be unacceptable. In the case of Cluster 1, any of the products listed below product 327 (Figure 8.2) could be used to illustrate unacceptable quality, with products 741 and 893 depicting the most unacceptable deviation of intensity and combination of sensory characteristics, especially regarding the key attributes. In practice the different shadings would only be used for assessor training and reporting purposes. The training process should build the assessor's competence to a level where they can confidently recognize the different grades of product; therefore the scales on the actual assessment ballots should be left unshaded to minimize assessment bias.

III.B.2. Quality criteria

The quality levels are defined under each sensory modality for all key attributes in the following manner. As a general guide for defects and off-flavors, a product demonstrating the slightest degree of a designated major defect or off-flavor would automatically be considered unacceptable and placed on hold, pending further decisions whether to re-work or scrap as appropriate.

From an internal business perspective it may be beneficial to include other sensory attributes in the specification, especially if they are associated with processing factors or raw materials that are closely allied to the key drivers. For example, referring back to Figure 8.6, which shows all product sensory attributes, it could be useful to monitor 'sticky' and/or 'firmness' of texture to further aid control of 'floury' texture, a key negative driver. The target range for any additional attributes would again be defined in the specification with reference to the intensities shown by the selected reference products, ensuring that the consumers' requirements are retained.

III.C. Maintaining reference material

It is important to maintain examples of all the quality grades but especially those illustrating the current target range and its boundary limits. The 'control' material is needed to calibrate and validate the assessors' performance and reliability both initially and on-going as appropriate. This is usually less of an issue for long shelf-life products, which can be either frozen or held at a low chill temperature to maintain stability for at least 6 months. Freezing and/or gas flushing in cans or foil packs can also prove useful as methods in maintaining material for some short shelf-life products and are well worth investigating. When the reference material (either a short or long shelf-life product) is replenished, it should be done using the expertise of those trained according to the current target standards. Sensory comparative tests such as Duo Trio or Paired Comparison are best used to ensure that the new control material is sufficiently similar to the old.

IV. APPLICATION OF THE SPECIFICATION

IV.A. On-line quality control (QC)

The specification has the flexibility to be used at two levels of detail. One would be to provide an overall grade for each sensory modality without grading each specific attribute. This is usually more practical for on-line QC applications where assessments are being conducted several times per shift per day and rapid decisions are required whether to release or hold the product.

IV.B. Quality assurance (QA)

The full detailed grading version is more applicable to QA and further assessment of any product placed on 'hold'. It should be noted that for the specification to be used reliably at either level assessors must be fully trained and validated. Both applications provide diagnostic information, but the detailed level provides greater focus in determining which specific sensory criteria are deviating out of specification and therefore greater insight into which process variables need attention to counteract such deviation.

V. CONCLUSION

This consumer-focused approach fully utilizes the product sensory learning obtained in the development process. It helps bridge the link between product development and production by promoting the need to connect the sensory development criteria with the sensory specifications from the initial stages of the process, ensuring that the resulting specification has much greater relevance to the needs of the end user while being practical commercially.

REFERENCES

Beckley, J. P., & Kroll, D. (1996). Searching for sensory research excellence. *Food Technology, 50*(2), 61–64.

Greenhoff, K., & MacFie, H. J. H. (1994). Preference Mapping in Practice. In H. J. H. MacFie, & D. M. H. Thomson (Eds.), *Measurement of Food Preferences* (pp. 137–165). London: Blackie Academic and Professional.

King, S., Gillette, M., Titman, D., Adams, J., & Ridgely, M. (2002). The Sensory Quality System: a global quality control solution. *Food Quality and Preference, 13*(6), 385–395.

Meilgaard, M., Vance Civille, G., & Carr, B. T. (1991). *Sensory Evaluation Techniques* (2nd Edition, pp. 280–284).. Boca Raton, FL: CRC Press.

Munñoz, A. M. (2002). Advances in Sensory Evaluation for Quality Control. *Special Issue: Food Quality and Preference, 13*(6), 327–328.

Stone, H., & Sidel, J. L. (1985). *Sensory Evaluation Practices,* (pp. 230–236). London: Academic Press.

CHAPTER 9

How We Consume New Products: The Example of Exotic Foods (1930–2000)

Faustine Régnier

Contents

Abstract

How do we consume new food products? How are they integrated? This presentation is based on the example of exotic foods and on the detailed analysis of 9,758 recipes taken from four German and French magazines between the 1930s and 2000. Magazines play an important role in the spread of food taste preferences, and they are a precious data source for studying domestic and daily cooking practices and their transformation. The first point concerns the definition and characteristics of 'exoticism,' which can be examined through an ordered list of foreign recipes from France and Germany that have evolved through time. This list highlights the historical and sociological context of interest in foreign and exotic

Global Issues in Food Science and Technology
© 2009 Elsevier Inc.

foods, and the relationship between the taste for exotic foods (i.e. exoticism) and the history of gastronomy, colonization, immigration streams, and tourism, which in turn reveal national food consumption models. The second point examines the reactions toward culinary novelty: fear as well as seduction. This chapter describes the various processes that encourage consumption of new products (selection of products and dishes similar to national tastes, and the importance of promoters of such food). At the same time, foreign recipes are seductive because they are different from the national food consumption habits (e.g. foreign foods make us travel during our dinner; we eat these foods because they are said to be good for our health). In consequence exotic foods may be a way to improve national cooking and food consumption habits.

I. INTRODUCTION

To understand how and why culinary innovation has been integrated into French cooking, this contribution uses the example of culinary exoticism in France and Germany between the 1930s and the end of the 1990s (Régnier, 2004). The word 'exotic' is taken here in a broad sense to mean foreign products and food consumption habits. This research analyzed a corpus of 9,758 cooking recipes taken from four women's magazines: *Marie Claire* and *Modes et Travaux* published in France, and *Brigitte* and *Burda* published in Germany. Women's magazines are not particularly well considered as tools of research, whereas they constitute a particularly rich data source. Indeed, women's magazines were selected because of the important role played by the media in taste diffusion (Besnard and Grange, 1993; Warde, 1997); magazines are at the origin of fashions, as well as reflecting these fashions (Bourdieu, 1984). Moreover, women's magazines are a precious data source for studying domestic daily cooking and its transformations. Indeed, a particular style of cooking is studied here, the same one we use for consumption in the home (not the restaurant). The magazines *Modes et Travaux*, *Marie Claire*, *Brigitte*, and *Burda* were selected because they have existed for a long time, they have a very wide circulation (some of the most widely read magazines in France and Germany), and they are different. The readers of *Marie Claire* and *Brigitte* generally live in urban areas, while the readers of *Modes et Travaux* and *Burda* generally live in rural areas. Consequently, these magazines were selected to compare the social distinctions between countries and periodicals.

Each edition of the four magazines since their creation up to 2000 was analyzed: 3,830 issues were systematically surveyed. Analysis followed two points of view. On the one hand, a quantitative and statistical analysis was made concerning the importance of each type of exoticism and its diffusion. On the other hand, focus was placed on the discourses (on recipes and magazine commentaries). It was a classical analysis of contents and also a work realized with special software for textual analysis (Hyperbase), which could be used in French as well as German.

French interest in foreign products and dishes is ancient, but 'exoticism' itself is much more recent; it was not until the beginning of the 20th century that the terms 'exoticism' and 'exotic' began appearing in French culinary books (Régnier, 2004). Nonetheless, the consumption of unknown products is not always easy in France; products tend to frighten consumers because consumption supposes close contact with a foreign foodstuff or dish. By buying, preparing, and eating foreign products, a person incorporates something strange. The consumer introduces a foreign thing (i.e. food) deeply into their body, abolishing in consequence the border between the body and the still unknown. Unknown aliments, first of all, can be regarded as dangerous, which makes the fear of incorporation all the more strong as the food is new (Augé, 1986; Fischler, 1990). Moreover, the cook does not know the exact way to prepare or consume exotic products. Foreign food habits may also disgust consumers when they consider the use of non-edible aliments in the culinary system of the consumer. For example the consumption of raw fish, which is popular in Japanese cooking, has been considered 'very strange' (i.e. non-edible) for a long time in France and Germany. Lastly, a foreign way of cooking may upset the traditional culinary system. However, respect for these traditions is really a respect for the 'order of the world' (Certeau et al., 1994), including that of the consumer. Exotic foods and dishes that mix for example sweet and salty flavors could then upset the traditional systems of food classification.

Consequently, this chapter will (1) highlight the historical and social context supporting the diffusion of new culinary practices; (2) stress the reactions concerning culinary innovation: the process of selection allowing a consumer to eat new products; and (3) highlight the desire for new foods and different dishes.

II. HISTORICAL AND SOCIAL CONTEXT

II.A. Rankings and diffusion of exotic foods

Classification of the 9,758 recipes studied by geographical area according to number of recipes can be viewed in Table 9.1. Recipes in France and Germany have several common features. Recipes from southern Europe are the most frequent, and more than 60% are Italian. Also noted is a similar high frequency of exoticism from Western Europe as well as a few recipes from sub-Saharan Africa.

Table 9.1 Number of exotic recipes in French and German magazines

Exoticism	France		Germany	
	Number of recipes	**Proportion (%)**	**Number of recipes**	**Proportion (%)**
Southern Europe	749	28.3	2213	31.1
Distant islands	277	10.5	58	0.8
Western Europe	206	7.8	1408	19.8
Recipes using exotic fruit	202	7.6	655	9.2
United States	135	5.1	210	3.0
Mixed exoticisms	133	5.0	264	3.7
India	132	5.0	341	4.8
Far East	129	4.9	421	5.9
Middle East	125	4.7	149	2.1
Eastern Europe	113	4.3	600	8.4
Northern Africa	103	3.9	30	0.4
Latin America	88	3.3	138	1.9
Recipes using spices	80	3.0	154	2.2
Scandinavia	69	2.6	279	3.9
Exotic	53	2.0	108	1.5
Tropics	18	0.7	6	0.1
Sub-Saharan Africa	17	0.6	5	0.1
Indonesia	17	0.6	73	1.0
Total	2646	100.0	7112	100.0

However, several differences clearly appear in the percentages of exotic recipes. In particular, Western European exoticism accounts for only 7.8% of the exotic recipes in France but 19.8% in Germany. This can be explained by the large number of French exotic recipes in Germany (55% from Western Europe in German magazines; whereas German exotic recipes in French magazines make up only 13% of Western European exoticism).

The fact remains that the German magazines are much more interested in foods from nearby European countries. Thus, the number of recipes coming from nearby countries, from a geographical or cultural point of view (Southern Europe, Western Europe, United States, Eastern Europe, and Scandinavia), is more significant in Germany (66.2%) than France (48.1%). France, on the other hand, is more interested in recipes from distant countries, for example from Northern Africa and the Middle East, whereas these recipes are negligible in Germany (Régnier, 2003). In other words, the exotic foods specific to France are from distant lands like Northern Africa, and to a lesser extent, from Sub-Saharan Africa and the Tropics (i.e. former colonies).

The diffusion of exotic foods through time highlights other national particularities. Throughout the period, the proportion of exotic recipes related to the total number of recipes – exotic and national – is higher in the German magazines than French ones. This suggests a much greater interest within Germany in the discovery of foreign culinary practices as shown in Figure 9.1.

The interest in exotic foods was very strong in the German magazines until the 1970s; after World War II Germany was very attracted to exotic foods. As noted in Table 9.1, this proportion diminished in the 1980s, although *Burda* shows a new taste for exoticism in the 1990s. In France, the diffusion of exotic recipes displays a more stable profile, except in the 1980s in *Marie Claire* (Table 9.1) where the proportion of exotic recipes is particularly significant. *Modes et Travaux* in particular throughout its history shows a low proportion of exotic recipes.

II.B. History of gastronomy

The stability of the diffusion of exotic foods in France, and on the contrary, the strong interest in Germany after World War II, can partly be explained

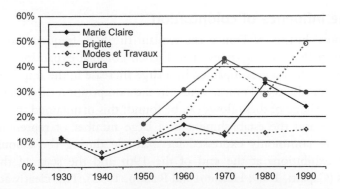

Figure 9.1 Comparison of proportion of exotic recipes in relation to total number of recipes (% per 10-year period) in French and German magazines.

by the differences in the history of gastronomy. The early establishment of the state with a strong royal court and centralization in France at the end of the 17th century has been the framework for formation of a high standard cuisine in which there is an established interest ('taste') for foreign cuisines and foodstuffs (Elias, 1939a, 1939b; Mennell, 1987). For some historians, the French interest in foreign food habits is even older, dating from the Crusades (Ketcham-Wheaton, 1984). This taste for foreign foods, which has been observed since the first major French cookbooks, is linked to a great extent to the numerous trips of French cooks abroad for work and pleasure, for example Vincent La Chapelle at the beginning of the 18th century who traveled in England, the Netherlands, Portugal and possibly the West Indies; or Antonin Carême at the beginning of the 19th century who traveled in England and Russia and visited many other courts of Europe (Carême, 1833). They transformed and enriched the French art of cooking with foreign products and methods of cooking discovered outside of France (Régnier, 2003).

This former interest could explain the much slower and more stable diffusion of exotic recipes in the French magazines of the 20th century (it is an ancient tradition). The strong interest in foreign recipes in Germany after World War II can be explained by the amplification and diversification of alimentation in the 1950s following years of privation and misery. Germany had experienced years of inflation, unemployment, economic crisis, and the disappearance of many foodstuffs (in particular imported food products), as well as the hunger and privations of World War II and the years just afterwards. The following years in which alimentary produce once more became accessible saw unrestrained alimentary consumption (Wildt, 1996). The diffusion of exotic foods during the 20th century is moreover related to three social phenomena: colonization, immigration, and tourism, all of which involve contact with a foreign people.

II.C. The influence of colonization

The colonial history of France can explain the high frequency of distant exoticism in French magazines. The French colonial expansion was long and extensive. It was accompanied by a large increase in the importation of alimentary products to France, where in the composition of several exotic foods the heritage of these flows can be found; this maintained the taste for exoticism – and the establishment of a large number of representations, which were commonly expressed for example during the colonial and universal exhibitions at the end of the 19th and the start of the 20th centuries (Capatti, 1989; Hassoun and Raulin, 1995). This clearly accounts for the specificity of the exoticisms of Northern Africa and the West Indies in France, developed at the time of the colonial expansion. By comparison, the German colonial expansion was much more limited in space and time, which can explain the smaller representation of distant exoticisms in Germany.

The decolonization, and more particularly, the wars of decolonization, also had an influence in the taste for exotic products in France; the disappearance of Algerian food in France takes place at the time of the war in Algeria, and the time of the war in Indochina coincides with a very low number of Asian recipes. In time of war, foreign foods are not attractive anymore, which proves that an unknown product is consumed only if it is perceived in a positive way, which is not the case when it is the food of one's enemy.

II.D. Immigration streams and prestige

Colonization also was at the origin of migratory flows. Some may be at the origin of the taste for foreign products. The interest in Asian foods can be explained to a greater degree by migratory streams, especially with the arrival of refugees from Vietnam in the 1970s. The economic activity of the Asian community is characterized by the importance of the alimentary sector. The restaurant sector is the most visible and well-known part of Asian commerce. Furthermore, Asian groceries and restaurants are 'exotic': before the 1970s these restaurants were devoted to French customers searching for typical Asian dishes and thus constituted an important factor in the discovery of Asian food. These restaurants were open to families and middle-class customers, which favored the discovery of Asian food practices (Ma Mung and Simon, 1990).

However, the relationship between immigration and culinary exoticism is usually difficult to establish. For example, North Africans are the biggest foreign community in France, whereas North African recipes account for only 3.4% of the total number published in French magazines, a fact related to various factors. First of all, North Africans in France have proximity food stores but they do not sell exotic products (except for stores selling 'harissa' or 'halal' meat, specifically for Muslim consumers). In addition, this weak correlation between demographic importance of an immigrant population and culinary exoticism is related to the ambivalent status of the immigrant. Immigrants are generally in marginal economic activities; they face more unemployment, unstable job conditions, and less income. Such images of the poor worker are without prestige and do not correspond with the taste for foreign food. Moreover, in situations of migration, proximity with immigrants is strong and the dream of exoticism can no longer be developed. In order to consume foreign foods, it is necessary that foreigners remain slightly distant and mysterious: when foreigners are too close, the differences are exacerbated and perceived negatively.

One can thus underline a significant relation between consumption of new products and prestige. Food practices and social prestige are closely linked, and alimentary consumption, particularly in the upper class, can be viewed as a 'conspicuous consumption' (Veblen, 1899). As Flandrin (1995) showed, certain exotic products are rapidly integrated in the field of alimentary innovation when traditional indigenous produce is replaced. More importantly, Flandrin emphasizes that a foreign product can be easily introduced when it confers symbolic benefits or carries a connotation of prestige: 'For a foreign foodstuff to be adopted its mere presence is not sufficient, other conditions must be fulfilled. It must be known that it is appreciated by a socially superior group or population of prestige such as the Italians during the Renaissance' (Flandrin, 1989). In fact, spices illustrate clearly the relation between food consumption and prestige. Their use was strong in Europe in the 14th and 15th centuries and they were a marker of social distinction. However, spices were less consumed during the 17th and 18th centuries as they did not have any distinctive value anymore, and in consequence, were abandoned by the aristocracy. Before this period, spices came from the East, a very prestigious place for Europeans at the time, but from the 17th century on, Europeans began affirming their superiority in this part of the world. This started the decline in the prestige of the East followed by a decrease in the prestige of spices, which also contributed to the social decline of the East.

In this context, during the 20th century the movement of foreign elites was at the origin of a strong taste for exotic and foreign foods. For example in Paris, the arrival of White Russians fleeing the Russian Revolution explains the taste for recipes from Eastern Europe in France. Consequently, culinary exoticism is particularly sensitive to prestigious events. Another example would be the Olympic Games: in Tokyo in 1964 they were at the origin of a Japanese fashion in food.

II.E. Tourism and gastronomic discovery

Strong equivalences between tourism and exoticism can be pointed out. The most visited places also coincide with the most significant exotic foods (Southern Europe and Western Europe in particular). The contrary is also true. For example, Sub-Saharan Africa, a part of the African continent still seldom visited, is poorly represented in the food field. Tourism no doubt constitutes a mode of discovering products, those unknown dishes that one discovers on holiday and then seeks to cook at home.

The more recent fashion of Italian cuisine and other forms of cooking from Southern Europe rely to a great extent on the influence of tourism in this area. Thus, Southern Europe, which is distinguished in exoticism by its high occurrence, is also the most popular area in the world for tourism; a link between the development of tourism in the Mediterranean and a developing taste for the cuisine of this region is quite clear. Thus, tourism explains to a large extent the success of Southern European dishes. At the same time Southern Europe constitutes the most familiar and nearest of places where consumers can find new flavors, and new products they are not used to eating.

However, it should be noted that Germany has a longer and more important history of foreign tourism than France. In the 1990s, the level of foreign departures was 60% for Germany compared to only 16% for France. Thus, tourism is a more relevant factor in Germany than in France; German magazines regularly refer to tourism and holidays from the 1970s onwards.

III. FOREIGN FOODS ARE LIKE NATIONAL PRODUCTS

What are the reactions towards culinary novelty? The first is fear. In order for foreign products to be eaten, they cannot appear to be too different from the familiar native products.

III.A. Substitution strategies

In the consumption of foreign foods and integration of new products into French (or German) cooking, various processes exist. These processes seek to replace unknown practices with more familiar food habits. Culinarily speaking, many processes rely on the substitution of different products, which is based on the analogy between exotic and autochthon products. This substitution has two different meanings. The first would be when local products can take the place of some exotic products in exotic recipes. For example, in 1937 *Marie Claire* proposed a recipe from the West Indies called *acras*. The magazine explained that *acras* was made originally with 'vegetable roots which do not exist under our climates' (*Marie Claire*, November 1937), mentioning the use of yam or sweet potato – still barely known at the time – and that the reader could replace these vegetables with Jerusalem artichokes. Substitution, in this case, is certainly related to the difficulty of getting still rare products, but it is also a manner of making a French dish from an unknown dish. Second, substitution also consists of the introduction of an exotic product in a French or German traditional recipe. *Marie Claire* mentioned, for instance, an 'exotic flognarde' (*Marie Claire*, 1974), a French traditional dessert; the flognarde was cooked with an exotic fruit at the time – the banana.

III.B. Conformity to national tastes

To integrate a culinary innovation, magazines also must underline that the foreign culinary practices are close to national culinary practices. Indeed culinary exoticism relies on the search for its proximity with national food practices and tastes. Thus, we select among exotic foods what is similar to national habits. Exoticism is an adaptation to typically French or German food practices and tastes. Parallels with French or German cooking are often made in the magazines to underline the proximity of exotic foods. Magazines mention the proximity of savors or products, and sometimes it is simply the proximity of speech explaining a foreign recipe through a comparison with the national practices. Moreover, magazines try to classify and insert within familiar food habits the unknown plants, products, or animals, replacing them within the French or German culinary culture; this preserves a familiar classification system and makes known a culinary practice that otherwise seems strange, explaining it by culinary analogies. For instance the evangelical expression, 'the daily bread of North Africa' (*Brigitte*, 1998) is employed for couscous.

In consequence the same exotic food is constructed differently in France and Germany. Thus, chili con carne is a marker of Mexican food in the two countries but it does not refer to the same dish. In France it is always a meat stew with beans, tomatoes, and chili powder, i.e. the combination of meat and a leguminous plant found in many other French dishes (e.g. lentils and morteau sausage). In Germany the dish refers to something else, mainly meatballs with red bean sauce, tomatoes, and chili powder, which is a typical German way of cooking, as meatballs are a very traditional dish.

This selection of what is most similar to one's national culinary practices explains to a very large part the success of Spanish exoticism in Germany. Indeed, the first marker of this exotic food in German magazines is the term *tapas*, which are numerous small dishes consumed in Spain with aperitif while waiting for the evening meal. *Tapas* perfectly fit with the German way of eating *Abendbrot*, the German word for dinner. *Abendbrot* is much simpler than in France; it is most often composed of salad, bread and varied delicatessen, and cheeses. As a result, *tapas* in Germany became a real dinner and represent a variant of *Abendbrot*.

III.C. Codification of exotic foods

Within this framework, the introduction of foreign foods involves the reduction and recombination of foreign culinary practices. This reconstruction is not due to hazard but refers to a logic of metonymy, the rhetorical figure that makes it possible to express a whole by one of its parts. Here, exoticism relies on the use of only one exotic product, and transforms a classical recipe into an exotic one: adding a slice of pineapple changes the traditional recipe to a Hawaiian one, such as the product 'Hawaiian toast,' which is toast with a slice of pineapple. The reduction of exoticism by combining a traditional recipe with Hawaiian pineapple relies on the history of the food industry. The first attempts to can pineapple were done in

Hawaii in the year 1880. It is also in Hawaii at the beginning of the 20th century that the first canning machine was developed. From that time onward, canned pineapples from Hawaii were produced and distributed by large American firms throughout the world, particularly in Europe where the product was especially successful (Péhaut, 1996).

Exoticism also relies on the recombining of foreign culinary practices around certain associations with the products, but associations that remain in conformity with national tastes, for example banana and rum in West Indian recipes, whereas in Germany the banana is generally associated with salted products, for example with ham and cheese in a 'toast with banana,' where the distinction between sweet and salty *flavors* is less strong than in France.

III.D. The importance of promoters

Even though they are at the origin of fashions, magazines also mention various promoters who can help introduce a culinary innovation. Thus, *Marie Claire,* a very Parisian magazine and very aware of fashion, often introduces culinary exoticism through cinema stars. In this manner, the first Japanese recipes were introduced in the magazine in May 1965 by actress Odette Laure (her cook was Japanese). Stars can thus appear via this mechanism of diffusion of exotic foods to be true innovators. Moreover, because of their high social status they are able to legitimize culinary innovation in a country (France) where gastronomy is closely related to prestige. In consequence, culinary innovations are first consumed by upper class members of society, and adopted in modest categories 10–20 years later; tastes and culinary novelties are then consumed from the top to the bottom of the social scale. This consumer behavior is linked to a process of distinction (where upper class members distinguish themselves from popular practices), to economic constraints (where exotic foods and new products are usually more expensive than native products), and to the influence of geographical area; for instance, a rural or urban location does not offer the same ability to buy new products. Consumer behavior may also involve different attitudes towards the innovation, such as attachment to a tradition in a rural environment or taste for a novelty in an urban environment.

Others promoters are medical doctors. *Modes et Travaux*, for instance, often refers to the medical discourse concerning a new product as being a guarantee of safety, making it possible to introduce new products with less fear since the medical arguments justify the health benefits of the exotic product. Thus, doctors authorize the pleasure of exotic food consumption.

We also can see in magazines the discourses on exotic food concerning the first reaction to new products, aimed at affirming the proximity of the exotic food with national foods; magazines therefore codify foreign culinary practices by transforming them into more similar national practices. At the same time, the curiosity, the research of something new, the desire for variety, and especially the seduction of something different, all lead consumers to go beyond their fear of the unknown.

IV. THE SEDUCTION OF FOREIGN FOODS

IV.A. Desire for new products

Indeed, two attitudes constitute what Fischler (1990) calls 'the paradox of the homnivore,' fear of and the desire for innovation – or 'néophobie' and 'néophilie.' We also find in women's magazines discourses valorizing the difference in exotic foods as compared to French or German foods. The origin of the seduction of exotic products relies much more on this difference. Consuming exotic foods means to experience charming and strange products, unexpected flavors, colorful dishes, and new ways of cooking. This is why, while conforming to French or German food habits, foreign culinary practices must remain a little mysterious, as expressed in *Modes et Travaux* (1987): 'Magic of the perfumes, savors and colors, the spices will blow on your kitchen a wind of exoticism, mystery and wonders.' In consequence, foreign foods are better than one's traditional food and here is the singularity of exoticism (Todorov, 1989).

Exotic foods are opposed to the national daily and familiar food practices, and appear to be in opposition to our common habits. In consuming exotic foods, it is possible to vary one's daily habits, to fight against the boredom (of eating the same foods), or to renew a food tradition. Exotic products make it possible to leave the ordinary behind. Consequently, exoticism is related to leisure, feasting, sensuality, and sensuousness (Todorov, 1989).

IV.B. Traveling during a meal

The consumption of exotic food is also seductive because it gives the illusion of travel during a meal, e.g. creating during a dinner the mood of the holidays and traveling abroad, or a way of anticipating the pleasure of a trip to come or of recalling memories of a past holiday (Chiva, 1993; Régnier, 2005). In this sense, culinary exoticism is a way to discover foreign countries, in contrast to regional foods, which are a way to discover the treasures of local (i.e. French) culinary practices (Csergo, 1996). Thus, the consumer can be transported on the road to India, for example, thanks to an exotic Indian meal prepared at home. Cooking an exotic product is therefore a form of tourism that is far easier to achieve by cooks who remain in their own kitchen.

The consumer of exotic food can also travel through time, thanks to a kind of 'historical exoticism' (Verdier, 1979), which is only anecdotal, but which mobilizes pleasant images, increasing the effect of exoticism. For example, Cortez the conqueror at the beginning of the 16th century was delighted by Mexican tamales – corn sheets of dough stuffed with meat; the conqueror was thus conquered by this culinary innovation.

Finally, consumption of exotic food is a trip through the 'words' that create exoticism because the simple name of a dish, and not its composition, can be enough to create the difference and the geographical distance. The terms in foreign language, often employed by the magazines, create the dream and increase the effect of exoticism. One may remember what Lévi-Strauss said: 'a food should not only be good to eat, but also good to think.' The terms in a foreign language increase the effect of exoticism; they make one dream, by the mystery they introduce, by their play on sonorities.

IV.C. Discovering new practices

Cooking new products is also a way to discover new culinary practices while remaining in the kitchen, and a way to discover new countries: 'We will make you discover the Gulf of Genoa and its gastronomy this summer,' wrote *Marie Claire* (1989) proposing to reveal still unknown and ignored culinary treasures. Women's magazines evoke rules of preparation and tasting as well. A Swedish buffet, for example, 'must be very abundantly furnished. It is necessary to taste the dishes according to a rigorous order . . . Between each dish . . . eat a steamed potato' (*Marie Claire,* 1970). The use of two expressions (of injunction), 'must be' and 'to be necessary,' as well as the imperative verbs suggest that many rules should be followed if one wants to be faithful to the tradition of origin.

Simplification and codification of foreign foods does not prevent the search for a certain authenticity; if we study culinary exoticism over a long period, we discover that the exotic recipes become more respectful of original practices. In France, culinary exoticism was perceived for a long time from a national point of view. For example, until the 1980s, the Italian recipes that came from 'Italy' were recognized globally. Progressively, with the diffusion of Italian exoticism, in particular from the 1980s, several Italian regional foods appeared (from Tuscany, Venetia, Milano, etc.), indicating an interest in the regional food habits of Italy. In the same way, the exotic foods are refined and specified progressively with their diffusion, allowing the integration of new products. For example, in the 1930s the characteristic spice of West Indian food was curry (not specifically West Indian in origin; curry is a Tamil word; the spice itself comes from East India); at the end of the 1950s, colombo (specifically a West Indian spice) finally appeared, illustrating a greater conformity with what was really consumed in the French West Indies. Consequently, foreign foods are a way to improve national cooking, especially because they encourage consumers to eat unknown products. The food repertoire of consumers therefore increases through the consumption of foreign foods.

IV.D. Good for one's health

Finally, exotic foods are indeed perceived as good for one's health and the exotic products as having significant therapeutic virtues, another factor that encourages consumers to eat new products. The link established between food and health is inherent to eating; by the incorporation of a foodstuff, the consumer eats its properties, in particular its virtues, real or imaginary, when the food is regarded as positive. The consumer indeed also incorporates the symbol of this foodstuff, in relation to the magic thought of incorporation and contagion (Frazer, 1911). Exotic products are regarded as particularly good for the health. Very ancient representations point out the strong correlation established between the mystery of a product and its therapeutic virtues, as related to the degree of social prestige of a food. The medicinal capacity of exotic products relies on their strangeness and scarcity; consequently, it also relies on their prestige, obviously in relation to social distinctions. This process is all the more important as the process of integration of food innovation relies on the relation between taste, food practices, and prestige. In medieval culture, spices, to which so many virtues were attributed, were also symbols of luxury. Consequently, they were consumed a lot: 'History shows that in a great number of cases, the medicinal virtues lent to a product justify its use. Then this medicine penetrates the menus' (Fischler, 1990).

During the 20th century, two categories of exotic products were regarded as particularly good for the health: spices, which have the global power to stimulate the body; and fruits, which are good for the health because of the vitamins they contain and the slimming capacities of some (e.g. the 'pineapple diet,' or the 'Hollywood diet' based on the consumption of various exotic fruits). Two exotic types of food (Mediterranean and Asian) are also thought to be particularly good for the health and to constitute proper diets; they are used in Southern Europe recipes (considered very healthy because of the balanced nature of the dishes) and in Far East recipes (particularly useful in the fight against obesity and heart conditions).

V. CONCLUSION

The consumption of new foodstuffs takes place in a particular national context, revealing the historical influence of gastronomy, as influenced by the colonization of a country, immigration flow, and tourist migrations, resulting in the introduction of foreign foods and a stream of new products.

The mechanism of integration of new products depends on several factors: a country's history, the economy, technical conditions (Bruegel, 1995; see for history of canned food use), human flow, and food product flow. However, the consumption of poorly known foods is not easy. In consequence, various processes can be used to introduce foreign culinary practices. These include, first of all, processes aimed at replacing these practices within a familiar framework: strategies of substitution and processes of selection and recombining of foreign practices, establishing proximity between the national food habits and foreign ones (in short, making edible what is unknown). It is through this assimilation of the unknown (Lévi-Strauss, 1962) that we can, in an almost paradoxical way, enjoy the consumption of new products. At the same time, the seduction of foreign products comes from their differences with ours; the desire for variety leads consumers to go beyond their fear of the unknown. Consequently, it is important to maintain a kind of mystery. Through exoticism, our national cooking becomes richer and, for albeit one time only, foreign habits may appear better than the national food. Culinary innovation is then integrated all the more easily but only if there are prestigious and positive connotations, and if it is neither too unknown nor too familiar.

REFERENCES

Augé, M. (1989). *Aimer, manger, mourir, in Noirot Paul (dir.), 1989, L'honnête volupté. Art culinaire, art majeur*. Paris: Michel de Maule, 6–9.

Besnard, P., & Grange, C. (1993). La fin de la diffusion verticale des goûts? *L'année sociologique, 43*, 269–294

Bourdieu, P. (1984). La métamorphose des goûts. In *Questions de sociologie* (pp. 161–172). Paris: Éditions de Minuit.

Bruegel, M. (1995). 'Un sacrifice de plus à demander au soldat': l'armée et l'introduction de la boîte de conserve dans l'alimentation française, 1872–1920. *Revue Historique, 596*, 259–284.

Capatti, A. (1989). *Le goût du nouveau. Origines de la modernité alimentaire*. Paris: Albin Michel.

Carême, A. (1833). *L'art de la cuisine française au XIXe siècle*. Paris: Payot.

Certeau, M., Giard, L., & Mayol, P. (1994). *L'invention du quotidien II, Habiter, cuisiner*. Paris: Gallimard.

Chiva, M. (1993). L'amateur de durian. In: C. N. Diayne (ed.) La gourmandise. Paris, Autrement, série Mutations 140: 90–96.

Csergo, J. (1996). L'émergence des cuisines régionales. In: Flandrin, J.L., Montanari, M. (eds.) Histoire de l'alimentation, (pp. 823–841). Fayard, Paris.

Elias, N. (1939a). *La civilisation des mœurs*. Paris: Calmann Lévy.

Elias, N. (1939b). *La dynamique de l'Occident*. Paris: Calmann Lévy.

Fischler, C. (1990). *L'homnivore, p. 165*. Paris: O. Jacob.

Flandrin, J. -L. (1989). Le lent cheminement de l'innovation alimentaire. In: Piault, F. (ed.) Nourritures. Plaisirs et angoisses de la fourchette. Autrement, Paris 108: 68–74, 97.

Flandrin, J.-L. (1995). L'innovation alimentaire du XIVe au XVIIIe siècle d'après les livres de cuisine. In N. Eizner (Ed.), *Voyage en alimentation* (pp. 19–36). Paris: A.R.F. Editions.

Frazer, J. G. (1911). *Le rameau d'or*. Paris: Robert Laffont.

Hassoun, J.-P., and Raulin, A. (1995). Homo exoticus. In: S. Bessis (ed.) Mille et une bouches. Cuisines et identités culturelles. Autrement, Paris, 154: 119–129.

Ketcham-Wheaton, B. (1984). *L'office et la bouche*. Paris: Calmann-Lévy.

Lévi-Strauss, C. (1962). *La pensée sauvage*. Paris: Plon.

Ma Mung, E., & Simon, G. (1990). *Commerçants maghrébins et asiatiques en France*. Paris: Masson.

Mennell, S. (1987). *Français et Anglais à table, du Moyen Age à nos jours*. Paris: Flammarion.

Péhaut, Y. (1996). L'invasion des produits d'outre-mer. In J. L. Flandrin, & M. Montanari (Eds.), *Histoire de l'alimentation* (pp. 747–766). Paris: Fayard.

Régnier, F. (2003). Spicing up the Imagination: Culinary Exoticism in France and Germany, 1930–1990. *Food and Foodways*, *11*(4), 189–214.

Régnier, F. (2004). *L'exotisme culinaire. Essai sur les saveurs de l'Autre*. Paris: Presses Universitaires de France.

Régnier, F. (2005). Le monde au bout des fourchettes: voyage dans l'exotisme culinaire. Website: http://www.lemangeur-ocha.com/fileadmin/Pdf_agenda_et_actus/Regnier_Exotisme_culinaire.pdf

Todorov, T. (1989). *Nous et les autres. La réflexion française sur la diversité humaine*. Paris: Seuil.

Veblen, T. (1899). *The Theory of the Leisure Class*. New York, London: Macmillan.

Verdier, Y. (1979). *Façons de dire, façons de faire, p. 277*. Paris: Gallimard.

Warde, A. (1997). Consumption. *Food and Taste*. London: Sage Publication.

Wildt, M. (1996). Vom kleinen Wohlstand. Eine Konsumgeschichte der fünfziger Jahre. Fischer Taschenbuch Verlag GmbH, Frankfurt-am-Main.

OTHER REFERENCES: FASHION MAGAZINES

Brigitte. Issues: Jan. 1952 through Dec. 1999. Grüner und Jahr, Hambourg. (Quoted: 8, Apr. 1998, p. 226).

Burda Moden. Issues: Jan. 1950 through Dec. 1999. Verlag Aenne Burda, Offenburg.

Marie Claire. Issues: Mar. 1937 through Dec. 1997. Paris. (Quoted: 38, Nov. 1937, p. 46; 257, Jan. 1974, p. 75; 443, Jul. 1989, p. 139; 220, Dec. 1970, p. 59).

Modes et Travaux. Issues: 1919 through Dec. 1997. Paris. (Quoted: 1042, Sept. 1987, p. 166).

CHAPTER 10

Consumer Response to a New Food Safety Issue: Food Terrorism

Jean Kinsey, Tom Stinson, Dennis Degeneffe, Koel Ghosh, *and* **Frank Busta**

Contents

Abstract

Deliberate contamination of some component of the food supply with the intention of doing physical or economic harm, or creating fear (terror), defines food terrorism. Food is one of several vectors used to induce intense prolonged fear with imagined or real future dangers, and has been used around the world. Witness such examples of mercury found in Jaffa oranges in five European countries in 1978, salmonellae in salad bars in the U.S. in 1984, and rat poisoning in school breakfasts in China in 2003. More than four thousand U.S. consumers were surveyed in 2005 to ascertain their level of concern about food defense measures relative to food safety, and their preferences for allocating resources to defend the food system compared to defending other infrastructure such as airlines, public transportation, or national monuments. Contrary to actual food defense expenditures, consumers indicated that 13% more should be allocated to food systems than to airlines. Twenty-six percent were not confident the food supply is safe; 55% were not confident the food supply is secure. Segmenting the representative sample into archetypes, based on attitudes, shows that those who have a high fear level as well

Global Issues in Food Science and Technology
© 2009 Elsevier Inc.

ISBN 9780123741240

as those who are risk averse allocated the most to defending food. Food defense has become another factor in the quest for safe food consumption. Consumers are more concerned about food terrorism than food safety.

I. INTRODUCTION

Deliberate contamination of some component of the food supply with the intention of doing physical or economic harm, or creating fear (terror), defines food terrorism. Food is one of several vectors used to induce intense prolonged fear with imagined or real future dangers, and has been used around the world. Witness such examples of mercury found in Jaffa oranges in five European countries in 1978, salmonellae in salad bars in the U.S. in 1984, and rat poisoning in school breakfasts in China in 2003 (Kennedy and Busta, 2006). In this century terrorism of all types has heightened awareness, spawned research into new technologies to detect and decontaminate, and intensified protection and defense protocols in food plants. Consumers are not only beneficiaries of increased security measures but also the victims of fear and potential physical harm.

Food defense refers to activities that protect an already safe food supply from deliberate contamination with the intent to cause catastrophic physical harm or death to many people, and/or to create an economic disaster within the food industry by causing it to stop production and sale of food for an extended period of time. Food safety traditionally deals with the accidental or natural contamination of food with an agent that has made people ill and may lead to death. The agents of concern are innate pathogenic microorganisms and their proliferation pathways. Food defense deals with the deliberate contamination of food with unfamiliar agents that typically cause high mortality rates.

Consumers in the United States as well as much of the rest of the world have come to expect that the food they purchase is safe. Good manufacturing practices, food safety regulations and inspections, and the protection of private brand names all conspire to provide food that can generally be considered safe to eat. Keeping food and the food supply chain safe is the business of those involved in food protection and defense. It is increasingly the business of every food producer, manufacturer, distributor, and retailer.

II. CONSUMER BEHAVIOR THEORY

Psychologists know that consumers willingly accept high levels of risk but are loath to being subjected involuntarily to risk and uncertainty (Lowrance,

1976; Kuchler and Golan, 2006). Classic examples of this behavior are the general acceptance of a high risk of injury and death from driving an automobile compared to outrage and anger over the low risk of a dreaded disease like cancer. Similarly, the low risk of harm from a terrorist attack on the food system, which is beyond the control of individual consumers and their loved ones, and something they could be subjected to involuntarily, leads to more fear and outrage than more familiar hazards.

Behavioral economic theories predict that consumers would willingly pay more to prevent a loss than to obtain a gain (Kaheman and Tversky, 1979). Consumers generally consider their food supply to be safe, something they take for granted in most countries. For their food to be contaminated deliberately would be considered a great loss. Therefore, we expect consumers would willingly pay more to prevent these types of losses than to gain an increment of food safety.

Becker and Rubinstein (2005) found that those who are frequent users of a product and/or have a large investment were less likely to be intimidated or to stop using the product after a terrorist attack. This is also consistent with the voluntary nature of product use and the differentiation between gains and losses. For example, airline pilots whose livelihood depends on flying are less likely to stop flying than the casual vacationer. The pilot has salary to gain; the vacationer perceives a potential loss and finds alternative modes of transportation. Likewise, consumers must eat food to live so the contamination risks that accompany this act are involuntarily received. Being harmed by eating food deliberately contaminated is an involuntary risk and therefore more fearsome than a risk taken voluntarily (like flying). When it comes to allocating funds for protection against terrorist attacks, one would expect consumers to allocate more to protect food than airlines on the basis of behavioral theory.

III. RESEARCH ON CONSUMER ATTITUDES AND PREFERENCES

A large survey of U.S. consumers' attitudes and concerns about terrorism in the United States was conducted over the internet in the first week of August, 2005. The survey, funded by the National Center for Food Protection and Defense, was designed to elicit the relationship between consumers' levels of fear and preferences for allocating funds to protect and defend the food system relative to the airlines and other potential terrorist targets. Responses were obtained from 4,260 U.S. residents over the age of 16, and then weighted by age, race and ethnic origin, sex, income, and geographic region to reflect the characteristics of the national population.

Respondents were asked how likely they believed each of six different types of attacks to be and about their perceptions of the physical, economic, psychological, and emotional damage each type of act would inflict on the country and on them personally. The separate terrorist acts covered by the survey included: another aircraft hijacking, an incident involving some other form of public transportation, destruction of a national monument, deliberate contamination of the food supply, disruption of the power grid, and release of a chemical or biological agent in a public area.

To provide a further indication of the relative concern U.S. residents attach to different types of terrorist attacks, respondents also were asked how they believed anti-terrorist spending should be allocated among the potential types of targets. The exact wording of the question was: 'For every $100 that you think should be spent to protect the country from terrorism, how would you divide it across the following types of attacks? Enter a dollar amount for each. The amounts must sum up to $100.' The choices given – another attack on a passenger aircraft, an attack on other public transportation, destruction of a national monument, deliberate chemical or biological contamination of a common food product, disruption of the electrical power grid, release of a biological or chemical agent in a crowded public area and 'other' – were randomized across survey respondents; the passenger aircraft and 'other' public transportation alternatives were always paired; 'other' was always the last option; thus an alternative terrorist target could be chosen by respondents. Asking respondents to divide $100 among specific types of terrorism risks, rather than to construct a contingent valuation (willingness-to-pay) question, has the advantage of being readily understood and simple to do. A preference ranking is obtained, indeed a percentage measure, for the allocation of finite resources. This type of question was designed after conducting four focus group studies to determine how consumers can readily rank and 'price' the value of reducing various types of terrorism risk.

It should be noted that the $100 in question was not specified as a public or private expense. Some of the total spending to protect and defend the food supply, as well as other potential terrorist targets, would be from private companies, some from individual consumers taking precautionary measures, and some from public (government) agencies. Nevertheless, the average percentage allocations obtained in this survey provide a measure of the relative intensity of concern over different types of terrorism. The percentage allocations can also be used to provide a crude estimate of the amount the public believes should be spent on anti-terrorist activity by comparing this amount against known levels of spending.

IV. CONSUMER PREFERENCES FOR ALLOCATIONS TO DEFENSE

Most respondents to the survey believed that further terrorist attacks would occur in the near future. At least one act of terrorism within the next four years (after August 2005) is expected by 95% of the public (Table 10.1). Differences in the perceived probabilities of attacks within four years are relatively small, but differences between pairs of terrorist acts were all statistically significant except for those between release of a chemical or biological agent in a public area and disruption of the power grid. After possible attacks on trains or subways are excluded, nearly 81% of the public expects at least one terrorist incident during the next four years. An attack on the food system is thought least likely, but 44% of U.S. residents still expect an attempt to introduce a toxin into the food supply chain within the four years following the survey.

Survey respondents were asked how the nation's anti-terrorism budget should be divided to protect against different types of terrorist attacks. Despite their belief that other types of terrorism are more likely in the near future, the public believes the implications of an attack on the food system are so serious that a greater percentage of anti-terrorism spending should be allocated to protecting the food supply than to defending other potential targets (Table 10.2).

On average, U.S. residents believe that more than 19% of the resources spent to protect against terrorism should be spent to defend the food supply chain. Protection against the release of a chemical or biological

Table 10.1 Percentage of U.S. residents expecting a terrorist attack within the next four years, by type of attack, August 2005

Attack target	Percent
Passenger aircraft	53
Public transportation	84
Destruction of a national monument	49
Deliberate chemical or biological contamination of a common food	44
Disruption of the power grid	51
Release of a chemical or biologic agent in a crowded public area	51

Table 10.2 Percentage of anti-terrorism spending in United States that residents believe should be allocated for protection against particular types of terrorist attacks, by type of terrorist attack, August 2005

Attack target	%	% Aircraft spending
Passenger aircraft	16.88	100.0
Public transportation	17.06	101.1
Destruction of a national monument	8.16	48.3
Deliberate chemical or biological contamination of a common food	19.13	113.3
Disruption of the power grid	14.97	88.7
Release of a chemical or biologic agent in a crowded public area	18.90	112.0
Other	4.91	29.1

agent in a public area is also seen as a high priority, receiving almost the same percentage allocated as protection against the nation's food supply.

The public believes that about 17% of the anti–terrorism budget should be spent to secure subways and railways, and slightly less than 17% to protect airline transportation. Preventing disruption of the power grid was allocated 15% of the anti-terrorism budget, while 8% would go to preventing destruction or damage to a national monument and 5% to preventing other forms of terrorism.

The percentage allocations chosen for protection against different types of attack are all statistically significant. Differences between the public's allocations for activities to protect the food supply chain or to protect against potential chemical and biological attacks versus the percentage that should go for protecting air transportation are also statistically significantly different at the 95% level.

When the public's percentage allocations for the homeland security budget are converted to spending levels as a percentage of the amount believed appropriate to provide secure air transportation, U.S. residents would allocate 13.3% more protection for the food supply than for airline travel. Preventing a chemical or biological attack was given 12% more, and protecting other transportation activities 1% more than preventing an aircraft hijacking.

V. FOOD TERRORISM ALLOCATIONS AFTER EDUCATION (POST SCENARIO)

Once respondents had made an initial distribution of funds for defending against possible terrorist activities a scenario describing the progression of events following a potential food terrorism incident was introduced.

> 'Emergency room visits and hospital admissions suddenly increase in the region where you live. A foodborne toxin is suspected to be the cause. The number of individuals affected continues to grow over the next several days and some of those hospitalized die. Similar patterns are seen in other metropolitan areas within the region. The number of fatalities associated with this problem grows. State and national agencies struggle to identify the source of the problem. Ten days after the first report a statement is issued by a government agency saying that there has been a deliberate attempt to contaminate the food system. By comparing the pattern of affected consumers and the distribution of various types of food products a single commonly used food product has been identified as the source. It is estimated that more than 50,000 units of the contaminated product have already been purchased. Consumers are instructed to bring all unused product to central collection sites for disposal. Ultimately the death toll from this incident reaches 1,500.'

After reading the above scenario, the survey respondents were again asked to allocate resources to combat terrorism. Differences between the naïve and post-scenario results are shown in Table 10.3.

As anticipated, the proportion of anti-terrorism spending that respondents believed should go to protect the food supply system increased substantially. Post-scenario responses call for programs protecting the food system to receive nearly 23% of all anti-terrorism spending, 3.75 percentage points more than before being informed of the potential consequences of an act of food terrorism. Protecting against release of a chemical or biological agent in a crowded public area remains the second highest priority; the proportion of the anti-terrorist budget that should be devoted to that mission remains almost constant, falling by just over 0.1 percentage points.

The proportion or resources devoted to protecting against airline hijacking fell by the largest amount, down 1.4 percentage points. Allocations to protecting other transportation systems, national monuments, the power grid, and other uses also fell, but by smaller amounts. All allocations in the post-scenario responses are significantly different from the airline allocation, and the allocations for each potential terrorist target changes significantly from the naïve responses, except for release of a chemical or

Table 10.3 Change in percentage of anti-terrorism spending in the United States that residents allocated for protection against particular types of terrorist attacks before and after reading a food defense scenario, by type of attack, August 2005

Attack target	% (Naïve)	% (Post-scenario)	Change in % age points
Passenger aircraft	16.88	15.48	(1.4)
Public transportation	17.06	16.39	(0.7)
Destruction of a national monument	8.16	7.84	(0.3)
Deliberate chemical or biological contamination of a common food	19.13	22.88	3.75
Disruption of the power grid	14.97	14.41	(0.6)
Release of a chemical or biologic agent in a crowded public area	18.90	18.77	(0.1)
Other	4.91	4.24	(0.7)

biological agent in a crowded public area. Post-scenario allocations primarily take funds away from airline security and allocate them to food security.

Women, African Americans, and individuals with a high school education or less show the largest increases in the proportion of spending they believed should be devoted to protecting the food supply after being exposed to additional information. African Americans and individuals residing in the east and west south central states reduced the proportion of spending believed necessary to secure the airways by more than the national average, while men and college graduates reduced their allocation for that activity by less than the national average. There were relatively small differences in the changes in resources thought appropriate for other anti-terrorist activities (Stinson *et al.*, 2006).

VI. FOOD DEFENSE AND FOOD SAFETY

Concerns and the relative allocation of funds to food safety were determined as well as food defense. Since there is a long history of food safety programs and expenditures, this provides a useful benchmark for the priority consumers give to food defense. Soliciting concern about both food safety and defense reveals that 49% of respondents were very concerned (checked five or six on a six-point scale) about food defense but only 29% were very concerned about food safety, even though there is a significant correlation between the two overall. In contrast, almost one-third were very confident about food safety and only 12% were very confident about food defense. Again there is a significant correlation between the two. In general, people are more confident about food safety and less concerned about it than food defense. This might be expected since we have become accustomed to safe food and since food defense is a new concept and new fear.

A key question asked all respondents was how they would allocate $100 between food safety and food defense. On average, they allocated $52 to food safety and $48 to food defense, while 2,160 respondents split the allocation fifty–fifty. Of the 28% who allocated *more* to food safety than food defense, compared to the 22% who allocated more to food defense, there was a tendency to have a higher level of education, to be from an Asian ethnic group, to be a middle-aged single person, and to have more confidence in food safety than food defense.

When asked who should be responsible for food safety and food defense there was little distinction across parties in the food supply chain from farmers to consumers, but 8% more thought the government should be responsible for food defense (53%) than food safety (45%). Food manufacturers and the government were selected as the parties that should bear the greatest share of cost for both food safety and food defense. Twenty-one percent said food manufacturers should bear the food safety cost and 23% said manufacturers should bear food defense costs. Twenty-four percent said government should bear the food safety cost and 28% said government should bear food defense cost. Clearly there is a belief and understanding that costs are borne by both public and private parties.

Communication with the public in a food safety or food defense crisis is an important part of planning for damage control. Most consumers (82–84%) want to receive information about how they can protect themselves during a food safety or food defense crisis. In a food defense crisis, three-quarters of the respondents want information about the scope and significance of the crisis and 65% want to know who is responsible. When asked the source and amount of information they would like to receive in a food safety crisis, the highest-ranked sources were a tie between television and medical/public health officials, followed by the internet, then another tie between newspapers, federal officials, and officials from private companies involved. These were followed by radio and state government officials. The information sources that ranked lowest were religious leaders and university experts.

VII. SEGMENTED ARCHETYPES

Segmentation has long been a marketing research method used by private industry to identify consumer segments. Marketers have deployed successful business strategies by focusing on the needs of specific groups of consumers in the development of meaningful new products, and highly effective advertising campaigns. In the context of food defense, the segmentation approach can provide food companies with an indication of the importance their consumers place on measures to protect their food from deliberate contamination and institute protective measures to meet consumer expectations. This same approach can be applied to public sector research, providing government agencies and policymakers with a tool to better understand the diversity of consumer needs and to develop more effective programs and communications.

The segmentation approach used a battery of 75 consumer attitude/value statements developed from a set of similar questionnaires used in the private sector by marketing research companies, and findings from a set of focus groups conducted for the purpose of questionnaire development for this study. The range of statements was intentionally very broad and general so as to enable the identification of fundamental consumer values that relate to an individual's sense of security/vulnerability in relationship to a potential terrorist attack. These statements included such dimensions as: lifestyle, outlook on life, aspirations, fears, views of authority, self image, health orientation, family focus, sense of social responsibility, and moral standards. In developing this battery of statements, each of these dimensions was believed to have some relevance to a person's concern over potential terrorist attacks. A balanced six-point Likert scale was used so as to enable some degree of discrimination without a neutral point. Respondents were asked their degree of agreement as to whether the statement described them with anchor points of 'strongly agree' (6) to 'strongly disagree' (1). The first five items on this list of statements are listed in Figure 10.1.

Other measures used in the segmentation analysis include concerns over different types of terrorist targets, expectations for the timing of potential attacks, and the allocation of defense spending by potential targets. These measures were used as a set of 'result' measures (dependent variables) in a factor analysis to identify relationships between individual attitude/value measures and orientation to terrorist attacks. The results of factor analysis provided the basis for collapsing the 75 attitudinal statements into a manageable number of clusters. These clusters (segments) and the dependent variables were used in canonical analysis to determine how each segment is related to dependent variables such as the allocation of $100 to various types of terrorist attacks. Once the segments were identified, all other questions in

Q- 14	Please indicate how much you agree or disagree with each of the following statements by using the 6 point scale where '1' means you 'Strongly Agree', and '6' means you 'Strongly Disagree'. Please circle the number you think best describes you.

Attitude/Values/Lifestyle Battery	Level of Agreement – Please Circle <u>one</u> for Each					
	Strongly Agree					Strongly Disagree
1 I like the challenge of doing something I have never done before	1	2	3	4	5	6
2 I like trying new things	1	2	3	4	5	6
3 I often crave excitement	1	2	3	4	5	6
4 I would like to spend a year or more in a foreign country	1	2	3	4	5	6
5 Everyone has the power to be successful if they just work hard	1	2	3	4	5	6

Figure 10.1   Example of attitude questions.

the questionnaire were used to describe the profile and characteristics of the people in each segment. These segments can be thought of as archetypes of consumers.

Indexes are used extensively to explain the findings and to make comparisons across archetypes relative to the total sample of citizens. For most scaled questions, these indexes are based on the percent of respondents who answered in the 'top two boxes' of the rating scale, either a five or a six. For example, the percent of respondents who marked a five or six on the attitude statements in Figure 10.1 would be the 'score' for that statement. The index is then computed as the 'score' for the archetype in segment one divided by the 'score' for the total sample on the same statement.

For all differences sited in the analysis, a significance test was done on the **mean** (not the 'top two box score' or the index) difference between the segment and the total sample. Unless otherwise noted all differences are significant at 0.95 or higher.

VII.A. Segments found

Six consumer segments were identified. These segments were studied with respect to the pattern of attitude/value statement responses and named based on the analyst's (D. Degeneffe) net impression about the attitude/value mindset expressed by each respective segment. (Other names could apply.) These segments vary in size from the largest, 'Predestinarians' at 19.9% of the general population over the age of 16, to the smallest, 'Principled & Self-Disciplined' at 13.6% of the general population over the age of 16. The segments and their relative sizes are shown in Figure 10.2.

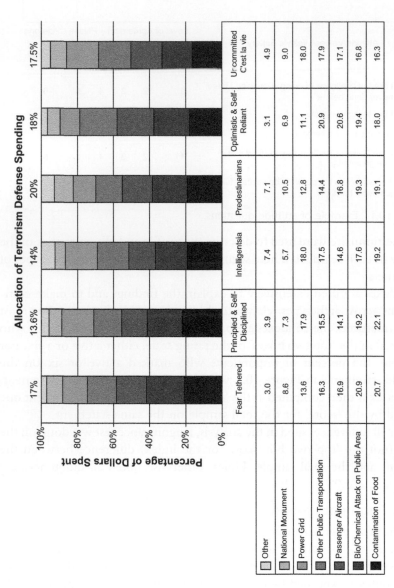

The image shows a chart titled "Allocation of Terrorism Defense Spending" with the y-axis labeled "Percentage of Dollars Spent" (0% to 100%). The data table below reads:

	Fear Tethered 17%	Principled & Self-Disciplined 13.6%	Intelligentsia 14%	Predestinarians 20%	Optimistic & Self-Reliant 18%	Ur-committed C'est la vie 17.5%
Other	3.0	3.9	7.4	7.1	3.1	4.9
National Monument	8.6	7.3	5.7	10.5	6.9	9.0
Power Grid	13.6	17.9	18.0	12.8	11.1	18.0
Other Public Transportation	16.3	15.5	17.5	14.4	20.9	17.9
Passenger Aircraft	16.9	14.1	14.6	16.8	20.6	17.1
Bio/Chemical Attack on Public Area	20.9	19.2	17.6	19.3	19.4	16.8
Contamination of Food	20.7	22.1	19.2	19.1	18.0	16.3

Figure 10.2 Names and size of population segments.

The differences in size are not dramatic, suggesting that each segment represents an appreciable proportion of the general population. Therefore, it is important to understand and address each segment equally in the development of policies and strategies, to mitigate the impact of potential terrorist attacks.

VII.B. Segment profiles described
VII.B.1. Segment 1

The *'Uncommitted C'est la vie'* segment tend not to worry about the unexpected and believe that the health threats they have heard about in the news are overblown. Relative to the general population (and other segments) they can be characterized as having low levels of concern over health and safety. The 'Uncommitted C'est la vie' attitude also carries over to social interactions. These consumers are less likely to worry about how others view them personally, and have yet to set any goals for their career or life in general.

The demographic profile of the 'Uncommitted C'est la vie' helps put some of these attitudes into perspective. Although this segment of the population includes people of all ages, there is a greater tendency of 'Uncommitted C'est la vie' consumers to be between 16 and 24 years of age, and male. Therefore, the attitude/value set likely reflects young adulthood, before the commitments of social, family, and career force one to consider the uncertainties and risks of life. The 'Uncommitted C'est la vie' are not likely to be concerned over the threat of terrorist attack, and are not likely to prepare for it. In the event of an attack they probably would be caught off guard, and experience a sense of helplessness. However they are more likely to have only themselves to protect.

VII.B.2. Segment 2

The *'Intelligentsia'* hunger for learning and experience. They value freedom of action and thought, and tend to question authority. The 'Intelligentsia' view themselves as more knowledgeable than most people and use this knowledge to gain the admiration of others. They are status seekers, but more from an intellectual standpoint than a material standpoint. The knowledge they accumulate gives the 'Intelligentsia' the sense that they are more in tune with reality. Therefore, they tend to be more concerned over health and security threats, as well as environmental risks.

The most notable demographic characteristic of the 'Intelligentsia' is education level. They are more likely to have a Bachelor's degree from college, or higher. They are nearly 50% more likely than the general population to have a post graduate degree. Also they tend to be 50 years or more of age, male, and live in one- to two-member households. Clearly the education levels attained by the 'Intelligentsia' are consistent with the attitudes and values they express.

When it comes to terrorism the 'Intelligentsia' will already have an awareness, if not a sense of fear over the potential for attacks. Their biggest concern is with the credibility of the information source.

VII.B.3. Segment 3

The *'Fear Tethered'* is the citizen segment with the greatest fear level in general, as well as with respect to terrorist attack. They have a much greater tendency than the general population (and other segments) to be frightened by the threat of disease, and threats to personal and family safety. This fear seems to emanate from a sense of powerlessness, such that the 'Fear Tethered' type senses little control or influence over future events. Still, they do have strong values and convictions. Family, religion, social consciousness, and the environment are all important. Further, they express a sense of ambition and set career/life goals. In this sense the 'Fear Tethered' manages their own affairs as best they can, but feels that they are at the mercy of dangerous forces beyond their control.

From a demographic standpoint, the 'Fear Tethered' tends to have family responsibilities. They have a higher tendency (than the general population) to be between the ages of 30 and 44, to be female, and have three or more members in households. They also tend to be moderately educated with high school diplomas to associate college degrees.

Clearly the attitudes/values of the 'Fear Tethered' emanate from their focus on the welfare of their family. They readily interpret the risks they hear about in the media to be risks to themselves and their family, and this results in fear and apprehension. In preparing this group for a real terrorist event, the need will be to provide objective factual information and a sense for how they can do what is most important, namely protect their family.

VII.B.4. Segment 4

The *'Principled & Self-Disciplined'* can be best characterized as risk avoiders. They deal with future uncertainty by planning and self discipline; they maintain a budget and set aside money for major purchases. They maintain a healthy and balanced diet, plan for the future, and have insurance policies in place. As a result, the 'Principled & Self-Disciplined' have an optimistic view of the future. Their principles are also reflected in other ways. They view others as inherently good, and have a strong social conscience, with a sense of responsibility for the welfare of society, and the natural environment. Still, from a moral standpoint they are conservative with strong personal integrity and religious convictions. 'Principled & Self-Disciplined' are practical and pragmatic people; they are less into the superficial trends and fashions, and adventure, or the need for the admiration of others.

From a demographic standpoint 'Principled & Self-Disciplined' people are generally average, but do tend to be older than the general population with many being over the age of 50. Additionally they have a slight tendency to be white, female, and to live in two-person households.

The 'Principled and Self-Disciplined' would likely be highly receptive to communications on how to prepare for the possibility of a terrorist attack. They would likely follow the advice of a credible spokesperson. In the event of an attack they would likely maintain a level head, and be willing to volunteer and help others.

VII.B.5. Segment 5

'Predestinarians' are generally supporters of the status quo, believing that future events are predestined to occur. They trust in the country's leadership and are optimistic toward the future, expecting that things will not be that different from the past. Therefore, they are less likely to be concerned over safety or sickness, and are less likely to plan for the future. 'Predestinarians' are the most morally conservative group, with a tendency to hold fundamental religious convictions and strong beliefs regarding gender roles. Still, they consider material wealth as being important, and are trend/fashion conscious. Demographically, 'Predestinarians' are much more likely than the general population to be under the age of 40, and to be moderately educated with high school diplomas to associate college degrees. They also tend to have incomes under $40,000, and live in rural and small metro areas. Also, relative to the general population, 'Predestinarians' have a higher incidence of minority groups: Hispanic, Black/African American, Asian, and Native American.

With 'Predestinarians' tending to trust the country's leadership, they are likely to rely on the government to protect and care for its citizens in the event of an attack.

VII.B.6. Segment 6

The **'Optimistic & Self-Reliant'** are likely to be highly involved in building careers and accumulating wealth. They are successful in life and prefer to assume leadership roles. They have financial plans and are optimistic toward the future. Therefore, they are fairly contented with life and are less likely to share the fears and anxieties of other segments.

The 'Optimistic & Self-Reliant' segment has the highest socio-demographic status. They are more likely to have incomes over $60,000 and are twice as likely to have incomes over $100,000. They tend to be more educated and live in larger population centers. Age (25 to 45) tends to reflect the career development life stage.

With respect to security from terrorism, the 'Optimistic & Self-Reliant' types are not likely to have invested much attention or thought. Careers are more likely to have taken a priority.

VII.C. Concern over terrorism by segment

Level of concern over terrorism varies dramatically across the segments. Respondents were asked to indicate how concerned they are about six alternative terrorist events. The event that received the highest level of concern in the overall sample was an attack on public transportation (other than airlines). For this type of attack, 88% of the 'Fear Tethered' segment indicated either a five or a six on the six-point scale, which contrasts sharply with a score of only 28% among the 'Uncommitted C'est la vie.' The top two box scores reflecting concern for an attack on public transportation (other than airlines) varied across the other segments as follows: 'Principled & Self-Disciplined' 76%, 'Intelligentsia' 69%, 'Predestinarians' 54%, and 'Optimistic & Self-Reliant' 52%. Similar patterns in levels of concern were seen across the other five types of terrorist events.

Allocation of $100 to defend the various types of terrorist targets against attack varied some across the archetypes. Figure 10.3 identifies the different allocations. The importance of these allocations is that they represent a surrogate of consumers' preferences and ranking of concerns. This information can be very useful to those making decisions about how to divide up scarce resources related to food protection and defense.

Allocation of defense spending does not vary greatly across archetypes – since the level of concern seems to pertain to terrorism in general. Still, some minor differences reflect different interest areas. For example, the 'Principled and Self-disciplined' allocated the most (22%) for food defense and the 'Uncommitted C'est la vie' allocated the least (16%). The 'Optimistic and Self-reliant' allocated considerably more to transportation modes.

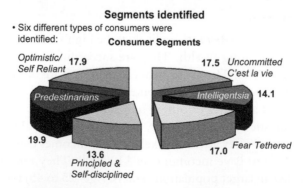

Figure 10.3 Percent of total sample in each segment.

VIII. CONCLUSION

Consumers in the United States in 2005 would allocate 13% more total dollars to defending the food system than the airline system against terrorist attacks. Based on current spending by the U.S. federal government to protect airline travel, this implies that $5.65 billion should be spent to protect the food system compared to the $93 million allocated to protect the food supply chain in the federal budget year 2007. Private companies are increasing their expenditures and vigilance related to food defense as well. In deciding how to defend the food system and to recover after a potential attack, understanding the preferences and behavior of consumers is important. Eliciting consumers' help and helping to educate them as to what they can do in self-defense can minimize the catastrophic impact of a potential attack. Sharing information and defense tactics is in everyone's best interest because defense of the nation's food supply is truly a public good from which we can all benefit.[1]

REFERENCES

Becker, G. S., & Rubinstein, Y. (2004). *Fear and the Response to Terrorism: An Economic Analysis. Unpublished manuscript.* Tel-Aviv University: University of Chicago Department of Economics and School of Economics.

Kahneman, D., & Tversky, A. (1979). Prospect Theory: An Analysis of Decision Under Risk. *Econometrica, 47*, 263–291.

Kennedy, S. P., & Busta, F. F. (2006). Biosecurity – Food Protection and Defense. In M. P. Doyle, & L. R. Beuchat (Eds), *Fundamentals and Frontiers* (3rd Edition). Washington D.C.: ASM Press. *Food Microbiology.*

Kuchler, F., & Golan, E. (2006). Where Should the Money Go? Aligning Policies with Preferences. USDA, ERS, Washington D.C. Amber Waves 4: 3, 31–37.

Lowrance, W. W. (1976). *Of Acceptable Risk: Science and Determination of Safety.* Los Altos, CA: William Kaufman.

Stinson, T., Kinsey, J., Degeneffe, D., & Ghosh, K. (2006). How Should America's Anti-Terrorism Budget Be Allocated? Findings from a National Survey of Attitudes of U.S. Residents. Working paper 01–06: The Food Industry Center, University of Minnesota, http://foodindustrycenter.umn.edu

[1] This research was conducted at The Food Industry Center, University of Minnesota. This research was supported by the U.S. Department of Homeland Security (Grant number N-00014-04-1-0659), through a grant awarded to the National Center for Food Protection and Defense at the University of Minnesota. Any opinions, findings, conclusions, or recommendations expressed in this publication are those of the authors and do not represent the policy or position of the Department of Homeland Security.

Food Safety

Rapid Methods and Automation in Food Microbiology: 25 Years of Development and Predictions

Daniel Fung

Contents

Abstract

Rapid methods and automation in microbiology is a dynamic area in applied microbiology dealing with the study of improved methods in the isolation, early detection, characterization, and enumeration of microorganisms and their products in clinical, pharmaceutical, food, industrial, and environmental samples. In the past 25 years this field has emerged as a sub-division of the general field of applied microbiology and is gaining momentum nationally and internationally as an area of research and application to monitor the numbers, kinds, and metabolites of microorganisms related to food spoilage, food preservation, food fermentation, food safety, and foodborne pathogens. Comprehensive reviews on the subject were made by Fung (2002, 2007) and are summarized in this chapter related to food safety and predictions of the future.

Global Issues in Food Science and Technology
© 2008 Elsevier Inc.

ISBN 9780123741240
All rights reserved.

I. INTRODUCTION

Medical microbiologists first became involved with rapid methods around the mid-1960s, and started to accelerate in the 1970s and continued to develop in the 1980s and 1990s, which continues up to the present today. Other disciplines such as environmental, industrial and pharmaceutical, cosmetic, water and food microbiology were lagging about 10 years behind. However, recently much progress has been made by all disciplines in microbiology in the field of rapid methods. Many symposia and conferences were held nationally and internationally to discuss the developments in this important applied microbiology topic. The most comprehensive hands-on workshop, the international workshop on Rapid Methods and Automation in Microbiology, was developed by author Daniel Y.C. Fung in 1981 at Kansas State University (Fung, 2002). To date about 4,000 people from 60 countries have participated in this international workshop.

II. ADVANCES IN VIABLE CELL COUNTS AND SAMPLE PREPARATION

II.A. Solid samples and liquid samples

The number of living organisms in a given product, on the surfaces in manufacturing environments, and within the air of processing plants is very important to the food industry. Colony-forming units (CFUs) are the standard way to express the microbial loads. In the past 25 years ingenious systems for 'massaging' solid and liquid foods via the Stomacher or 'pulsifying' food via the Pulsifier have been developed, wherein samples are homogeneously distributed in a disposable bag after 1–2 minutes of operation. There are more than 40,000 Stomacher instruments used worldwide. The Pulsifier instrument is especially suitable for 'shaking' microorganisms from the surface of the food without breaking the food matrix. As a result the 'pulsified' samples are a lot clearer with much less debris so that further microbiological manipulations are much more efficient.

Typically after 'Stomaching' or 'Pulsifying' the samples in the conventional method, one ml (after dilutions) is placed into melted agar to encourage microorganisms to grow into discrete colonies for counting (CFU/ml). These colonies can be isolated and further identified as pathogenic or non-pathogenic organisms. In the past 25 years convenient systems, such as (1) housing nutrients in films (via 3M Petrifilm, St. Paul, MN), (2) using a mechanical instrument (e.g. a Spiral Plater) to spread a sample over the surface of a preformed agar plate, (3) trapping microorganisms on a bacteriological membrane (via IsoGrid System, Neogen), and (4) looking for growth of target microorganisms on selective and non–selective culture media using an agar-less solidifying system (Redigel), have greatly helped to reduce labor time in performing viable cell counts, which is so important in the food industry. Chain and Fung (1991) made a comparative study of aerobic plate counts by comparing the effectiveness of the Petrifilm, Spiral plate, Redigel, and Isogrid systems on the enumeration of seven different foods. They prepared 20 samples each and found that the alternative systems and the conventional method (CFU/ml; or g) were highly comparable at an agreement of r = 0.95. In the same study they also found that the alternative systems cost less than the conventional system for making viable cell counts.

According to the authors (Fung *et al.*, 1980), the following Fung Scale is for viable cell counts in common food particles, edible food surfaces, and liquid foods:

0–2 log CFU/ml, g, and cm^2	Low count	Acceptable
3–4 log CFU/ml, g, and cm^2	Intermediate count	Caution
5–6 log CFU/ml, g, and cm^2	High count	Corrective actions
7 log CFU/ml, g, and cm^2	Index of spoilage	
8 log CFU/ml, g, and cm^2	Odoriferous	
9 log CFU/ml, g, and cm^2	Slime formation	
10 log CFU/ml, g, and cm^2	Discard immediately	

The scale of viable cell count is useful for ascertaining spoilage potential of most foods. This scale is not for pathogenic bacteria in food. For cooked foods, no pathogens are allowed. For raw ground beef no *Escherichia coli* O157:H7 is allowed. Of course, for fermented solid and liquid foods high levels of desirable organisms are needed to make excellent products and this scale does not apply.

II.B. Air samples and surface samples

The food industry also needs to ascertain the air quality as well as environmental surfaces for product manufacturing, storage, and transportation. An active air sampling instrument such as the SAS system of pbi (Milan, Italy) can 'suck' a known volume of air and deposit microorganisms on an agar surface (impaction). After incubation of the agar the number of organisms per cubic meter of air can be obtained. Another method is to suck a known volume of air and pass it through a tube of liquid (impingement) to capture the microorganisms in the liquid and later obtain CFU/$meter^3$ or ml. A variety of swabs, tapes, sponges, and contact agar methods have been developed to obtain the surface count in the food manufacturing environment, expressed as CFU/cm^2.

Fung Scale for airborne microorganism counts in food manufacturing plants (Al-Dagal and Fung, 1993):

0–100	$CFU/meter^3$	Acceptable
100–300	$CFU/meter^3$	Caution
>300	$CFU/meter^3$	Corrective actions

Fung Scale for food contact surfaces (knives, spoons, forks, chopping boards, table tops, etc.) (Fung *et al.*, 1995):

0–10	CFU/cm^2	Acceptable
10–100	CFU/cm^2	Caution
>100	CFU/cm^2	Corrective actions

The 3-tube or 5-tube Most Probable Number (MPN) system has been in use for more than 100 years in food and water microbiology. These MPN methods are used widely in Public Health Laboratories around the world. However, the methods are very time consuming, labor intensive, and use large quantities of glassware and expensive culture media. In 2007 a completely mechanized, automated, and hands-off TEMPO (bioMerieux, Hazelwood, MO) 16-tube MPN system was recently developed and marketed for ease in operating this tedious yet powerful viable cell count procedure. The operator only needs to make a 1:10 dilution of the food or water and then place a predetermined quantity of liquid (e.g. 1 ml) into a vial containing dehydrated medium for specific organisms (e.g. coliform). Other dilution factors can be made accordingly. A small delivery tube is then placed into the vial with the sample and nutrient. The delivery tube is attached to a sterile plastic card (4 inch × 3 inch × 1/8 inch), which has three series of 16 chambers. The top 16 chambers are very small holes in size and the second 16 chambers are 10 times larger, while the third series of 16 are again 10 times larger in volume. This creates a 3-dilution, 16-tube MPN test protocol. Several of these cards, each containing one sample in the 3, 16-tube series (48 chambers total), are placed in a rack, which is then inserted into a chamber. In the chamber the sample for each separate card is then 'pressurized' into all 48 chambers to distribute exactly one ml of the sample into the 16-tube MPN system. After, the filled cards with the delivery tube cut and sealed automatically are placed into an incubator (e.g. 35°C) and incubated overnight. After incubation a series of cards are placed in an instrument to read the presence of fluorescence in each of the 48 chambers per card due to biochemical reactions of the target organisms (total count, coliform count, etc.) in the liquid. The instrument automatically calculates the 16-tube 3-dilution series MPN of the sample. The entire procedure is highly automated with no need for a technician to transfer samples into a large number of tubes, and after incubation to read the color change or turbidity in each tube, as in the conventional 3-tube or 5-tube MPN methods to determine the viable cell count. The savings of time of operation and materials by TEMPO compared to the conventional MPN method are truly impressive. This author predicts that the system will be well received by food and water microbiologists.

II.C. Aerobic, anaerobic, and 'real time' viable cell counts

By using the correct gaseous environment or suitable reducing compounds one can obtain aerobic, anaerobic, facultative anaerobic microbial counts in products. Typically, microbial counts were obtained in 24–48 hours. Several methods have been developed and tested in recent years that provide 'real'-time viable cell counts, such as (1) the use of 'vital' stains (Acridine Orange) to report living cells under the microscope to count fluorescing viable cells (Direct Epifluorescent Filter Technique), (2) scanning cells interacted with Fluorassure (a vital dye in Chemunex Scan RDI system), and (3) measuring ATP of microcolonies trapped in special membranes (MicroStar System of Millipore), among other methods. These real-time tests can give viable cell counts in about 4 hours. A simple Fung Double Tube system using appropriate agar and incubation conditions was developed more than 20 years ago by the author for food. This system was evaluated for water testing in 2007 (Fung et al., 2007) and can provide a *Clostridium perfringens* count in about 6 hours. Hawaii is the only U.S. state to use *C. perfringens* as an indicator of fecal contamination. This information can be used to assess suitability for recreation and swimming in various beaches around the world.

III. ADVANCES IN MINIATURIZATION AND DIAGNOSTIC KITS

Identification of normal flora, spoilage organisms, clinical and foodborne pathogens, starter cultures, etc., in many specimens, is an important part of microbiology. This author systematically miniaturized biochemical reactions to identify and characterize large numbers of microbes in food and water in the Microtiter system in the late 1960s and early 1970s, which increased the efficiency of handling large numbers of bacterial and yeast isolates in the 96-well Microtiter system and used a multiple inoculator to transfer large numbers of cultures into a variety of biochemical tests and agar plates (Fung and Hartman, 1975). In the past 25 years many miniaturized diagnostic kits have been developed and widely used to conveniently introduce pure cultures into the system and to obtain reliable identification results in as short as 2–4 hours. Some systems can handle several or even hundreds of isolates at the same time. These diagnostic kits have no doubt saved many lives by rapidly and accurately, and conveniently, identifying important pathogens, so that treatments can likewise be made correctly and rapidly. Some of the more common miniaturized systems on the market today to identify pathogens, ranging from enterics (*Salmonella*, *Shigella*, *Proteus*, *Enterobacter*, etc.) to non-fermentors, anaerobes, Gram-positives, and even yeasts and molds, include API, MicroID, Enterotube, Crystal ID systems, Biolog, Vitek, Vitek-2, etc.

IV. ADVANCES IN IMMUNOLOGICAL TESTING

Antigen and antibody reaction has been used for decades for detecting and characterizing microorganisms and their components in medical and diagnostic microbiology. This is the basis for serotyping bacteria such as *Salmonella*, *Escherichia coli* O157:H7, *Listeria monocytogenes*, etc. Both polyclonal antibodies and monoclonal antibodies have been used extensively in applied food microbiology. The most popular format is the 'sandwiched' ELISA test. Recently some companies have completely automated the entire ELISA procedure and can complete an assay from 45 minutes to 2 hours after overnight incubation of the sample with suspect target organisms. VIDAS system (bioMerieux) is a totally automated system for identification of pathogens such as *Listeria*, *Listeria monocytogenes*, *Salmonella*, *E. coli* O157:H7, Staphylococcal enterotoxins and *Campylobacter*. There are more than 13,000 units in use worldwide. BioControl markets an Assurance EIA system that can be adapted to automation for high-volume testing.

Lateral Flow Technology (similar to a pregnancy test with three detection areas on a small unit) offers a simple and rapid test for target pathogens (e.g. *E. coli* O157) after overnight incubation of food or allergens (e.g. wheat gluten). The entire procedure takes only about 10 minutes with very little training necessary. The BioControl VIP, Neogen Reveal, and Merck KgaA systems are some of the commercial systems available on the market for *Salmonella*, *E. coli* O157, *Listeria*, etc.

A truly innovative development in applied microbiology is the immuno-magnetic separation (IMS) system. Very homogenized paramagnetic beads have been developed that can be coated with a variety of molecules, such as antibodies, antigens, DNA, etc., to capture target cells such as *E. coli* O157, *Listeria*, *Cryptosporidium*, *Giardia*, etc. These beads can then be immobilized, captured, and concentrated by a magnet stationed outside a test tube. After clean-up, the beads with the captured target molecules or organisms can be plated on agar for cultivation or used in ELISA, polymerase chain reaction (PCR), microarray technologies, biochips, etc., for detection of target organisms. Currently many diagnostic systems (ELISA test, PCR, etc.) are combining an IMS step to reduce incubation time and to increase sensitivity of the entire protocol. Dynal company in Norway produces very homogeneous paramagnetic beads, which can be used for coating antigens, antibodies, DNA, and even cells for use in the IMS system. A new and very efficient IMS system is the Pathatrix system by Matrix MicroScience. This system can circulate the entire preincubated sample (e.g. 25 g of food in 225 ml of broth for a few hours) over surfaces containing immobilized beads coated with specific antibodies for specific pathogens, such as *E. coli* O157:H7. After capturing the target pathogens, the entire broth liquid is discarded and the beads with pathogens can be released; then the detection of the pathogens can be made using the conventional ELISA test, agar plates, PCR, etc. This system greatly increases detection efficiency and time. In the author's laboratory as little as 1–10 *E. coli* O157:H7 CFU/25 g of ground beef can be detected in 5.25 hours from the time of placing the 25 g of ground beef into 225 ml of broth for growth (4.5 hours), circulating the broth in the Pathatrix system (0.5 hour) to capture target pathogens, and then to completing the ELISA test (0.25 hour). Many other laboratories have since used the Pathatrix for detecting pathogens in food, water, fresh produce, etc.

V. ADVANCES IN INSTRUMENTATION AND BIOMASS MEASUREMENTS

Instruments can be used to automatically monitor microbial growth kinetics and dynamics changes (e.g. ATP levels, specific enzymes, pH, electrical impedance, conductance, capacitance, turbidity, color, heat, radioactive carbon dioxide, etc.) of a population (pathogens or non-pathogens) in a liquid and semi-solid sample. It is important to note, that for the information to be useful, these parameters must be related to viable cell counts in the same sample series. In general, the larger the number of viable cells in the sample, the shorter the detection time of these systems. A scatter gram is then plotted and used for further comparison of unknown samples. The assumption is that as the number of microorganisms increases in the sample, these physical, biophysical, and biochemical events will also increase accordingly. The detection time is inversely proportional to the initial population of the sample. When a sample has 5-log or 6-log organisms/ml, detection time can be achieved in about 4 hours. Some instruments can handle hundreds of samples at the same time. Instruments to detect ATP in a culture fluid include Lumac, BioTrace, Ligthning, Hy-Lite, Charm, Celsis, Zyluz Profile 1, etc. The Bactometer and Rapid Automated Bacterial Impedance Technique can monitor impedance changes in the test samples, and the Malthus system can monitor conductance changes and relate the changes to microbial populations. Biosys (Neogen) is a convenient system that can monitor the change in color of the sample when microbial growth occurs. The uniqueness of this system is that it monitors the color changes in an agar plug below the liquid sample, so that there is no interference of food particles during monitoring of color intensity related to growth of organisms in a specially designed culture broth (e.g. a special broth for color change in the presence of *E. coli* growth). Basically, any type of instrument that can continuously and automatically monitor turbidity and color changes of a liquid in the presence of microbial growth can be used for rapid detection of the presence of microorganisms.

VI. ADVANCES IN GENETIC TESTING

Phenotypic expression of cells is subject to growth conditions such as temperature, pH, nutrient availability, oxidation-reduction potentials, etc. Genotypic characteristics of a cell are far more stable. Hybridization of DNA and RNA by known probes has been used for more than 30 years. More recently the polymerase chain reaction (PCR) is now an accepted method to detect viruses, bacteria, and even yeast and molds by amplification of the target DNA and detecting the target PCR products. By use of reverse transcriptase, target RNA can also be amplified and detected. After a DNA (double-stranded) molecule is denatured by heat (e.g. 95°C), proper primers will anneal to target sequences of the single-stranded DNA of the target organism, for example *Salmonella* at a lower temperature (e.g. 37°C). A polymerase (e.g. TAQ enzyme) will extend the primer at a higher temperature (e.g. 70°C) and complete the addition of complement bases to the single-stranded denatured DNA. After one thermal cycle one piece of DNA will become two pieces. After 21 and 31 cycles one piece will become 1 million and 1 billion copies, respectively. At the beginning PCR products are detected by gel electrophoresis, which is very time-consuming. Now there are ingenious ways to detect the occurrence of the PCR procedure either by fluorescent probes or special dyes, or by actually reporting the presence of the PCR products by molecular beacon. Since these methods generate fluorescence, the PCR can be monitored over time and provide 'real-time' PCR results. Some systems can monitor four different targets in the same sample (multiplexing) by using different reporting dyes. These methods are now standardized and easy to use and interpret.

To further characterize closely related organisms, detailed analysis of the DNA molecule can be made by obtaining the patterns of DNA of specific organisms by pulse field gel electrophoresis (DNA finger-printing) or by 'Riboprinting' of the ribosomal genes in the specific DNA fragment. Since different bacteria exhibit different patterns (e.g. *Salmonella* versus *E. coli*), and even the same species can exhibit different patterns (e.g. *Listeria monocytogenes* with 49 distinct patterns), this information can be used to compare closely related organisms for accurate identification of target pathogens (e.g. comparing different patterns of *E. coli* O157:H7 isolated from different sources in an outbreak) in epidemiological investigations.

VII. ADVANCES IN BIOSENSOR, MICROCHIPS, AND BIOCHIPS

Biosensor is an exciting field in applied microbiology. The basic idea is simple but the actual operation is quite complex and involves much instrumentation. Basically, a biosensor is a molecule or a group of molecules of biological origin attached to a signal recognition material. When an analyte comes in contact with the biosensor the interaction will initiate a recognition signal, which can be reported in an instrument. Many types of biosensors have been developed. Sometimes whole cells can be used as biosensors. Analytes detected include toxins, specific pathogens, carbohydrates, insecticides and herbicides, ATP, antibiotics, etc. The recognition signals used include electrochemical (e.g. potentiometry, voltage changes, conductance and impedance, light addressable, etc.), optical (e.g. UV, bioluminescence, chemiluminescence, fluorescence, laser scattering, reflection and refraction of light, surface phasmon resonance, polarized light, etc.) and miscellaneous transducers (e.g. piezoelectric crystals, thermistor, acoustic waves, quartz crystal, etc.).

Recently, much attention has been directed to 'biochips' and 'microchips' development to detect a large variety of molecules including foodborne pathogens. Due to the advancement in miniaturization technology as many as 50,000 individual spots (e.g. DNA microarrays), with each spot containing millions of copies of a specific DNA probe, can be immobilized on a specialized microscope slide. Fluorescent labeled targets can be hybridized to these spots and be detected. Biochips can also be designed to detect all kinds of foodborne pathogens by imprinting a variety of antibodies or DNA molecules against specific pathogens on the chip for simultaneous detection of pathogens such as *Salmonella*, *Listeria*, *Escherichia coli*, *Staphylococcus aureus*, etc. The market value is estimated to be as high as $5 billion at this moment. This technology is especially important in the rapidly developing field of proteomics and genomics, which require massive amounts of data to generate valuable information. Advanced bioinformatics are necessary to interpret these data. The potential of biochips and microarrays for microbial detection and identification in pharmaceutical and related samples is great but at this moment much more research is needed to make this technology a reality in applied and food microbiology.

Certainly, the developments of these biochips and microchips are impressive for obtaining a large amount of information for biological sciences. As for foodborne pathogen detection there are several important issues to be considered. These biochips are designed to detect minute quantities of target molecules. The target molecules must be free from contaminants before being applied to the biochips. In food microbiology, the minimum requirement for pathogen detection is 1 viable target cell in 25 grams of a food. Biochips will not be able to seek out such a cell from the food matrix without extensive cell amplification or sample preparation, separation, filtration, centrifugation, absorption, etc. Another concern is viability of the pathogens. Monitoring the presence of cell components in food does not necessarily indicate that the cells are alive or dead. Certainly, human beings regularly consume killed pathogens such as *Salmonella* in cooked chicken without becoming sick. To ensure a system is detecting living cells some form of culture enrichment or vital dyes are necessary to indicate that the pathogens to be detected are alive.

VIII. TESTING TRENDS AND PREDICTIONS

There is no question that many microbiological tests are being conducted nationally and internationally on pharmaceutical and food products, environmental samples, medical specimens, and water samples. The most popular tests are total viable cell count, coliform/*E. coli* count, and yeast and mold counts. A large number of tests are also performed on pathogens such as *Salmonella*, *Listeria* and *Listeria monocytogenes*, *E. coli* O157:H7, *Staphylococcus aureus*, *Campylobacter*, and other organisms. According to Strategic Consulting Inc. (Woodstock, VT, 2005) in 1998, the number of worldwide microbiological tests was estimated to be 755 million with a market value of $1 billion in the U.S. The newest Global Industrial Microbiological Testing projection for 2008 for food is 716.6 million tests (47.5% total tests); beverages, 137.0 million tests (9.1%); pharmaceuticals, 311.1 millions tests (20.7%); personal care products, 249.1 million tests (16.5%); environmental, 55.9 million tests (3.7%); and material processing, 36.5 million tests (2.4%). The total number of tests will be about 1.5 billion with a market value of $5 billion in the U.S. Presently, in 2008 about one third of the tests are being performed in North America (U.S. and Canada), another third in Europe, and the last third in the rest of the world. This author predicts in 20 years that the rest of the world will perform 50% of the tests with North America and Europe performing 25% each. This is due to rapid economic development and food and health safety concerns of the world in the years ahead.

VIII.A. Future predictions and conclusions

Following are **ten predictions** made by the author in 1995 as Divisional lecturer for the American Society for Microbiology Food Microbiology in Washington D.C. Many predictions have been correct in 2008; (+) is a good prediction; (?) is an uncertain prediction.

1. Viable cell counts will still be used in the next 25 years (+).
2. Real-time monitoring of hygiene will be in place (+).
3. PCR, Ribotyping, and genetic tests will become a reality in food laboratories (+).
4. ELISA and immunological tests will be completely automated and widely used (+).
5. Dipstick technology will provide rapid answers (+).
6. Biosensors will be in place for Hazard Analysis Critical Control Point Programs (?).
7. Biochips, microchips, and microarrays will greatly advance in the field (+).
8. Effective separation and concentration of target cells will assist rapid identification (+).
9. Microbiological alert systems will be in food and pharmaceutical packages (?).
10. Consumers will have rapid alert kits for pathogens at home (?).

In conclusion, it is safe to say that the field of rapid methods and automation in microbiology will continue to grow in the number and kinds of tests in the future due to the increased concern over food safety and public health nationally and internationally. The future looks very bright indeed for the field of rapid methods and automation in microbiology. The potential is great and many exciting developments will certainly unfold in the near and far future in the areas of food safety and security.

REFERENCES

Al-Dagal, M., & Fung, D. Y. C. (1993). Aeromicrobiology: An assessment of a new meat research complex. *J. Environmental Health, 56*(1), 7–14.

Chain, V. S., & Fung, D. Y. C. (1991). Comparison of Redigel, Petrifilm, Spiral Plate System, Isogrid and aerobic plate count for determining the numbers of aerobic bacteria in selected food. *J. Food Protect, 54*, 208–221.

Fung, D.Y.C. (2002). Rapid methods and automation in microbiology. *Comprehensive Reviews in Food Science and Food Safety. Inaugural Issue* 1(1): 3–22. Website: www.ift.org

Fung, D. Y. C. (2007). Rapid methods and automation in microbiology in pharmaceutical samples. *American Pharmaceutical Review, 10*(3), 82–86.

Fung, D. Y. C., Fujioka, K., Vijayavel, D., & SatoBishop, D. (2007). Evaluation of Fung Double Tube Test for *Clostridium perfringens* and Easyphage Test for F-Specific RNA Coliphages as Rapid Screening Tests for Fecal Contamination in Recreational Waters of Hawaii. *J. Rapid Methods Automat. Microbiol., 15*(3), 217–229.

Fung, D. Y. C., & Hartman, P. A. (1975). Miniaturized microbiological techniques for rapid characterization of bacteria. In C. G. Heden, & T. Illeni (Eds.), *New Approaches to the Identification of Microorganisms, Chapter 21*. New York: John Wiley and Son.

Fung, D. Y. C., Kastner, C. L., Lee, C. Y., Hunt, M. L., Dikeman, M. E., & Kropf, D. (1980). Mesophilic and psychrotrophic populations on hot boned and conventionally processed beef. *J. Food Protection, 43*, 547–550.

Fung, D. Y. C., Phebus, R., Kang, D. H., & Kastner, C. L. (1995). Effect of alcohol-flaming on meat cutting knives. *J. Rapid Methods Automat. Microbiol., 3*, 237–243.

Strategic Consulting Inc (2005). Global Review of Microbiology Testing in the Food Processing Market. Strategic Consulting, Inc., Woodstock VT. Email: weschler@stragetic-consult.com/.

CHAPTER 12

The Role of Standardization Bodies in the Harmonization of Analytical Methods in Food Microbiology

Bertrand Lombard *and* Alexandre Leclercq

Contents

Abstract

Several scientific reasons are given in the introduction of this chapter to explain why the harmonization of analytical methods through standardization is important in the field of food microbiology. This is mainly due to the nature of the analyte considered, a living microorganism, and the principle of the methods used, based on conventional microbiology. This chapter defines the notion of *standard* and *standardization*, and then introduces the structures in charge of standardization at European and international levels: the European Committee for Standardization (CEN) and the International Organization for Standardization (ISO), as well as their committees dedicated to standardization in food microbiology.

The different types of standards elaborated in the field of food microbiology are thus presented in this chapter, mainly standards defining reference methods and standards dealing with general aspects.

Global Issues in Food Science and Technology
© 2009 Elsevier Inc.

ISBN 9780123741240

The status of novel technologies in food microbiology standardization is then envisaged in more details. The principles for introducing these technologies in these standards are also defined, and a set of standards being developed on the use of PCR in food microbiology is quoted.

Further justifications for this standardization effort are finally envisaged: use of the standards developed for checks performed by food business operators, for official food controls by public control authorities, for accreditation of laboratories in the field of food microbiology.

I. INTRODUCTION

The harmonization of analytical methodology is crucial in food microbiology, as it is well known that food microbiological analysis is highly variable. This is mainly due to two factors: (1) the intrinsic nature of the analyte, which is a living microorganism with a large capability for genetic variation and mutation, and (2) the principles of detection/enumeration of microorganisms by conventional culture techniques, which include the use of biological substances that are also variable, such as culture media, antibodies, or biochemical reagents. In addition, the analytical results are closely linked to the method's principle, for conventional techniques, in particular for enumeration procedures: the choice of different steps (enrichment, isolation, confirmation), the selection of temperature/time of incubation, and the composition of culture media will directly impact the number of colonies growing on a plate (Lombard et al., 1996; Lombard, 2004a). The enumeration result also depends on the physiological initial state of the bacteria, which may be in a growth phase, or be altered or stressed (Leclercq et al., 2000). In the presence of stressed cells, if any resuscitation step is not conducted prior to the enumeration procedure, the counts of target microorganisms will most likely be underestimated. There is therefore a clear need to define common reference methods in food microbiology, so as to minimize as much as possible the variability of microbiological analyses, and to define in a less ambiguous way the analytical target: a given microorganism (or group of microorganisms) to be detected or a given number of microorganisms (or group of microorganisms) to be enumerated.

In addition, the analytical methods will need to be highly reliable since some of them enable detection or quantification of foodborne pathogens, which represent direct hazards to consumer health. Their presence at very low numbers in a given food product (a few cells in a test portion of 25 g) may be responsible for the illnesses of consumers of that food, ranging from the most common gastroenteritidis to severe renal affections, meningitis, and septicemia, which may if not properly treated lead to death.

Standardization has proven to be the preferred way (based on consensus) to harmonize reference methods at the international level or regional level (e.g. in Europe), and to facilitate trade between or within countries. It is also a means to ensure the validity of methods since they are selected by expert microbiologists from several countries, and their choice is based on a well-established set of scientific criteria.

The principles of standardization, the European and international bodies, and the structures active in food microbiology will first be introduced. The different types of standards developed in the field, covering bacteria, yeasts and molds and soon viruses, will then be presented before focusing on the status of novel technologies in standardization.

II. STANDARDIZATION: PRINCIPLES AND STRUCTURES – THE CASE OF FOOD MICROBIOLOGY

II.A. What is a standard?

Before introducing the standardization bodies, it seems necessary to define 'standard.' According to ISO (International Organization for Standardization) (ISO/IEC, 2004), the term means 'a document established by consensus, and approved by a recognized body, that provides, for common and repeated use, guidelines or characteristics for activities or their results, aimed at the achievement of the optimum degree of order in a given context.'

This general definition, which is designed to cover a very broad scope of standardization activities, underlines that a standard is primarily a text of reference established by a recognized body: (1) for a national standard, the national standardization body (one per country), (2) for a European standard, CEN (*Comité Européen de Normalisation* – European Committee for Standardization), and (3) for an international standard, ISO. A standard represents a *consensus* of all concerned parties, at the national, European, or international level; this is by far the main characteristic of a standard. Therefore, a standard cannot be developed to meet an individual need or immediate use; it must respond to a common interest as it is intended for repeated use.

A standard, as a reference document, can be used in:

- trade within and between countries ('to speak the same language')
- relationships between customers and suppliers (terms of reference for contracts)
- official food controls
- providing consumer information (for clarification of product's properties, quality).

Figure 12.1 represents different types of reference documents, ordered by a decreasing degree of flexibility and detail, and on an inverse sense, by their degree of legal validity. Standards are located between (1) business internal

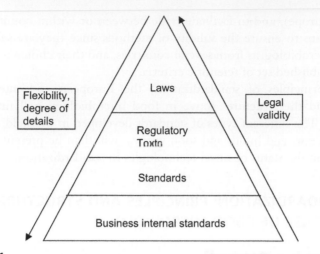

Figure 12.1

standards, which are very detailed, easy to modify but of limited validity outside the business, and (2) regulatory texts. Standards can be distinguished from regulatory texts by (1) their elaboration process (they are developed by all concerned parties instead of public authorities) and (2) their voluntary use, an important difference. Meanwhile, the use of a standard can become mandatory in some cases if they are referenced in a regulatory text, such as from the European Commission.

One of the main added values of standardization is the bringing together of partners often in competition with each other or with diverging interests, making them interact to produce a common document. In the field of food microbiology, the experts elaborating the standards come from public and private laboratories, reference laboratories, research laboratories, universities, accreditation/certification bodies, the food industry, and public authorities in charge of food hygiene (Leclercq *et al.*, 2000).

II.B. Standardization structures at international and European levels

Before presenting the standardization structures in food microbiology, let us introduce the international and European standardization bodies, ISO and CEN.

II.B.1. ISO – International Organization for Standardization

ISO, created in 1947, is an international non-governmental federation of national standardization bodies worldwide (158 to date). The central structure of ISO, called ISO Central Secretariat, is located in Geneva; it is a small structure in charge of coordinating the whole standardization process and of publishing the ISO standards and related texts.

The standardization process is conducted within technical committees, sub-committees, and working groups, which are settled per area of expertise. Each one is under the responsibility of a chair and a secretariat of the country that takes the lead. The secretariat is usually undertaken by the standardization body of the country taking the lead, at least for the technical committees and sub-committees. The standardization body with the secretariat manages the whole standardization process (circulation of documents, preparation and follow-up of meetings, editorial drafting of standards, etc.). The national ISO members and international organizations in liaison choose membership in a given committee/sub-committee, depending on their interest and willingness to participate in the standardization field considered. The technical development of standards is therefore not centralized in Geneva. The ISO Standards can optionally be adopted as national standards by ISO members.

II.B.2. CEN – European Committee for Standardization

CEN was set up more recently than ISO, in 1961. The central structure of CEN, called CEN Management Centre, is located in Brussels, and has a role similar to the ISO Central Secretariat. CEN members are the national standardization bodies of Europe at large: 30 countries are full members and six are associated members. CEN therefore corresponds more to a geographical Europe, and not only to the European Union. CEN has a privileged link with the European Commission, which can make specific requests for standardization.

CEN standardization functions in a similar way to ISO. The main difference is that CEN standards must be taken over and included in the collection of national standards of CEN Members, and that conflicting national standards along the same scope must be withdrawn. This ensures a higher degree of harmonization at the European level, made possible by the regional scale involved. Meanwhile, the use of these standards remains voluntary, except when cited in European regulatory texts.

II.C. Standardization structures in food microbiology

Most of the standardization works in food microbiology are conducted by two structures, one at the ISO level and the other at the CEN level:

- ISO/TC 34/SC 9: Sub-Committee 9 'Microbiology' of Technical Committee 34 'Food products' of ISO (in short, SC 9)
- CEN/TC 275/WG 6: Working Group 6 'Microbial contaminants' of Technical Committee 275 'Food analysis – Horizontal methods' of CEN (in short, WG 6).

ISO/TC 34/SC 9 is under the responsibility of France, chaired by Bertrand Lombard and with a secretariat provided by AFNOR (*Association Française de Normalisation* – French Standardization Association), namely Sandrine Espeillac. Forty-five ISO member countries are SC 9 members, and 11 international organizations are in liaison. Several working groups have been established by SC 9 to deal with specific topics before endorsement by SC 9 (Table 12.1). The main objective is to develop and to update ISO Standards for the microbiological analysis of a wide spectrum of sample types (Section III.A).

CEN/TC 275/WG 6 is also under French responsibility, chaired by Alexandre Leclercq and with the same AFNOR secretariat (Sandrine Espeillac). Thirty CEN Members and 14 international organizations in liaison are represented in WG 6. Similarly to SC 9, several task forces (TAG) have been established by WG 6 on given topics (Table 12.2). The objectives are shared with SC 9, and within the framework of the CEN/ISO cooperation agreement called the 'Vienna Agreement,' most of the standards developed by SC 9 are adopted as CEN Standards, and inversely, the standards developed by WG 6, essentially in the areas covered by its task groups, are adopted as ISO Standards. Up until recently, the WG 6 scope was restricted to foodborne pathogens; thus the ISO Standards dealing with food hygiene indicators were not adopted as CEN Standards. But the WG 6 scope was enlarged in 2006 to become identical with the SC 9 one, so now all standards have this scope in common. To illustrate how wide this scope is in covering the whole food chain, it has been said to be the 'Standardization of horizontal microbiological analysis methods for all food and animal feeding stuffs and for any other sample that can be the source of microbiological contamination of food products' (WG 6 Resolution 88, 2006 meeting; SC 9 Resolution 337, 2007 meeting).

Table 12.1 Working groups of ISO/TC 34/SC 9 'Food Products – Microbiology'

Working group number	Title	Convenor
WG 1	Meat and meat products	Enne De Boer (Netherlands)
WG 2	Statistics	Bertrand Lombard (France)
WG 3	Methods validation	Paul in't Veld (Netherlands)
WG 4	Proficiency testing schemes	Sue Passmore (UK)
WG 5	Culture media	Melody Greenwood (UK)

Table 12.2   Task groups of CEN/TC 275/WG 6 'Food Analysis, Horizontal Methods – Microbial Contaminants'

Task group number	Title	Convenor
TAG 1	Culture media	None (dormant group)
TAG 2	Validation of alternative methods	None (dormant group)
TAG 3	PCR	Kornelia Berghof-Jäger (Germany)
TAG 4	Viruses	David Lees (UK)
TAG 5	Primary production stage	Rosine Tanguy des Deserts (France)
TAG 6	Sampling technique	Martine Poumeyrol (France)
Ad hoc group	Shiga toxin producing *Escherichia coli*	Stephano Morabito (Italy)

III. DIFFERENT TYPES OF STANDARDS DEVELOPED IN FOOD MICROBIOLOGY

To date, 70 standards have been published in food microbiology (see CEN and ISO websites). The microorganisms considered are mainly bacteria, but yeasts and molds have also been covered; methods are currently standardized for viruses and works have been recently launched on foodborne parasites. Globally, the standards either consist of reference methods or deal with the general aspects related to microbiological analysis of food.

Examples of three general aspects:

1. A working group was established in 2003 by SC 9, known as WG 2 'Statistics,' to provide any statistical expertise requested during the standardization process, with the objective of giving sound scientific basis for the standards developed.
2. A working group was established in 1996 by WG 6 and transferred to SC 9, known as WG 3 'Validation of methods,' to define in particular the minimum validation requirements for a standardized method.
3. The willingness to harmonize SC 9/WG 6 standards (a) with different fields of microbiology, first dairy microbiology, then water microbiology, and (b) worldwide, especially with North America, in order to build a unique reference system in food microbiology, through a complete harmonization of reference methods.

III.A. Standards on reference methods

III.A.1. Presentation

Standards developed by SC 9 and WG 6 cover the main microorganisms of interest in food microbiology: food hygiene indicators, foodborne pathogens, microorganisms of technological interest, as well as microorganisms responsible for food alteration or for nutritive or commercial value; in dairy microbiology (Table 12.3).

For each target microorganism, detection (presence/absence tests) methods and/or enumeration methods, as well as semi-quantitative methods (in some cases) are standardized. These methods are mostly based on conventional microbiology, that is, the ability of bacteria to grow in enrichment broths or on/in solid media. Indeed, this choice enables one to respect the major characteristics of a standard, that is, the largest consensus and the broadest applicability in food microbiological laboratories at a worldwide level, in particular in developing countries but also in the developed world.

Some of these standards are divided into several parts. In most cases, one part deals with detection and enumeration at low levels (<10 or 100 colony-forming units (CFUs) per g or ml) using a most probable number technique, and the other part deals with enumeration at higher levels using a colony-count technique. In the case of *Listeria monocytogenes* (EN ISO 11290), one part deals with its detection and the other with its enumeration. For the enumeration of coagulase-positive staphylococci (EN ISO 6888), the choice between part 1 and part 2, each using a different culture medium, depends on the type of food analyzed, with either a poor or high-level background flora. For the enumeration of *Escherichia coli* (ISO 16649), the use of a membrane in part 1 is adapted to the recovery of stressed cells, whereas part 2 without membrane can be used when it is expected that the target bacteria are not stressed.

As indicated earlier, these standard methods are developed as applicable to a wide range of sample types, that is, all foods and animal feeding stuffs, the environment of food production and handling, and recently, with possibly some adaptations, the starting point of the food chain – the animal's primary production (i.e. breeding environment). The standard methods having such a large scope, applicable to all food families, are called 'horizontal.' The rationale for developing such 'horizontal' standards is that the principle of the method (i.e. culture media, incubation times and temperatures, confirmation reactions) is mainly designed to recover and to characterize the target microorganism, whatever the food matrix is, given the fact that the initial steps of the analysis (preparation of test sample and initial suspension) are dealt with specifically, per type of matrix (Section III.B). From a practical point of view, horizontal methods are preferred by

Table 12.3 List of reference methods in food microbiology standardized by CEN and ISO[*]

Microorganisms detected/enumerated[a]	Standard reference number[b]
Food hygiene indicators/Process microorganisms	
Enumeration of microorganisms – CCT at 30°C	EN ISO 4833: 2003
Enumeration of microorganisms in milk – Plate-loop technique at 30°C	ISO 8553: 2004
Enumeration of coliforms – MPN technique	ISO 4831: 2006
Enumeration of bacteria, yeasts and molds in cereals, pulses and derived products	ISO 7698: 1990
Enumeration of coliforms – CCT	ISO 4832: 2006
Enumeration of mesophilic lactic acid bacteria – CCT at 30°C	ISO 15214: 1998
Enumeration of psychrotrophic microorganisms – CCT at 6.5°C	ISO 17410: 2001
Enumeration of psychrotrophic microorganisms in milk – CCT at 6.5°C	ISO 6730: 2005
Estimation of psychrotrophic microorganisms in milk – CCT at 21°C (rapid method)	ISO 8552: 2004
Enumeration of contaminating microorganisms in butter, fermented milk and fresh cheese – CCT at 30°C	ISO 13559: 2002
Enumeration of characteristic microorganisms in yoghurt – CCT at 37°C	ISO 7889: 2003
Identification of characteristic microorganisms in yoghurt (*Lactobacillus delbrückii* subsp. *bulgaricus* and *Streptococcus thermophilus*)	ISO 9232: 2003
Enumeration of citrate-fermenting lactic acid bacteria in milk, milk products and mesophilic starter cultures – CCT at 25°C	ISO 17792: 2006
Enumeration of presumptive *Lactobacillus acidophilus* on a selective medium – CCT at 37°C	ISO 20128: 2006

(*continued*)

Table 12.3 List of reference methods in food microbiology standardized by CEN and ISO*—Cont'd

Microorganisms detected/enumerated[a]	Standard reference number[b]
Enumeration of yeasts and molds – CCT at 25°C	ISO 7954: 1987
Enumeration of yeasts and molds in milk and milk products – CCT at 25°C	ISO 6611: 2004
Enumeration of yeasts and molds in meat and meat products – CCT at 25°C	ISO 13681: 1995
Enumeration of sulfite-reducing clostridia growing under anaerobic conditions	ISO 15213: 2003
Enumeration of *Pseudomonas* spp. in meat and meat products	ISO 13720: 1995
Enumeration of *Brochothrix thermosphacta* in meat and meat products	ISO 13722: 1996
Detection and enumeration of presumptive *Escherichia coli* – MPN technique	ISO 7251: 2005
Enumeration of β-glucuronidase positive *Escherichia coli* – Part 1: CCT at 44°C using membranes and BCIG	ISO 16649-1: 2001
Enumeration of β-glucuronidase positive *Escherichia coli* – Part 2: CCT at 44°C using BCIG	ISO 16649-2: 2001
Enumeration of β-glucuronidase positive *Escherichia coli* – Part 3: MPN technique using BCIG	ISO/TS 16649-3: 2005
Enumeration of presumptive *Escherichia coli* in milk and milk products – Part 1: MPN technique using MUG	ISO 11866-1: 2005
Enumeration of presumptive *Escherichia coli* in milk and milk products – Part 2: CCT at 44°C using membranes	ISO 11866-2: 2005
Detection and enumeration of *Enterobacteriaceae* – Part 1: MPN technique with pre-enrichment	ISO 21528-1: 2004
Detection and enumeration of *Enterobacteriaceae* – Part 2: CCT	ISO 21528-2: 2004

Table 12.3 List of reference methods in food microbiology standardized by CEN and ISO*—Cont'd

Microorganisms detected/enumerated[a]	Standard reference number[b]
Food-borne pathogens	
Detection of *Salmonella* spp.	EN ISO 6579: 2002/ Amendment 1: 2007[c]
Detection of *Salmonella* spp. in milk and milk products	EN ISO 6785: 2001
Enumeration of coagulase-positive staphylococci – Part 1: Technique using Baird Parker agar	EN ISO 6888-1: 1999
Enumeration of coagulase-positive staphylococci – Part 2: Technique using Rabbit Plasma Fibrinogen agar	EN ISO 6888-2: 1999
Enumeration of coagulase-positive staphylococci – Part 3: Detection and MPN technique for low numbers	EN ISO 6888-3: 2003
Detection and enumeration of coagulase-positive staphylococci in milk and milk products – MPN technique	ISO 5944: 2001
Detection of thermonuclease produced by coagulase-positive staphylococci in milk and milk products	ISO 8870: 2006
Enumeration of presumptive *Bacillus cereus* – CCT at 30°C	EN ISO 7932: 2004
Detection and enumeration of presumptive *Bacillus cereus* in low numbers – MPN technique	EN ISO 21871: 2006
Enumeration of *Clostridium perfringens* – CCT	EN ISO 7937: 2004
Detection of potentially enteropathogenic *Vibrio* spp. – Part 1: *Vibrio parahaemolyticus and Vibrio cholerae*	ISO/TS 21872-1: 2007
Detection of potentially enteropathogenic *Vibrio* spp. – Part 2: Other species than *Vibrio parahaemolyticus and Vibrio cholerae*	ISO/TS 21872-2: 2007
Detection of *Campylobacter* spp.	EN ISO 10272-1: 2006

(continued)

Table 12.3 List of reference methods in food microbiology standardized by CEN and ISO*—Cont'd

Microorganisms detected/enumerated[a]	Standard reference number[b]
Enumeration of *Campylobacter* spp. – CCT	ISO/TS 10272-2: 2006
Detection of presumptive pathogenic *Yersinia enterocolitica*	EN ISO 10273: 2003
Detection of *Listeria monocytogenes*	EN ISO 11290-1: 1996/ Amd 1: 2004[c]
Enumeration of *Listeria monocytogenes*	EN ISO 11290-2: 1998/ Amd 1: 2004[c]
Detection of *Escherichia coli* O157	EN ISO 16654: 2001
Detection of *Shigella* spp.	EN ISO 21567: 2004
Detection of *Enterobacter sakazakii* in milk and milk products	ISO/TS 22964: 2006

a In the absence of indication the standard method cited is 'horizontal,' which is broadly applicable to all foods, animal feeding stuffs, and food production environment. 'MPN technique' means Most Probable Number technique; 'CCT' means a Colony-Count Technique.

b 'EN ISO' means that a common CEN/ISO Standard is concerned, 'ISO' that it is only an ISO Standard. 'ISO/TS' means ISO Technical Specification. The year of publication of the last version is indicated after the reference number.

c The standards were amended in 2007/2004.

* Data obtained from CEN website: http://www.cenorm.be; and ISO website: http://www.iso.org/.

laboratories for analyzing a broad range of food products, since they avoid using food-type-specific methods, requiring the use and maintenance of different culture media, incubators, etc. (Lombard, 2004b). The policy is to develop a 'sectorial' standard applicable to a given family of food, in only one of the following cases:

- The corresponding horizontal standard does not apply, based on sound justifications.
- A horizontal method is not needed for the microorganism considered (e.g. detection of *Brochothrix thermosphacta* in meat).
- A method applicable to all foods could not be developed (e.g. enumeration of *Pseudomonas* in meat and dairy products).

This policy is currently implemented in entirety for meat and meat products, since the standards specific to these types of food are dealt with under the responsibility of ISO/TC 34/SC 9, within its WG 1. A harmonization of horizontal standards with milk microbiological standards developed at the

ISO level by another sub-committee of ISO/TC 34, known as SC 5 'Milk & Milk Products,' in collaboration with the International Dairy Federation (IDF), and at the CEN level by another technical committee, known as CEN/TC 302 'Milk & Milk Products,' is under way. The objective is to have only one horizontal method for a given determination, recognized by ISO/TC 34/SC 5, IDF, and CEN/TC 302, that is fully applicable to milk and dairy products. The harmonization process has progressed to a large extent, although it is not yet completed, as illustrated in Table 12.3 in several cases of duplicated standards for the same determination. For another family of food (cereals, pulses, and derived products), there is a sectorial Standard, ISO 7698, that deals with specific sample preparation procedures and enumeration methods. It will be deleted and specific aspects will be found in an amendment to the standard dealing with preparation of samples for this family of food (EN ISO 6887-4, see Section III.B).

Regarding samples from the primary production stage (e.g. animal feces), most likely their analysis will require an adaptation of the food methods. The first case dealt with is the detection of *Salmonella*: an adapted method has been included in an amendment to EN ISO 6579.

In some cases (topics where a broad consensus cannot be reached for several reasons, e.g. a recent topic still under development), it is considered impossible to publish a full standard, and another temporary status is adopted, an ISO/CEN Technical Specification (TS), which is published for a limited period of time (3 years). At the end of this period and based on the experience gained from the implementation of the TS, consideration is needed to either confirm it as a full standard 'as is,' to revise it, or to cancel it. Finally, the year of publication given in Table 12.2 indicates that SC 9 and WG 6 regularly update the set of standards under their responsibility.

III.A.2. Current developments

Horizontal standards are currently being developed for the following microorganisms:

- *Enterobacter sakazakii,* under the lead of WG 6
- *Clostridium botulinum* by TAG 3 of WG 6 (Table 12.2 and Section IV)
- STEC (Shiga-toxin producing *E. coli*), by an ad hoc group of WG 6 (Section IV)
- The major foodborne viruses (noroviruses and hepatitis A viruses), by TAG 4 of WG 6 (Table 12.2 and Section IV)
- Selected foodborne parasites, either under SC 9 or WG 6 leadership.

Regarding the detection of *Enterobacter sakazakii*, the purpose is to prepare a transformation of the ISO Technical Specifications developed for milk products (ISO/TS 22964) into a full standard also applicable to other types of products. As regards to STEC, a foodborne pathogen causing a rare but severe kidney illness in young children (hemolytic and uremic syndrome – HUS), the first and most difficult step has been to define the precise target of the future reference method. The target (i.e. the main five STEC serogroups involved in foodborne illnesses) was defined at the 2008 WG 6 meeting and the principle of the method, based on real-time PCR, was determined: a first step consists in STEC PCR screening, and if positive, a second step of PCR detection of the main five serogroups involved in foodborne illnesses, especially HUS (O26, O103, O111, O145 and O157); and if positive, a third step of confirmation, by colony isolation and characterization. The development of standard reference methods for the detection of foodborne viruses is of utmost importance: despite the significance of viruses in public health linked to food (a high number of gastroenteritis cases due to foodborne viruses), no regulatory controls have been established due to the absence of standardized reference methods. In that field, analytical development is still required before standards can be written.

It was recently decided (2007) to launch standardization works in the field of foodborne parasites, on *Trichinella* under WG 6 leadership, to meet the needs of European regulation, on *Giardia* and *Cryptosporidium* under SC 9 leadership, based on methods already developed and validated in the UK.

A criticism often heard, especially in North America, is that these standard methods have not been fully validated by a proper collaborative study. Due to the living nature of the analyte, the organization of such studies is particularly difficult and costly in food microbiology, thus this delay. To overcome it, a project funded by the European Commission, Standard, Measurement and Testing Programme was conducted in 1996–2000, during which seven main standard methods were fully validated: enumeration of *Bacillus cereus*, enumeration of coagulase-positive staphylococci (two methods), enumeration of *Clostridium perfringens*, detection and enumeration of *Listeria monocytogenes*, and detection of *Salmonella* (Lahellec, 1998). In addition, the European Commission recently gave CEN/TC 275/WG 6 an ambitious mandate to further validate 15 standard horizontal methods. This mandate is to be signed in the coming months before validation studies can be launched.

III.B. Standards on general aspects

There is a set of CEN/ISO standards that deals with the general aspects related to microbiological analysis of food, as shown in Table 12.4.

The following topics are covered, also taking into account the current standardization works:

1. *Good laboratory practices (GLP) in food microbiology (ISO 7218).* This standard gathers common aspects of the horizontal standards developed

Table 12.4 List of CEN/ISO standards on general aspects of food microbiological analysis[*]

Topic[a]	Standard reference number[b]
General rules and recommendations for microbiological analysis of food	EN ISO 7218: 2007
Preparation of samples, initial suspension and decimal dilutions – Part 1: General rules for the preparation of initial suspension and decimal dilutions	EN ISO 6887-1: 1999
Preparation of samples, initial suspension and decimal dilutions – Part 2: Specific rules for preparation of meat and meat products	EN ISO 6887-2: 2003
Preparation of samples, initial suspension and decimal dilutions – Part 3: Specific rules for preparation of fish and fish products	EN ISO 6887-3: 2003
Preparation of samples, initial suspension and decimal dilutions – Part 4: Specific rules for preparation of products other than milk, meat and fish products	EN ISO 6887-4: 2003
Preparation of samples, initial suspension and decimal dilutions for milk and milk products	EN ISO 8261: 2001
Carcass sampling techniques for microbiological analysis	ISO 17604: 2003
Surface sampling techniques, with contact plates and swabs	ISO 18593: 2004
Preparation and production of culture media – Part 1: Quality assurance for laboratory preparation of culture media	EN ISO/TS 11133-1: 2000
Preparation and production of culture media – Part 2: Performance testing of culture media	EN ISO/TS 11133-2: 2003
Estimation of measurement uncertainty for quantitative determinations	ISO/TS 19036: 2006
Quality control in dairy microbiological laboratories – Part 1: Analyst performance assessment for colony counts	ISO 14461-1: 2005
Quality control in dairy microbiological laboratories – Part 2: Reliability of colony counts of parallel plates	ISO 14461-2: 2005

(continued)

Table 12.4 List of CEN/ISO standards on general aspects of food microbiological analysis*—Cont'd

Topic[a]	Standard reference number[b]
Protocol for the validation of alternative methods	EN ISO 16140: 2003
Quantitative determination of milk bacteriological quality – Conversion relationship between routine method and anchor method	EN ISO 21187. 2004

a In the absence of indication the standard method cited is 'horizontal,' which is broadly applicable to all foods, animal feeding stuffs and food production environment.

b 'EN ISO' means that a common CEN/ISO Standard is concerned, 'ISO' that it is only an ISO Standard. 'ISO/TS' means ISO Technical Specification. The year of publication of the last version is indicated after the reference number.

* Data obtained from CEN website: http://www.cenorm.be; and ISO website: http://www.iso.org/.

by SC 9 and WG 6, and, in defining GLP, represents the basis for laboratory accreditation in food microbiology. It recently underwent a deep revision/update and the revised version was published in 2007, and also adopted as CEN.

2. *Quality assurance/performance testing of culture media (EN ISO/TS 11133).* The rationale to establish this standard was that the description in the detection/enumeration horizontal standards of a culture medium and its composition is not enough to assure users it will be fit for use and have appropriate performance. Initially developed by TAG 1 of WG 6, the EN/ISO TS 11133 is being converted into a full standard by WG 5 of ISO/TC 34/SC 9 (Table 12.1), working jointly with water microbiology to prepare a common standard for food and water microbiology.

3. *Preparation of test samples, initial suspension, and decimal dilutions (EN ISO 6887).* Part 1 of this standard deals with the first steps of the enumeration methods (preparation of initial suspension and decimal dilutions). The four other parts of the standard cover the initial step of microbiological analysis, i.e. the preparation of test sample (or taking of test portion from the laboratory sample), in a specific way, that is, per type of food. The main families of food and feed are dealt with (meat and meat products, fish and fish products, animal feeding stuffs, egg products, etc.), milk and milk products being covered by a separate standard (EN ISO 8261). The standards EN ISO 6887 and EN ISO 8261, for the following steps of the microbiological analysis, enable use of a 'horizontal' method applicable to all foods but specific to a given bacterium (Section III.A). Currently, the specific Standard EN ISO

8261 is being incorporated into EN ISO 6887 series as an additional part 5, and a new part will be dedicated to the sample preparation from the primary production stage.

4. *Measurement uncertainty (MU) for quantitative determinations (ISO/TS 19036).* This ISO Technical Specification, published at the beginning of 2006, provides a guide on a topic that is currently a main concern for food microbiological laboratories. For laboratory accreditation and according to EN ISO 17025 (ISO, 2005), laboratories have to report the uncertainty associated with their results, if requested by their customers. The guide recommends a 'global' or 'top down' approach for the MU estimation, based on the experimental reproducibility standard deviation. An amendment to ISO/TS 19036 is currently being developed to cover cases of low count. MU for qualitative determinations is investigated by WG 2 of ISO/TC 34/SC 9 (Table 12.1).

5. *Method validation.* A set of standards on method validation, for the community of microbiological laboratories and users of analytical results, is under development by the new WG 3 of SC 9 to deal with:
   - Validation of proprietary kits: the Standard EN ISO 16140, providing a technical protocol for the organization and interpretation of validation studies, is being revised, as to update the whole standard and to improve in particular the statistical treatment for quantitative methods.
   - Intra-laboratory (or in-house) method validation.
   - Intermediate validation (between in-house study and a full collaborative study).
   - Method verification: in the particular context of accreditation, to verify that a method already developed and validated is correctly operated in a given laboratory.
   - Technical minimum requirements for the standardization of new or revised reference methods.

6. *Proficiency testing (PT) schemes.* This topic is covered by WG 4 of SC 9 (Table 12.1). The purpose is to develop a guide dealing with aspects of PT organization and interpretation that are specific to food microbiology, linked in particular to (i) the living nature of the analyte and its consequences on the homogeneity and stability of the samples, and (ii) the importance of qualitative tests (Leclercq, 2006). This specific guide would be used in addition to the general standard on PT schemes' organization under development by ISO (ISO 17043).

IV. STATUS OF NOVEL TECHNOLOGIES

This section will focus on the status of novel technologies in the standardization of methods. In this context, it is first necessary to mention one important characteristic of any standard: the requirement/recommendation to use a commercial product should be avoided as far as possible in order to not give commercial exclusivity to a given manufacturer and to not link the use of a standard with the commercial availability of the product. At the CEN level, one way being investigated to avoid this problem is to define the performance criteria of the standard without mentioning any specific commercial method.

IV.A. Principles

Two aspects on the use of novel technologies in a standardized framework can be distinguished:

- introduction of novel technologies in specific standards development, and
- standardization of general guidelines on use of PCR in food microbiology (Section IV.B).

Regarding the first aspect, the principles for introduction of novel technologies in developing specific standards (detection/enumeration of a given microorganism) have been clarified at the joint meetings of ISO/TC 34/SC 9 and CEN/TC 275/WG 6 in Parma (April 2004). In particular, the SC 9 Resolution Nr 233 states the following:

1. *Introduction of novel technologies* (such as PCR) in specific standardized reference methods. Two cases are distinguished to allow for such an introduction:
   - Introduction is allowed if conventional microbiology is recognized as not satisfactory to detect/enumerate the given target microorganism at the level of interest. To date, examples include the detection of STEC, *Clostridium botulinum*, viruses, and partly, *E. coli* O157. In the first three cases, the PCR methods target the virulence genes of the bacteria considered (note: presence of genes does not automatically mean that virulence is expressed). For viruses, the PCR methods could also target some RNA/DNA sequences specific to the targeted viruses. For *E. coli* O157, the Standard EN ISO 16654 is based on conventional microbiology but introduces an extraction step based on a novel technology, immuno-magnetic separation.
   - In addition to the standard reference method for a given microorganism, based on conventional microbiology and with a taxonomical target, another standard based on novel technology can be developed with a different target than taxonomy, such as the pathogenicity of the given microorganism. The two first cases will be the detection of pathogenic *Vibrio* and pathogenic *Yersinia enterocolitica*.

2. *Use of commercial products (e.g. PCR systems and ELISA kits).* Given the characteristics of any standard mentioned in this section, these products should not be mentioned in specific standards, but need to be validated against the corresponding standard reference method to show they give at least equivalent results, in accordance with the validation protocol of the Standard EN ISO 16140.

IV.B. Use of PCR in food microbiology

Use of PCR in food microbiology is covered by the TAG 3 of CEN/TC 275/WG 6 (Table 12.2). This group mainly develops general guidance standards on the use of PCR in this field, and several of them have been published recently in 2005/2006 (Table 12.5).

These general guidance standards deal with:

- general requirements and definitions
- for qualitative methods, general guidelines on sample preparation, amplification and detection, and
- performance criteria for thermocyclers.

Another standard is currently being prepared to give general guidance on real-time PCR. TAG 3 has been also recently entrusted by WG 6 to develop specific standard methods based on PCR. Currently, three standards are being prepared: one for the detection of neurotoxin genes of *Clostridium botulinum*, a second for the detection of pathogenic genes of *Yersinia enterocolitica*, and another for the detection of pathogenic genes of *Vibrio*.

Table 12.5 List of CEN/ISO standards on the use of PCR for the detection of pathogenic microorganisms in food[*]

Topic	Standard reference number[a]
General requirements and definitions	EN ISO 22174: 2005
Sample preparation for qualitative detection	EN ISO 20837: 2006
Amplification and detection for qualitative detection	EN ISO 20838: 2006
Performance testing for thermocyclers	EN ISO/TS 20836: 2005

a 'EN ISO' means that a common CEN/ISO Standard is concerned, 'ISO' that it is only an ISO Standard. 'ISO/TS' means ISO Technical Specification. The year of publication of the last version is indicated after the reference number.

* Data obtained from CEN website: http://www.cenorm.be; and ISO website: http://www.iso.org/.

V. CONCLUSION

One might question the large amount of effort and money being spent to standardize methods in food microbiology, when it is well known that every 'good' microbiologist is convinced they must have the best method available, adapted to their laboratory and objectives. Several reasons of a scientific nature have been already envisaged in the introduction, but there are additional reasons of a different nature that can also be mentioned. For instance, the use of standard methods favors the recognition of analytical results in the framework of the supplier/customer relationship. Further, the use of standard methods improves the recognition of analyses conducted in the framework of conducting official food controls at the national level, as well as in the context of international trade.

At the European level, for example, a legislative text was published at the end of 2005, Regulation 2073/2005 (EC, 2005), that gathered and updated microbiological criteria on food. This text is formally applicable to food business operators for their own checks of food products, but it is also indirectly applicable to official controls performed by Member States of the European Union. In the annex of the text defining certain criteria, a reference method is specified for each criterion, consisting in almost all cases in the CEN/ISO Standards presented in this chapter. The status of the CEN/ISO Standards as reference methods for both food business operator checks and official controls thus becomes clear from a European perspective. The text mentions that alternative methods, including PCR, can be used if validated according to the Standard EN ISO 16140, thus it is in line with the policy defined by SC 9 and WG 6.

The set of standards developed is also a sound and common basis for the accreditation of laboratories in food microbiology. ISO and CEN play a direct role in preparing reference documents that afterwards can be used for accreditation purposes in this field. The standardization bodies, either at the national, regional (CEN for Europe), or international level (ISO), play a key role in developing and maintaining a complete and updated set of standards, harmonizing reliable analytical methods in food microbiology.

REFERENCES

CEN (2007). European Committee for Standardization. Website, http://www.cenorm.be
EC (2005). Commission Regulation (EC) No 2073/2005 of 15 November 2005 on microbiological criteria for foodstuffs. Official Journal of the European Communities 22/12/2005: 1–26.
EN ISO/IEC 17025 (2005). General requirements for the competence of testing and calibration laboratories. International Organization for Standardization/International Electrotechnical Commission; Geneva.

ISO (2007). International Organization for Standardization. Website, http://www.iso.org Guide 2: Standardization and related activities General vocabulary. International Organization for Standardization/International Electrotechnical Commission.

Lahellec, C. (1998). Development of standard methods with special reference to Europe. *Int. J. Food Microbiol., 45*(1), 6–13.

Leclercq, A., Lombard, B., & Mossel, D. A. (2000). Normaliser les méthodes d'analyse dans le cadre de la maîtrise de la sécurité microbiologique française des aliments: atout ou contrainte (To standardize analytical methods in the framework of microbiological food safety control in France: pros and cons). *Science des Aliments, 20*(2), 179–202.

Leclercq, A. (2006). Development of a standard for the organization of the interlaboratory comparisons in microbiology of food. *Accreditation and Quality Assurance, 11*, 367–369.

Lombard, B., Gomy, C., & Catteau, M. (1996). Microbiological analysis of foods in France: standardized methods and validated methods. *Food Control, 7*(1), 5–11.

Lombard, B. (2004a). Les essais inter-laboratoires en microbiologie des aliments (Inter-laboratory trials in food microbiology). PhD, Institut National Agronomique de Paris-Grignon; Paris.

Lombard, B. (2004b). Microbiological analysis of foods and animal feeding stuffs. *ISO Focus, 1*(8), 16–17.

CHAPTER 13

Harmonization and Validation of Methods in Food Safety – FOOD-PCR: A Case Study

Martin D'Agostino *and* **David Rodríguez-Lázaro**

Contents

Abstract

Harmonized and validated methods related to food safety, especially with regard to molecular microbiology, are rarely found in the scientific literature. Commercial molecular-based systems that have been approved by standardization bodies do exist, but they have the disadvantage of being very expensive and do not afford the flexibility of non-proprietary molecular methods, many of which can be found in the scientific literature. To benefit from the advances in molecular microbiology, especially polymerase chain reaction (PCR) and real-time (quantitative) PCR, it is a priority that these advances be made useable as tools for diagnostic laboratories, rather than for exclusive use by the researcher. For years, diagnosticians have used traditional culture-based methods. Although these methods are predominantly labor-intensive and very time-consuming, the thought of introducing molecular techniques into the daily routine can be extremely daunting to the 'traditional' analyst. However, a recent project endeavored to make the process of adapting these techniques from a researcher's tool to a useable method much more straightforward, using a harmonized and standardized approach.

Global Issues in Food Science and Technology
ISBN 9780123741240

I. INTRODUCTION

Promoting a high level of food safety is a major policy worldwide (European Commission, 2000). Moreover, guaranteeing the safety and quality of food products along the food chain is the principal demand of consumers, as they expect their food to be tasty and wholesome as well as safe (European Commission, 2005). Concern for the safety of food products has been increased considerably during the last decades by the rapid globalization of the food market, and profound changes in food consumption habits (Käferstein *et al.*, 1997; Mead *et al.*, 1999; European Commission, 2000). It is no longer unusual, for example, to go into a shop and buy a selection of summer fruits at any time of the year, most of which have been imported from many different countries around the world. This has given consumers a huge choice in selection, which years ago would not have been available to them; however it has also brought with it new challenges in the area of food safety.

The primary objective in the area of food safety is to improve the health and well-being of citizens through a higher quality of their food, which can be achieved through the improved control of food production and related environmental factors, thus giving priority to the consumer's demand and rights for high-quality and safe food (European Commission, 2000). The innocuousness of foodstuffs is a concept inherent within the principles of food safety, and is related to many aspects of agrarian production technologies as well as to manipulation and elaboration of foods (FAO, 2001). Hygienic practices may well vary between countries and be less transparent than those currently employed. Under the World Trade Organization Sanitary and Phytosanitary (WTO SPS) Agreements and the codes of practice issued by the Codex Alimentarius Commission, a benchmark for international harmonization now exists that guarantees the trade of safe food. Inevitably, food safety is still mainly the responsibility of the consumer (Aruoma, 2006). However, the ability to detect foodborne pathogens quickly and effectively in produce before it is sold in retail outlets still presents a challenge, which is an ongoing concern for both research scientists and diagnostic food laboratories. Classical microbiological methods for detection of foodborne pathogens can be laborious, often requiring a lengthy process to obtain conclusive results. Therefore, the development and optimization of novel rapid alternatives for monitoring, characterization, and enumeration of foodborne pathogens is one of the priorities in food safety. However, to date there is a clear lack of harmonization in the development of this kind of rapid alternative, and a need for validation of new methods. This chapter will focus on the harmonization of methods employed in a 3-year European FP5-funded project, 'Validation and Standardization of Diagnostic Polymerase Chain Reaction for Detection of

Foodborne Pathogens' EC Framework Five Project QLK1-CT-1999-00226 ('FOOD-PCR'), which resulted in the development of molecular methods to detect five major human foodborne pathogens, these being enterohemorrhagic *Escherichia coli* (EHEC), *Salmonella* spp., *Campylobacter* spp., *Listeria monocytogenes*, and *Yersinia enterocolitica*. The project involved a harmonized and standardized approach that led to the publication of several papers, which are now regularly cited in the scientific literature.

II. CURRENT CHALLENGES AND THE 'FOOD-PCR' APPROACH

II.A. Where are we?

The number of new molecular microbiological methods used in research has increased rapidly over the past few years. Some commercially available diagnostic kits and their performance characteristics are non-proprietary and some are patented without revealing their performance characteristics, so today's researcher/diagnostic scientist is presented with a perplexing array of options that basically profess to being able to do the same thing, but are either quicker, or more specific or sensitive than others. Further details on molecular microbiology diagnostics can be found in Chapter 14. One of the main problems that still exists, however, is the transference of technological advances in molecular microbiology that have been made over the years, particularly with regard to non-commercial and non-proprietary methods, to food testing laboratories. Of course, traditional microbiological methods, or 'gold standards,' are still widely used, even though many of the methods have been around for decades, are labor-intensive, and take much longer to perform than newer methods. The question is which one should be chosen and why? Will the traditional method be used, or can the newer molecular method or kit be used routinely and the results accepted by the customer? Before making this decision a number of factors naturally should be considered, such as the number of samples, complexity of matrix, time, and cost. The main factor is in deciding which method will be most suitable, although it will always come down to whether or not the method is a recognized standard, i.e. one that has been accepted by standardization bodies such as CEN (European Committee for Normalisation) or ISO (International Organization for Standardization). Further details required by standardization bodies and their objectives can be found in Chapter 12. One of the main benefits of prompting the idea of using molecular methods in diagnostic laboratories is the method's ability to produce results much more quickly, and consequently fulfilling the time requirements of food producers and/or providing immediate results at the time of production. Therefore these methods are cheaper in cost than traditional methods as they save time in analysis of a foodstuff and the amount of time it is stored. It is true that the initial cost to buy equipment and reagents, etc., may be high; however, turn around time, along with the possibility of performing the tests using high-throughput technology, would in theory soon outweigh any negative cost issues. So, if a diagnostic laboratory decides that a particular kit or molecular method would be a suitable alternative to traditional methods, it would be prudent and appropriate to consider the steps needed, if any, to validate the kit/method, and thus confirm its ability to provide results that are at least as good as the 'gold standard' traditional method. In order to do this suc-cessfully, a harmonized approach must be taken to validate these methods. This is what the FOOD-PCR project aimed to do. We will now focus on the structure, development, and organization of this project.

II.B. The FOOD-PCR project – aims

The original objective envisioned by the FOOD-PCR consortium, which consisted of 35 institutes from 14 EU countries and seven associate and applicant states, was to facilitate implementation of diagnostic PCR for both verification and detection of foodborne pathogens through harmonization and standardization of methods in Europe (Hoorfar, 1999). Some of the participants involved in the project were end-user laboratories that were keen to see the implementation of the methods in a diagnostic setting. It was hoped that once these five new methods were developed, there would be an opportunity to validate the methods against the traditional 'gold standard' method, but this has not yet been realized, mainly due to lack of funding available for such a study, as it would require a substantial amount of money due to the logistical complexities involved. Although the methods were validated using interlaboratory ring trials, validation was purely to check the reproducibility and the robustness of the methods using pre-enriched samples prepared by individual laboratories. For these methods to be accepted and widely used in a diagnostic laboratory setting, a subjective comparison between the traditional standard methods and the PCR-based methods would have to be carried out. In other words, the repeatability and reproducibility would have to be compared. In order to progress toward this step, a series of standardized approaches to the development of the methods was performed by the FOOD-PCR consortium.

II.C. The structure

Since the project's aims were to develop PCR-based methods to detect the five most common foodborne pathogens in a variety of food matrices, five working groups were formed to focus on each pathogen. Each group had a task leader whose role was to coordinate the work of the scientists involved with the pathogen in question. Each working group had approximately

10–12 participants, some of whom had responsibilities in other working groups if they had an interest in more than one pathogen. A dedicated website was created for the project and regularly updated with the latest developments; it also included all information required by the participants, including SOPs, contact names, e-mail addresses, etc. This proved to be an invaluable resource for the project and it is now very common to see projects with their own dedicated website.

The project was conducted in three phases, each requiring cooperative activity with all EU partners concerned (Hoorfar and Cook, 2003), with a total of six work packages, involving 12 main tasks (Table 13.1).

II.C.1. Phase 1. Method development

The development of the PCR-based methods involved following a series of harmonized steps. For instance, criteria and terminologies were agreed upon for each stage (Table 13.2): standardized DNA solutions were prepared and sent to the laboratories; standardized ring trial materials were prepared and couriered to the laboratories; standardized SOPs were prepared, etc. These were then used by everyone in the project, thus ensuring that all the participants used 'identical' materials and methods.

The final PCR-based methods produced also had identical performance criteria. Importantly, a consensus was agreed upon by all five working groups collectively, not on a group-by-group basis. Following these working principles resulted in the successful harmonization of the methods as a whole, so when it came to performing a particular PCR, many potential problems were effectively avoided. As an example, imagine no harmonization existed with

Table 13.1 Work packages involving 12 tasks

Work packages (WP)		Project milestones
WP1	Construction of DNA banks	Stabilized DNA ampoules available
		Protocols for sample treatment methods prepared
WP2	Validation of thermocyclers	Validated PCR for thermophilic *Campylobacter* spp.
WP3	Pre-PCR treatment	Validated PCR for *Salmonella* spp.
		Validated PCR for EHEC
WP4	Ring trials (WP leader)	Validated PCR for *Yersinia enterocolitica*
		Validated PCR for *Listeria monocytogenes*
WP5	Automated detection	Diagnostic accuracy of each PCR established
WP6	Guidelines and workshops	Guideline for first-time implementation of PCR prepared
		Draft proposals submitted to CEN/TC 275
		Phase 3 ring trials performed
		Pre-standard guidelines prepared

Table 13.2   Definition of terms used in the FOOD-PCR project

Stage	Term	Definition
Diagnostic accuracy of PCR assay	Selectivity	Inclusivity (detection of targets) Exclusivity (non-detection of non-targets)
	Detection limit	Lowest number of cells that can be detected with 99% probability in a single reaction
Statistical analysis of PCR assay ring trials	Accordance	Repeatability (odds of getting same results from two identical samples sent to one laboratory)
	Concordance	Reproducibility (odds of getting same results from two identical samples sent to two different laboratories)
	Concordance Odds Ratio (COR)	Odds of getting same results from two identical samples sent to one laboratory or two different laboratories
PCR-based method ring trial evaluation	Sensitivity	Correct identification of inoculated samples
	Specificity	Correct identification of uninoculated samples

regard to performing PCRs. What would be the consequences? There would be perhaps upwards of 20 laboratories across the EU choosing their own preferred primers, their own choice of enzyme, their own form of controls, their own choice of cycling parameters, their own interpretation of results, etc. Performing the experiments in this way would make the task of statisticians performing statistical analysis impossible, but by following an agreed upon set of performance criteria, such pitfalls could be avoided. These criteria include integration of an Internal Amplification Control (IAC) into each PCR assay (Hoorfar *et al.*, 2003, 2004), the use of Hot Start enzyme, a defined selectivity, and a defined detection probability. Importantly, the methods also should be compatible with the standard ISO methodology available (Malorny *et al.*, 2003).

II.C.2. Phase 2. Assay ring trials

Ring trials to test the reproducibility and repeatability of the PCR assays were performed according to identical criteria chosen by the working groups. For example, the experts within the relevant pathogen group agreed upon the selection of target and non–target strains used in the ring trials (see Table 13.3). A standard operating procedure was drafted and distributed to all the group members, amended accordingly and agreed upon collectively. Finally the results of the ring trials were sent to one laboratory expert in PCR for analysis, preventing any differences in interpretation of results, which through previous experience had been highly variable. Setting these criteria across all five working groups was very effective indeed and led to a much more simplified analysis and dissemination of results.

II.C.3. Phase 3. Collaborative trial of the PCR-based methods

A similar pattern was set when testing the PCR–based method as a whole. Again, all the working groups followed an agreed–upon set of criteria, ensuring complete harmonization. Following the structure set out in phase 2, the individual task groups determined which expert laboratory would inoculate the 12 test samples, the level of inoculation required, and the means by which samples would be sent to the ring trial participants. The criteria agreed upon were that there would be at least 12 participants in each trial; the samples would be inoculated at low, medium, and high levels, three each per level, and three uninoculated control samples would be included; the expert laboratory would enrich all the samples. One ml of enriched sample was sent out to the participants who then had to perform the next stage of treatment, followed by PCR. Once again a detailed SOP accompanied the samples, with results sent back to the expert laboratory for analysis.

II.D. The outcomes of the FOOD-PCR project and beyond

The cooperative activity resulted in the publication of several robust, repeatable, and reproducible PCR–based methods to detect foodborne bacterial pathogens (Abdulmawjood et al., 2003, 2004; Lübeck et al., 2003;

Table 13.3 Number of isolates used to determine the selectivity of each PCR method in expert laboratories

Pathogen	Number of target strains	Number of non-target strains
Campylobacter spp.	121	24
EHEC	32	91
L. monocytogenes	38	51
Salmonella spp.	242	122
Y. enterocolitica	117	58

D'Agostino *et al.*, 2004; Heuvelink *et al.*, 2004; Josefsen *et al.*, 2004; Malorney *et al.*, 2004). In all, the FOOD-PCR project produced more than 30 publications, many of which are now frequently cited in the scientific literature. The evidence shows that this work is highly valued and, in particular, that the need for robust validated methods is recognized by the scientific community. Indeed, there has been a lot of interest shown in the methods by various organizations who wish to incorporate these methods into their routine sampling strategies.

Furthermore, many members of the FOOD-PCR consortium have contributed to the activities of the standardization body (TAG 3 of CEN/TC 275/WG 6), and the standards now published (ISO/TS 20836, 2005; ISO 20837, 2006; ISO 20838, 2006; ISO 22174, 2005) were based on the findings of the FOOD-PCR project. The publication of these standards highlights the requirement that PCR-based methods should be implementable for food analysis.

One of the visions of the CEN/TAG3 group was to have PCR-based methods used routinely alongside conventional standard methods in end-user laboratories by the year 2010. To our knowledge, the FOOD-PCR methods developed thus far are not used routinely alongside conventional methods for food analysis, although there is enormous diagnostic potential. Part of the problem in progressing their use is technical, however a lack of international validation, the need for training of analysts, and difficulties in the transfer of technology also play major roles. Why are these problems seemingly insurmountable? Perhaps one reason is that many analysts still regard the conventional 'gold standard' as the only completely trustworthy method, and as mentioned earlier, any new method should be shown to work at least as well as and reliably as conventional methods. This really is what is required to convince the potential end-users of the effectiveness of FOOD-PCR methods and little if any progress has been made to reach this goal, although the technology has been around for many years now. This is why the focus of the FOOD-PCR project was the removal of these barriers through establishment of a research group whose goal was development, harmonization, and validation of methodology.

II.E. Implementation – how will it be achieved?

A direct comparison between the traditional standard method and the PCR-based method where the performance criteria for both methods are known is essential. This like other ring trials would ideally involve 10–12 partners. The two specific elements of both methods, which would be determined and compared, are diagnostic specificities (related to number of false positives) and diagnostic sensitivities (related to number of true positives). This direct comparison should be performed according to a standard protocol such as ISO 16140 (Anon, 2003). This document defines the general principle and the technical protocol for the validation of alternative methods in the field of microbiological analysis of food, animal feeding stuffs, and environmental and veterinary samples. To be fully effective and widely accepted, the validation exercise should be performed at an international level. Within the EU Framework 6 Programme there is no apparent funding for this type of study, and although a FOOD-PCR 2 project has been set up, there is no substantial funding available for research on this scale, however it has meant that active communication still exists within the FOOD-PCR consortium. This continued open communication allows the foundations of a very successful consortium to remain in place and provides the basis for accepting opportunities for this work in the future should opportunity arise in the FP7 framework.

III. CONCLUDING REMARKS

A search of publications in the scientific literature relating to PCR-based detection methods for various foodborne pathogens will reward the researcher with dozens, maybe hundreds, of different papers on the subject, showing a significant amount of work being carried out today in research laboratories all over the world. One may still ask though, are these methods being used routinely in analytical laboratories, or are they simply a researcher's tool? The answer is most likely the latter due to the majority of these methods not being taken to the next stage, which is validation against the current standard method. Unless this is done, these methods will never be considered as reliable and robust alternatives that can stand alongside the 'gold standard' method. It will be impossible for them to reach full potential as useful tools in the constant battle against the ever-present threat of foodborne pathogens. It is hoped that the valuable work performed by the FOOD-PCR consortium can provide a model for those wishing to embark on such a step, towards international validation of their own methods.

ACKNOWLEDGMENTS

David Rodríguez-Lázaro was a research fellow of the European Union's Marie Curie program (Contract MEIF-CT-2005-0011564). The authors would like to thank Dr. N. Cook, Central Science Laboratory, York, UK for checking the manuscript.

REFERENCES

Abdulmawjood, A., Cook, N., Bülte, M., Roth, S., Schönenbrücher, H., & Hoorfar, J. (2003). Toward an international standard for PCR-based detection of *E. coli* O157. Part 1.

Assay development and multi-center ring-trial validation. *Journal of Microbiological Methods*, *55*, 775–786.

Abdulmawjood, A., Bülte, M., Roth, S., et al. (2004). Toward an international standard for PCR-based detection of foodborne *Escherichia coli* O157: validation of the PCR-based method in a multicenter collaborative Trial. *Journal of AOAC International*, *87*, 856–860.

Anon (2003). Microbiology of food and animal feeding stuffs – Protocol for the validation of alternative methods. ISO 16140. International Organization for Standardization; Geneva.

Aruoma, O. I. (2006). The impact of food regulation on the food supply chain. *Toxicology*, *221*(1), 119–127.

D'Agostino, M., Wagner, M., Vazquez-Boland, J. A., et al. (2004). A validated PCR-based method to detect *Listeria monocytogenes* using raw milk as a food model – towards an international standard. *Journal of Food Protection, 67*, 1646–1655.

European Commission (2000). White Book on food safety. Brussels.

European Commission (2005). Special Eurobarometer: risk issues. Survey. Directorate-General Press and Communication.

FAO. (2001). CFS: 2001/Inf7/Add.1/Rev1. New Challenges to the Achievement of the World Food Summit Goals. *Committee on World Food Security.* Rome: Twenty-seventh Session, 2001.

Heuvelink, A. E., Tassios, P. T., Lindmark, H., Kmet, V., Barbanera, M., Fach, P., Loncarevic, S., & Hoorfar, J. (2004). Validation of a PCR-based method for detection of foodborne thermotolerant campylobacters in a multi-center collaborative trial. *Applied and Environmental Microbiology, 70*, 4379–4383.

Hoorfar, J. (1999). EU seeking to validate and standardize PCR testing of food pathogens. *ASM News, 65*(12), 799.

Hoorfar, J., & Cook, N. (2003). Critical aspects of standardization of PCR. In K. Sachse, & J. Frey (Eds.), *PCR Detection of Microbial Pathogens* (pp. 51–64). Totowa: Humana Press, Methods in Molecular Biology.

Hoorfar, J., Cook, N., Malorny, B., Rådström, P., De Medici, D., Abdulmawjood, A., & Fach, P. (2003). Diagnostic PCR: Making internal amplification control mandatory. *Journal of Clinical Microbiology, 41*(12), 5835.

Hoorfar, J., Malorny, B., Abdulmawjood, A., Cook, N., Wagner, M., & Fach, P. (2004). Practical considerations in design of internal amplification control for diagnostic PCR assays. *Journal of Clinical Microbiology, 42*(5), 1863–1868.

ISO/TS 20836 (2005). Microbiology of food and animal feeding stuffs – Polymerase chain reaction (PCR) for the detection of food-borne pathogens – Performance testing for thermal cyclers. International Organization for Standardization; Geneva.

ISO 20837 (2006). Microbiology of food and animal feeding stuffs – Polymerase chain reaction (PCR) for the detection of food-borne pathogens – Requirements for sample preparation for qualitative detection. International Organization for Standardization; Geneva.

ISO 20838 (2006). Microbiology of food and animal feeding stuffs – Polymerase chain reaction (PCR) for the detection of food-borne pathogens – Requirements for amplification and detection for qualitative methods. International Organization for Standardization; Geneva.

ISO 22174 (2005). Microbiology of food and animal feeding stuffs – Polymerase chain reaction (PCR) for the detection of food-borne pathogens – General requirements and definitions. International Organization for Standardization; Geneva.

Josefsen, M. H., Cook, N., D'Agostino, M., et al. (2004). Validation of a PCR-based method for detection of food-borne thermotolerant campylobacters in a multicenter collaborative trial. *Appl. Environ. Microbiol., 70*(7), 379–383.

Kaferstein, F. K., Motarjemi, Y., & Bettcher, D. W. (1997). Foodborne disease control: a transnational challenge. *Emerg. Infect. Dis, 3*(4), 503–510.

Lübeck, P. S., Cook, N., Wagner, M., Fach, P., & Hoorfar, J. (2003). Towards an international standard for PCR-based detection of food-borne thermotolerant campylobacters. Part 2: Validation of the PCR assay in a multi-center collaborative trial. *Applied and Environmental Microbiology, 69*, 5670–5672.

Malorny, B., Tassios, P. T., Rådström, P., Cook, N., Wagner, M., & Hoorfar, J. (2003). Standardization of diagnostic PCR for the detection of foodborne pathogens. *International Journal of Food Microbiology, 83*, 39–48.

Malorny, B., Cook, M., D'Agostino, M., et al. (2004). Multicenter collaborative trial validation of a PCR-based method for detection of *Salmonella* in chicken and pig samples. *Journal of AOAC International, 87*, 861–866.

Mead, P. S., Slutsker, L., Dietz, V., McCaig, L. F., Bresee, J. S., Shapiro, C., Griffin, P. M., & Tauxe, R. V. (1999). Food-related illness and death in the United States. *Emerging Inf. Dis., 5*, 607–625.

CHAPTER 14

Current Challenges in Molecular Diagnostics in Food Microbiology

David Rodríguez-Lázaro, Nigel Cook,
Martin D'Agostino, *and* Marta Hernández

Contents

Abstract

Foodborne diseases produced by microbial pathogens are one of the most serious worldwide public health concerns. Consequently, the application of microbiological controls within the quality assessment programs of the food industry is a way to minimize the risk of infection. Classical microbiological methods are laborious, time consuming, and not always reliable. A number of alternative, rapid, and sensitive molecular-based methods for the detection, identification, and quantification of foodborne pathogens have been developed to overcome these drawbacks. They have a number of advantages including rapidity, excellent analytical sensitivity and selectivity, and potential for quantification. However, the use of expensive equipment and reagents, the need for qualified personnel, and the lack of standardized protocols are impairing their practical implementation for monitoring and microbiological control.

I. INTRODUCTION

Incidences of foodborne diseases have increased considerably over the last few decades and are a major cause of morbidity (Wallace *et al.*, 2000).

Global Issues in Food Science and Technology
© 2009, Elsevier Inc. and British Crown Copyright.

Reasons for this include rapid globalization of the food market, the increase in personal and food transportation, and the profound changes in food consumption habits (Käferstein *et al.*, 1997). More than 250 known diseases are transmitted through food, with symptoms ranging from mild gastro-enteritis to life-threatening syndromes, with the possibility of chronic complications or disability (Mead *et al.*, 1999). The causes of foodborne illness include pathogens, toxins, and metals. More than 40 different foodborne pathogens are known to cause human illness (CAST, 1994), among which over 90% of confirmed human illness cases and deaths caused by foodborne pathogens reported to the Center for Disease Control and Prevention (CDC) have been attributed to bacteria (Bean *et al.*, 1990).

The impact of foodborne pathogens in public health systems is considerable. Foodborne pathogens cause 14 million illnesses, 60,000 hospitalizations and 1,800 deaths per year in the USA (Mead *et al.*, 1999), with annual medical and productivity losses around 6,500 million dollars due only to the five major foodborne pathogens (Crutchfield and Roberts, 2000). In England and Wales, foodborne pathogens cause 1.3 million illnesses, 20,759 hospitalizations and 480 deaths each year (Adak *et al.*, 2002).

The development and optimization of novel alternatives for the monitoring, characterization and enumeration of foodborne pathogens is one of the key aspects of food microbiology (Stewart, 1997), and has become increasingly important in the agricultural and food industry (Malorny *et al.*, 2003). Classical microbiological methods for the presence of microorganisms in foods involve, in general, pre-enrichment and isolation of presumptive colonies of bacteria on solid media, and final confirmation by biochemical and/or serological identification. Thus, they are laborious, time-consuming and not always reliable (e.g. viable but non-culturable VBNC forms) (Rollins and Colwell, 1986; Tholozan *et al.*, 1999). The adoption of molecular techniques in microbial diagnostics has become a promising alternative to overcome these disadvantages (Scheu *et al.*, 1998; Fung, 2002).

II. CURRENT CHALLENGES

The inherent advantages of amplification techniques (e.g. shorter turn-around, improved detection limits, specificity and potential for automation) can foster its implementation in food laboratories that routinely test for foodborne pathogens. The polymerase chain reaction (PCR) has become the most extensively used molecular technique, and is predicted to be established as a routine reference within the next 10 years (Hoorfar and Cook, 2003). Other techniques have also been developed such as nucleic acid sequence based amplification (NASBA), although until now, they have had limited practical relevance for food monitoring and control. However further developments are needed for effective implementation of amplification techniques in food microbiology. The main issues that must be addressed for the effective adaptation of molecular techniques in food laboratories are: the development of rational and easy-to-use strategies for pre-PCR treatment of food samples; the design and application of analytical controls; the inability to unambiguously detect viable targets; the development of strategies for the quantitative use of real-time PCR for food samples and greater automation of the whole analytical process.

II.A. Pre-amplification processing of samples

The efficiency and performance of molecular methods can be negatively affected by the presence of inhibitory substances generally found in foods, growth media, and nucleic acid extraction reagents (Rossen et al., 1992; Wilson et al., 1997; Rådström et al., 2003). Amplification inhibitors may interfere with the cell lysis, degrade or capture nucleic acids, and/or inhibit the amplification reaction (Wilson, 1997). Consequently they can reduce or even block amplification reactions, leading to the underestimation of the bacterial load or production of false-negative results. Usual amplification inhibitors include components of body fluids and reagents present in food components (e.g. organic and phenolic compounds, milk proteinases, glycogen, fats, and Ca^{2+}); environmental substances (e.g. phenolic compounds, humic acids, and heavy metals); constituents of bacterial cells; 'contaminant' (non-target) nucleic acids; and laboratory products such as glove powder, plasticware, and cellulose. Thus, PCR-friendly sample preparation prior to the amplification reaction is crucial for the robustness and performance of amplification-based methods, and is a priority for the implementation of molecular methods as diagnostic tools in food microbiology laboratories.

The purpose of sample preparation is to homogenize the sample to be amplified, increase the concentration of the target organism to the practical operating range of a given assay, and reduce or exclude amplification-inhibitory substances. Hence, pre-amplification treatment aims to convert biological samples into amplifiable samples (Rådström et al., 2003). As food samples vary in homogeneity, consistency, composition, and accompanying microbiota, pre-amplification procedures should be adapted for each food matrix. A large range of pre-amplification procedures has been developed, but many of these are laborious, expensive, and time-consuming (Jaffe et al., 2001). Procedures can either be biochemical, immunological, physical, or physiological (Rådström et al., 2003), or a combination thereof (Table 14.1).

Table 14.1 Sample preparation procedures used for different types of samples[1]

Category	Subcategory	Sample preparation procedure	Sample
Biochemical	Adsorption	Lectin-based separation	Beef meat
		Protein adsorption	Blood
	Nucleic acids extraction	Nucleic acid purification procedures	Diverse matrixes
		Lytic procedures	Diverse matrixes
Immunological	Adsorption	Immunomagnetic capture	Diverse matrixes
Physical		Aqueous two-phase systems	Soft cheese
		Buoyant density centrifugation	Minced meat
		Centrifugation	Diverse matrixes
		Dilution	Diverse matrixes
		Filtration	Diverse matrixes
		Mechanical disruption by ceramic spheres	Diverse matrixes
		Grinding by mortar and pestle	Diverse matrixes
		Boiling	Diverse matrixes
		Other heat treatments	Diverse matrixes
Physiological		Enrichment	Diverse matrixes

1 Source: Adapted from Rådström, P., Knutsson, R., Wolffs, P., Dahlenborg, M., and Löfström, C. (2003).

II.B. Analytical controls

Molecular-based detection techniques, as with any instrumental techniques, can produce false-negative and false-positive results (Knowk and Higuchi, 1989). Contamination is one of the principal concerns in food analysis laboratories. The main causes of production of false-positive results are accidental contamination of the samples or the reagents with positive samples (cross-contamination) or with amplification products and plasmid clones (carryover contamination). In addition, the efficiency of amplification-based techniques can be negatively influenced by several conditions including malfunction of equipment, incorrect reaction mixture, poor enzyme activity, or the presence of inhibitory substances in the original sample matrix. This can result in weak or negative signals and lead to underestimation of the amount of microbial load in the sample. The potential presence of amplification inhibitors in the reaction is a serious problem that can compromise the applicability of the amplification-based techniques in food analysis. Therefore, adequate control of the efficiency of the reaction is a fundamental aspect in such assays (Hoorfar and Cook, 2003; Hoorfar et al., 2004). A series of controls are recommended to correctly interpret the results of molecular techniques (Table 14.2).

II.B.1. Internal amplification controls (IAC)

Other fundamental aspects rely on the adequate control of the amplification reaction efficiency. In this sense, the application of internal amplification

Table 14.2 Analytical controls for molecular-based techniques[1]

<u>P</u>rocessing <u>P</u>ositive <u>C</u>ontrol (PPC): A negative sample spiked with a sufficient amount of target (e.g. pathogen, species, etc.), and processed throughout the entire protocol. A positive signal should be obtained indicating that the entire process (from nucleic acids extraction to amplification reaction) was correctly performed.

<u>P</u>rocessing <u>N</u>egative <u>C</u>ontrol (PNC): A negative sample spiked with a sufficient amount of a non-target sample or water, and processed throughout the entire protocol. A negative signal should be obtained indicating the lack of contamination along the entire process (from nucleic acids extraction to amplification reaction).

Premise Control or <u>E</u>nvironmental <u>C</u>ontrol (EC): A tube containing the master mixture or water left open in the PCR setup room to detect possible contaminating nucleic acids in the environment.

<u>A</u>mplification <u>P</u>ositive <u>C</u>ontrol (PC): A template known to contain the target sequence. A positive amplification indicates that amplification was performed correctly.

No <u>T</u>emplate <u>C</u>ontrol or Reagent Control or Blank (NTC): Includes all reagents used in the amplification except the template nucleic acids. Usually, water is added instead of the template. A negative signal indicates the absence of contamination in the amplification assay.

Internal <u>A</u>mplification <u>C</u>ontrol (IAC): Chimerical non-target nucleic acid added to the master mixture in order to be co-amplified by the same primer set as the target nucleic acid but with an amplicon size visually distinguishable or different internal sequence region from the target amplicon. The amplification of IAC both in presence and absence of target indicates that the amplification conditions are adequate.

1 Source: Adapted from Dieffenbach, C.W., Lowe, T.M.J., and Dveksler, G.S. (1995); Hoorfar, J., and Cook, N. (2003); Stirling, D. (2003).

controls allows the assessment and interpretation of the diagnostic results of the molecular techniques. An internal amplification control or 'IAC' is a non-target nucleic acid sequence, which is co-amplified simultaneously with the target sequence (Cone *et al.*, 1992; Rodríguez-Lázaro *et al.*, 2004d, 2005b). In a reaction without an IAC, a negative response (no signal) can mean that there was no target sequence present in the reaction. But, it could also mean that the reaction was inhibited. In a reaction with an IAC, a control signal will always be produced when there is no target sequence present. When no control signal is observed, this means that the reaction has failed, and the sample must be reanalyzed. In an amplification-based assay, an IAC should be based on flanking nucleic acid sequences (DNA for PCR assays, and RNA for RT-PCR and NASBA assays) with the same primer recognition sites as the target, with non-target internal sequences (Rodríguez-Lázaro *et al.*, 2004d, 2005b). The principal requirements of an optimal internal amplification control (IAC) for use in food diagnostic assays are reviewed in Hoorfar *et al.* (2004).

There are two main strategies for using an IAC in diagnostic assays: competitive IAC and non-competitive IAC. Both strategies are useful, although the first is recommended to avoid the risk of undesired interactions of multiple primers, and, especially, to get comparable amplification efficiency of the IAC and the target sequence. In the competitive IAC strategy, the target and the IAC are co-amplified in a single reaction tube with the same primer set, and, in consequence, there is always some competition between target nucleic acid and IAC. Thus, the most critical parameter to consider is the optimal initial number of IAC copies in the diagnostic assay as it directly affects the target detection limit (Abdulmawjood *et al.*, 2002). If used at too high a concentration, the IAC might not allow detection of weak inhibition, which could cause false-negative results if the target is present in low concentrations (Rosenstraus *et al.*, 1998). However, having too low a concentration of initial IAC leads to substantial variations in IAC amplification, indicating poor reproducibility. The initial IAC copy number in the reaction must therefore be determined at a compromise level that allows reproducible IAC detection and avoids inhibition of the target-specific reaction. In the non-competitive IAC strategy, the target and the IAC are amplified by two different primer sets. The IAC primer set can target a synthetic nucleic acid (DNA for PCR assays or RNA for RT-PCR or NASBA assays) or a gene different from the target.

II.C. Determination of viable bacteria

The determination of bacterial viability is a key issue for the application of food risk management, and thus a rational approach to detect only viable bacterial cells by using molecular-based methods is necessary. However,

PCR-based methods detect DNA, which survives cell death. For this purpose the use of mRNA as template for amplification can be a promising solution (Klein and Juneja, 1997), though this requires removing any trace of bacterial DNA in the reaction in order to avoid false-positive results in viability assays (Cook, 2003).

Nucleic acid sequence-based amplification (NASBA) technique is a promising diagnostic tool for the analysis of viable microorganisms, since it is based on amplification of RNA rather than DNA. This technique is a sensitive transcription-based amplification system specifically designed for the continuous amplification of RNA in a single mixture at isothermal conditions (Compton, 1991; Deiman et al., 2002). Thus, this technique has the potential for detection of viable cells through selective amplification of messenger RNA, even in a background of genomic DNA, which PCR does not possess. NASBA employs a battery of three enzymes (AMV reverse transcriptase, RNase H, and T7 RNA polymerase) to amplify sequences from an original single-stranded RNA template, leading to a main amplification product of single-stranded RNA. The reaction also includes two oligonucleotide primers, complementary to the RNA region of interest. One of the primers also contains a promoter sequence that is recognized by T7 RNA polymerase at the 5'-end. The reaction is performed at a single temperature, normally 41°C for 1 to 2 h in a self-sustained manner. At this temperature, the genomic DNA from the target remains double-stranded and does not become a substrate for amplification. The principal characteristics of NASBA are summarized in Table 14.3.

The application of NASBA for detection of foodborne pathogens is at around the same stage as PCR was a decade or so ago, with a few methods being sporadically published in the scientific press (Table 14.4) (Cook et al., 2003; Rodríguez-Lázaro et al., 2006). Hence, considerable further development is required before NASBA can follow in PCR's footsteps to realize its potential for routine use. However, since NASBA can equal the rapidity and accuracy of PCR and has the additional potential advantage of unambiguous detection of viable pathogens, NASBA is a very promising diagnostic tool for food and clinical microbiology, and could become a reference in future decades.

Another approach has recently been devised to distinguish viable bacterial cells: the staining of cells with ethidium monoazide bromide (EMA) prior to DNA extraction and PCR to inhibit the amplification of DNA from dead cells (Nogva et al., 2003; Rudi et al., 2005a). This strategy combines the use of viability (live–dead) discriminating dye with the speed, specificity, and selectivity of amplification-based techniques such as real-time PCR. EMA is a phenanthridinium nucleic acid-intercalating agent (Waring, 1965), and photolysis of EMA with visible light

Table 14.3   Characteristics of NASBA[1]

• A single step isothermal amplification reaction at 41°C.
• Especially suited for RNA analytes because of the integration of RT into the amplification process.
• The single-stranded RNA product is an ideal target for detection by various methods including real-time detection using molecular beacons.
• The fidelity of NASBA is comparable to other amplification processes that use DNA polymerases lacking the 3' exonuclease activity.
• The use of a single temperature eliminates the need for special thermocycling equipment.
• Efficient ongoing process results in exponential kinetics caused by production of multiple RNA copies by transcription from a given cDNA product.
• Unlike amplification processes such as PCR, in which the initial primer level limits the maximum yield of product, the amount of RNA obtained in NASBA exceeds the level of primers by at least one order of magnitude.
• NASBA RNA product can be sequenced directly with a dideoxy method using RT and a labeled oligonucleotide primer.
• The intermediate cDNA product can be made double-stranded, ligated into plasmids, and cloned.
• Three enzymes are required to be active at the same reaction conditions.
• Low temperature can increase the non-specific interactions of the primers. However, these interactions are minimized by the inclusion of DMSO.
• A single melting step is required to allow the annealing of the primers to the target.
• The NASBA enzymes are not thermostable and thus can only be added after the melting temperature.
• The length of the target sequence to be amplified efficiently is limited to approximately 100 to 250 nucleotides.

1 Source: Modified from Deiman, B., van Aarle, P., and Sillekens, P. (2002).

produces a nitrene that can form stable covalent links to DNA (Hixon *et al.*, 1975; Coffman *et al.*, 1982). The unbound EMA, remaining free in solution, is simultaneously photolysed and converted to hydroxylamine, and is no longer capable of covalent attachment to DNA (DeTraglia *et al.*, 1978). Thus, the application of EMA prior to bacterial DNA extraction can lead to selective removal of DNA from dead cells. This approach has already been tested with different foodborne pathogens such as *Escherichia coli* 0157:H7 (Nogva *et al.*, 2003; Guy *et al.*, 2006; Nocker *et al.*, 2006),

Table 14.4   NASBA methods for detection of pathogenic microorganisms in food[1]

Target	Food	Reference
Bacillus spp.	Milk	Gore *et al.*, 2003
Campylobacter jejuni, C. coli, C. lari	Poultry products	Uyttendaele *et al.*, 1994
C. jejuni	Poultry products	Uyttendaele *et al.*, 1995a
C. jejuni	Poultry products	Uyttendaele *et al.*, 1995b
C. jejuni	Poultry products	Uyttendaele *et al.*, 1997
C. jejuni	Poultry products, meat products, dairy products	Uyttendaele *et al.*, 1999
Hepatitis A virus	Blueberries and lettuce	Jean *et al.*, 2001
Hepatitis A virus and norovirus	Deli sliced turkey and lettuce	Jean *et al.*, 2004
Listeria monocytogenes	Poultry products, meat products, seafood, vegetables, dairy products	Uyttendaele *et al.*, 1995c
L. monocytogenes	Egg products, dairy products	Blais *et al.*, 1997
M. paratuberculosis	Milk, water	Rodríguez-Lázaro *et al.*, 2004e
Salmonella enteritidis	Liquid egg	Cook *et al.*, 2002
Salmonella enteritidis	Fresh meats, poultry, fish, ready-to-eat salads and bakery products	D'Souza *et al.*, 2003

1 Source: Adapted from Rodríguez-Lázaro et al., 2006.

Salmonella (Nogva *et al.*, 2003; Guy *et al.*, 2006; Nocker *et al.*, 2006), *Listeria monocytogenes* (Nogva *et al.*, 2003; Rudi *et al.*, 2005a,b; Guy *et al.*, 2006; Nocker *et al.*, 2006), *Campylobacter* (Rudi *et al.*, 2005a), and *Vibrio vulnificus* (Wang and Levin, 2005). However, it has been reported that EMA can also penetrate the membrane of viable bacterial cells and co-valently cross-linked with the DNA during photolysis, resulting in loss of a percentage of the genomic DNA of viable cells and PCR inhibition

Table 14.5 Real-time quantitative PCR methods for principal foodborne pathogens

Target	Sequence	Matrix	Quantification limit	Detection limit	Reference
Campylobacter	VS1	Pure culture	10 CFU/reaction	1 CFU/reaction	Yang et al., 2004
	16S rRNA	Fecal samples	3×10^3 CFU/g	3×10^3 CFU/g	Inglis and Kalischuk, 2004
	C. jejuni fragment	Fecal samples	20 CFU/reaction	2 CFU/reaction	Rudi et al., 2004
	VS1	Poultry, milk, and environmental water	150 CFU/reaction	6–15 CFU/reaction	Yang et al., 2003
	C. jejuni fragment	Pure culture	10 CFU/reaction	1 CFU/reaction	Negva et al., 2000a
E. coli	Stx	Fecal samples	1×10^3 CFU/g	1×10^3 CFU/g	Sekse et al., 2005
	sfmD	Culture medium	1×10^2 CFU/ml	1×10^2 CFU/ml	Kaclikova et al., 2005
	16S rRNA	Fecal samples	4×10^3 CFU/ml	4×10^3 CFU/ml	Penders et al., 2005
	uspA	Urine samples	– –		Hinata et al., 2004
	stx1, eae	Environmental samples	2.6×10^4 CFU/g	2.6×10^4 CFU/g	Ibekwe and Grieve, 2003
	eae, stx1	Fecal samples	1×10^3 CFU/ml	1×10^3 CFU/ml	Sharma, 2002
	lacZ	Water	10 CFU/ml	10 CFU/ml	Foulds et al., 2002
	stx1, stx2, eae	Dairy wastewater	3.5×10^4 CFU/ml	6.4×10^3 CFU/ml	Ibekwe et al., 2002
Listeria monocytogenes	actA	Pure culture	10^{-2} CFU/ml	10^{-2} CFU/ml	Oravcova et al., 2006
	hly	Several foods	10^{-3} CFU/g	10 CFU/g	Rodríguez-Lázaro et al., 2005b
	LightCycler L. monocytogenes Detection kit	Salad	10 CFU/100g	10 CFU/100g	Berada et al., 2006

Organism	Target	Sample			Reference
	hly	Raw salmon and smoked salmon	10^{-3} CFU/g	10 CFU/g	Rodríguez-Lázaro et al., 2005a
	hly	Biofilm	6×10^2 CFU/cm^2	6×10^2 CFU/cm^2	Guilbaud et al., 2005
	hly	Cheese	100 CFU/g	10 CFU/g	Rudi et al., 2005a
	hly	Meat and meat products	10^{-3} CFU/g	100 CFU/g	Rodríguez-Lázaro et al., 2004c
	hly	Pure culture	30 CFU/reaction	1 CFU/reaction	Rodríguez-Lázaro et al., 2004b
	hly	Pure culture	30 CFU/reaction	1 CFU/reaction	Rodríguez-Lázaro et al., 2004a
	hly	Cabbage	140 CFU/25g	9 CFU/reaction	Hough et al., 2002
	iap	Milk	60 CFU/reaction	6 CFU/reaction	Hein et al., 2001
	hly	Pure culture	60 CFU/reaction	6 CFU/reaction	Nogva et al., 2000b
Salmonella	invA	Culture medium, clinical sample	—	—	Fujikawa et al., 2006
	invA	Chicken	7.5×10^2 CFU/100 ml	2.2 CFU/100 ml	Wolffs et al., 2006
	flaA	Blood samples	—	—	Massi et al., 2005
	fimC	Culture medium	1×10^3/ml	1×10^3/ml	Piknova et al., 2005
	sefA	Ice cream	1×10^3/ml	1×10^3/ml	Seo et al., 2006
	TaqMan Salmonella PCR Amplification kit	Culture medium	25 CFU/reaction	25 CFU/reaction	Nogva and Lillehaug, 1999
Yersinia enterocolitica	16S rRNA	Pork meat	4.2×10^3 CFU/ml	4.2×10^3 CFU/ml	Wolffs et al., 2004

(Rueckert *et al.*, 2005; Nocker and Camper, 2006). This drawback can be overcome using a similar staining strategy with a more selective molecule such as propidium monoazide (PMA). PMA is a modification of propidium iodide that does not penetrate the membrane of viable cells, but is efficiently taken up by permeabilized cells (Nocker *et al.*, 2006).

II.D. Application of real-time PCR in food samples for quantitative purposes

PCR is a simple, versatile, sensitive, specific, and reproducible assay (Scheu *et al.*, 1998; Hoorfar and Cook, 2003; Malorny *et al.*, 2003). However, in its conventional format this technique does not allow the quantification of the initial sample. Enumeration of foodborne pathogens is a main aspect of molecular microbiological diagnostics, especially if it wants to be used for quantitative risk assessment. Although several methodological approaches have been described based on conventional PCR (Scheu *et al.*, 1998; Rijpens and Hermann, 2002), the most promising alternative is the application of adequate real-time (Q-)PCR assays (Norton, 2002). Q-PCR represents a significant advance as it allows monitoring of the synthesis of new amplicon molecules in real time during the PCR process by using fluorescence (Heid *et al.*, 1996), and not only at the end of the reaction, as occurs in conventional PCR. Other major advantages of RTi-PCR for its application in diagnostic food laboratories include rapidity and simplicity to perform analysis, the closed-tube format that avoids risks of carryover contamination, the extremely wide dynamic range of quantification (more than eight orders of magnitude) (Heid *et al.*, 1996), and the significantly higher reliability of the results compared to conventional PCR. Currently, different scientific contributions have been done in which novel Q-PCR-based methods have been developed for accurate quantification of food microbial contaminants (Table 14.5).

III. CONCLUDING REMARKS

The guarantee of microbiological safety and quality of foods, and the possession of means to meet the challenges posed by potential emerging microbial risks, requires the development of novel methods, and refinement of existing analytical methodology. Molecular methods, as reviewed above, are a promising alternative that could complement or substitute the current reference methods in food microbiology. More suitable and reliable results can be achieved in terms of speed and precision, and have the potential to detect and enumerate specifically viable microorganisms. There is, however, much work to be done before these methods can be implemented into the fabric of existing standard methods, the most important being full-scale multicenter validation. This is discussed in some detail in Chapter 17. In short, the current standard methods are still viewed as the benchmark by which all other methods are measured, and so any new method should be shown to be at least as effective as the standard method already in use within a laboratory. This is not a minor undertaking, but it is essential if the new method is to be trusted as a truly robust alternative to the standard. In addition, new methods and techniques are regularly contributed to the scientific community in the area of molecular diagnostics and many claims are made as to how effective they are, but very often the performance characteristics and parameters used to develop the method are undisclosed, preventing them from being optimized and tailored to suit a diverse range of organisms and which are compatible with current ISO standards. Ultimately, it is unlikely that molecular methods will completely replace traditional methods anytime soon. Rather they can provide an efficient means to screen for target microorganisms and complement the existing standard, by providing more detailed and relevant information more rapidly than can be obtained by the standard methods currently available. This should encourage their adoption in food microbiology laboratories.

ACKNOWLEDGMENTS

David Rodríguez-Lázaro was a research fellow of the European Union's Marie Curie program (Contract MEIF-CT-2005-0011564) and Marta Hernández holds a contract from the Spanish Instituto Nacional de Investigación y Tecnología Agraria y Alimentaria (INIA).

REFERENCES

Abdulmawjood, A., Roth, S., & Bülte, M. (2002). Two methods for construction of internal amplification controls for the detection of *Escherichia coli* O157 by polymerase chain reaction. *Mol. Cell. Probes, 16*, 335–339.

Adak, G. K., Long, S. M., & O'Brien, S. J. (2002). Trends in indigenous foodborne disease and deaths, England and Wales: 1992–2000. *Gut, 51*, 832–841.

Bean, N. H., Griphin, P. M., Goulding, J. S., & Ivey, C. B. (1990). Foodborne disease outbreaks, 5 year summary, 1983–1987. *J. Food Prot, 53*, 711–728.

Berrada, H., Soriano, J. M., Pico, Y., & Manes, J. (2006). Quantification of *Listeria monocytogenes* in salads by real-time quantitative PCR. *Int. J. Food Microbiol., 107*, 202–206.

Blais, B., Turner, G., Sooknanan, R., & Malek, L. (1997). A nucleic acid sequence-based amplification system for detection of *Listeria monocytogenes* hlyA sequences. *Appl. Environ. Microbiol., 63*, 310–313.

CAST (1994). CAST Report: Foodborne Pathogens: Risks and Consequences. Task Force Report No. 122, Washington, DC: Council for Agricultural Science and Technology.

Coffman, G. L., Gaubatz, J. W., Yielding, K. L., & Yielding, L. W. (1982). Demonstration of specific high affinity binding sites in plasmid DNA by photoaffinity labeling with ethidium analog. *J. Biol. Chem., 257*, 13205–13297.

Compton, J. (1991). Nucleic acid sequence-based amplification. *Nature, 350*, 91–92.

Cone, R. W., Hobson, A. C., & Huang, M. L. (1992). Coamplified positive control detects inhibition of polymerase chain reactions. *J. Clin. Microbiol., 30*, 3185–3189.

Cook, N. (2003). The use of NASBA for the detection of microbial pathogens in food and environmental samples. *J. Microbiol. Methods, 53*, 165–174.

Cook, N., Ellison, J., Kurdziel, A. S., Simpkins, S., & Hays, J. P. (2002). A NASBA-based method to detect *Salmonella enterica* serotype Enteritidis strain PT4 in liquid whole egg. *J. Food Prot., 65*, 1177–1178.

Crutchfield, S., & Roberts, T. (2000). Food safety efforts accelerate in 1990s. *USDA, Econ. Res. Service Food Rev., 23*, 44–49.

D'souza, D. H., & Jaykus, L. A. (2003). Nucleic acid sequence based amplification for the rapid and sensitive detection of *Salmonella enterica* from foods. *J. Appl. Microbiol., 95*, 1343–1350.

Deiman, B., van Aarle, P., & Sillekens, P. (2002). Characteristics and applications of nucleic acid sequence-based amplification (NASBA). *Mol. Biotechnol, 20*, 163–179.

DeTraglia, M. C., Brand, J. S., & Tometski, A. M. (1978). Characterization of azido-benzamidines as photoaffinity labeling for trypsin. *J. Biol. Chem., 253*, 1846.

Dieffenbach, C.W., Lowe, T.M.J., & Dveksler, G.S. (1995). General concepts for PCR primers design. In: C. W. Dieffenbach, & G. S. Dveksler (eds) PCR primers: a laboratory manual, (pp. 133–142. Cold Spring Harbor, NY, USA: Cold Spring Harbor Laboratory Press.

Foulds, I. V., Granacki, A., Xiao, C., Krull, U. J., Castle, A., & Horgen, P. A. (2002). Quantification of microcystin-producing cyanobacteria and *E. coli* in water by 5'-nuclease PCR. *J. Applied Microbiol., 93*, 825–834.

Fujikawa, H., Shimojima, Y., & Yano, K. (2006). Novel method for estimating viable *Salmonella* cell counts using real-time PCR. *Shokuhin Eiseigaku Zasshi, 47*, 151–156.

Fung, D. Y. C. (2002). Predictions for rapid methods and automation in food microbiology. *J. AOAC Int., 85*, 1000–1002.

Gore, H. M., Wakeman, C. A., Hull, R. M., & McKillip, J. L. (2003). Real-time molecular beacon NASBA reveals hblC expression from *Bacillus* spp. in milk. *Biochem. Biophys. Res. Commun, 311*, 386–390.

Guilbaud, M., de Coppet, P., Bourion, F., Rachman, C., Prevost, H., & Dousset, X. (2005). Quantitative detection of *Listeria monocytogenes* in biofilms by real-time PCR. *Appl. Environ. Microbiol., 71*, 2190–2194.

Guy, R. A., Kapoor, A., Holicka, J., Shepherd, D., & Horgen, P. A. (2006). A rapid molecular-based assay for direct quantification of viable bacteria in slaughterhouses. *J. Food Prot., 69*, 1265–1272.

Heid, C. A., Stevens, J., Livak, K. J., & Williams, P. M. (1996). Real-time quantitative PCR. *Genome Res., 6*, 986–994.

Hein, I., Klein, D., Lehner, A., Bubert, A., Brandl, E., & Wagner, M. (2001). Detection and quantification of the iap gene of *Listeria monocytogenes* and *Listeria innocua* by a new real-time quantitative PCR assay. *Res. Microbiol., 152*, 37–46.

Hinata, N., Shirakawa, T., Okada, H., Shigemura, K., Kamidono, S., & Gotoh, A. (2004). Quantitative detection of *Escherichia coli* from urine of patients with bacteriuria by real-time PCR. *Mol. Diagn, 8*, 179–184.

Hixon, S. C., White, W. E., & Yielding, K. L. (1975). Selective covalent binding of an ethidiumanalog tomitochondrial DNA with production of petite mutants in yeast by photoaffinity labeling. *J. Mol. Biol., 92*, 319–329.

Hoorfar, J., & Cook, N. (2003). Critical aspects in standardization of PCR. In K. Sachse, & J. Frey (Eds), *Methods in Molecular Biology PCR detection of microbial pathogens* (pp. 51–64). Totowa, USA Humana Press.

Hoorfar, J., Malorny, B., Abdulmawjood, A., Cook, N., Wagner, M., & Fach, P. (2004). Practical considerations in design of internal amplification control for diagnostic PCR assays. *J. Clin. Microbiol., 42*, 1863–1868.

Hough, A. J., Harbison, S. A., Savill, M. G., Melton, L. D., & Fletcher, G. (2002). Rapid enumeration of *Listeria monocytogenes* in artificially contaminated cabbage using real-time polymerase chain reaction. *J. Food Prot, 65*, 1329–1332.

Ibekwe, A. M., & Grieve, C. M. (2003). Detection and quantification of *Escherichia coli* O157: H7 in environmental samples by real-time PCR. *J. Applied Microbiol., 94*, 421–431.

Ibekwe, A. M., Watt, P. M., Grieve, C. M., Sharma, V. K., & Lyons, S. R. (2002). Multiplex fluorogenic real-time PCR for detection and quantification of *Escherichia coli* O157:H7 in dairy wastewater wetlands. *Appl. Environ. Microbiol., 68*, 4853–4862.

Inglis, G. D., & Kalischuk, L. D. (2004). Direct quantification of *Campylobacter jejuni* and *Campylobacter lanienae* in feces of cattle by real-time quantitative PCR. *Appl. Environ. Microbiol., 70*, 2296–2306.

Jaffe, R. I., Lane, J. D., & Bates, C. W. (2001). Real-time identification of *Pseudomonas aeruginosa* direct from clinical samples using a rapid extraction method and polymerase chain reaction (PCR). *J. Clin. Lab. Anal., 15*, 131–137.

Jean, J., Blais, B., Darveau, A., & Fliss, I. (2001). Detection of hepatitis A virus by the nucleic acid sequence-based amplification technique and comparison with reverse transcription-PCR. *Appl. Environ. Microbiol., 67*, 5593–5600.

Jean, J., D'Souza, D. H., & Jaykus, L. A. (2004). Multiplex nucleic acid sequence-based amplification for simultaneous detection of several enteric viruses in model ready-to-eat foods. *Appl. Environ. Microbiol., 70*, 6603–6610.

Kaclikova, E., Pangallo, D., Oravcova, K., Drahovska, H., & Kuchta, T. (2005). Quantification of *Escherichia coli* by kinetic 5'-nuclease polymerase chain reaction (real-time PCR) oriented to sfmD gene. *Lett. Applied Microbiol., 41*, 132–135.

Käferstein, F. K., Motarjemi, Y., & Bettcher, D. W. (1997). Foodborne disease control: a transnational challenge. Emerging Infect. *Dis, 3*, 503–510.

Klein, P. G., & Kuneja, V. J. (1997). Sensitive detection of viable *Listeria monocytogenes* by reverse transcription-PCR. *Appl. Environ. Microbiol., 63*, 4441–4448.

Knowk, S., & Higuchi, R. (1989). Avoiding false positives with PCR. *Nature, 339*, 237–238.

Malorny, B., Tassios, P. T., Rådström, P., Cook, N., Wagner, M., & Hoorfar, J. (2003). Standardization of diagnostic PCR for the detection of foodborne pathogens. *Int. J. Food Microbiol., 83*, 39–48.

Massi, M. N., Shirakawa, T., Gotoh, A., Bishnu, A., Hatta, M., & Kawabata, M. (2005). Quantitative detection of *Salmonella enterica* serovar Typhi from blood of suspected typhoid fever patients by real-time PCR. *J. Med. Microbiol., 295*, 117–120.

Mead, P. S., Slutsker, L., Griffin, P. M., & Tauxe, R. V. (1999). Food-related illness and death in the United States. *Emerging Infectious Diseases, 5*, 607–625.

Nocker, A., & Camper, A. K. (2006). Selective removal of DNA from dead cells of mixed bacterial communities by use of ethidium monoazide. *Appl. Environ. Microbiol., 72*, 1997–2004.

Nocker, A., Cheung, C. Y., & Camper, A. K. (2006). Comparison of propidium monoazide with ethidium monoazide for differentiation of live vs. dead bacteria by selective removal of DNA from dead cells. *J. Microbiol. Methods, 67*, 310–320.

Nogva, H. K., & Lillehaug, D. (1999). Detection and quantification of *Salmonella* in pure cultures using 5'-nuclease polymerase chain reaction. *Int. J. Food Microbiol., 51*, 191–196.

Nogva, H. K., Dromtorp, S. M., Nissen, H., & Rudi, K. (2003). Ethidium monoazide for DNA-based differentiation of viable and dead bacteria by 5'-nuclease PCR. *Biotechniques, 34*, 804–813.

Nogva, H.K., Bergh, A. Holck, A., & Rudi, K. (2000a). Application of the 5'-nuclease PCR assay in the evaluation and development of methods for quantitative detection of *Campylobacter jejuni*. *Appl. Environ. Microbiol., 66*, 4029-4036.

Nogva, H.K., Rudi, K., Naterstad, K., Holck, A., & Lillehaug, D. (2000b). Application of 5'-nuclease PCR for quantitative detection of *Listeria monocytogenes* in pure cultures, water, skim milk, and unpasteurized whole milk. *Appl. Environ. Microbiol.*, 66: 4266–4271.

Norton, D. M. (2002). Polymerase chain reaction-based methods for detection of *Listeria monocytogenes*: toward real-time screening for food and environmental samples. *J. AOAC Int, 85*, 505–515.

Oravcova, K., Kaclikova, E., Krascsenicsova, K., Pangallo, D., Brezna, B., Siekel, P., & Kuchta, T. (2006). Detection and quantification of *Listeria monocytogenes* by 5'-nuclease polymerase chain reaction targeting the actA gene. *Lett. Appl. Microbiol., 42*, 15–18.

Penders, J., Vink, C., Driessen, C., London, N., Thijs, C., & Stobberingh, E. E. (2005). Quantification of *Bifidobacterium* spp., *Escherichia coli* and *Clostridium difficile* in faecal samples of breast-fed and formula-fed infants by real-time PCR. *FEMS Microbiol. Lett., 243*, 141–147.

Piknova, L., Kaclikova, E., Pangallo, D., Polek, B., & Kuchta, T. (2005). Quantification of *Salmonella* by 5'-nuclease real-time polymerase chain reaction targeted to fimC gene. *Curr. Microbiol., 50*, 38–42.

Rådström, P., Knutsson, R., Wolffs, P., Dahlenborg, M., & Löfström, C. (2003). Pre-PCR processing of sampling. In K. Sachse, & and J. Frey (Eds), *Methods in Molecular Biology PCR detection of microbial pathogens* (pp. 31–50). Totowa, USA: Humana Press.

Rijpens, N. P., & Herman, L. M. (2002). Molecular methods for identification and detection of bacterial food pathogens. *J. AOAC Int., 85*, 984–995.

Rodríguez-Lázaro, D., Hernandez, M., Scortti, M., Esteve, T., Vazquez-Boland, J.A., & Pla, M. (2004a). Quantitative detection of *Listeria monocytogenes* and *Listeria innocua* by real-time PCR: assessment of hly, iap, and lin02483 targets and AmpliFluor technology. *Appl. Environ. Microbiol., 70*, 1366–1377.

Rodríguez-Lázaro, D., Hernandez, M., & Pla, M. (2004b). Simultaneous quantitative detection of *Listeria* spp. and *Listeria monocytogenes* using a duplex real-time PCR-based assay. *FEMS Microbiol. Lett., 233*, 257–267.

Rodríguez-Lázaro, D., Jofre, A., Aymerich, T., Hugas, M., & Pla, M. (2004c). Rapid quantitative detection of *Listeria monocytogenes* in meat products by real-time PCR. *Appl. Environ. Microbiol., 70*, 6299–6301.

Rodríguez-Lázaro, D., D'Agostino, M., Pla, M., & Cook, N. (2004d). A construction strategy for an internal amplification control (IAC) for real-time NASBA-based diagnostic assays. *J. Clin. Microbiol., 42*, 5832–5836.

Rodríguez-Lázaro, D., Lloyd, J., Herrewegh, A., Ikonomopoulos, J., D'Agostino, M., Pla, M., & Cook, N. (2004e). A molecular beacon-based real-time NASBA assay for detection of *Mycobacterium avium* subsp. paratuberculosis in water and milk. *FEMS Microbiol. Lett., 237*, 119–126.

Rodríguez-Lázaro, D., Jofre, A., Aymerich, T., Garriga, M., & Pla, M. (2005a). Rapid quantitative detection of *Listeria monocytogenes* in salmon products: evaluation of pre-real-time PCR strategies. *J. Food Prot., 68*, 1467–1471.

Rodríguez-Lázaro, D., Pla, M., Scortti, M., Monzó, H.J., & Vazquez-Boland, J.A. (2005b). A novel real-time PCR for *Listeria monocytogenes* that monitors analytical performance via an internal amplification control. *Appl. Environ. Microbiol., 71*, 9008–9012.

Rodríguez-Lázaro, D., Hernández, M., D'Agostino, M., & Cook, N. (2006). Application of nucleic acid sequence based amplification (NASBA) for the detection of viable foodborne pathogens: progress and challenges. *J. Rapid Methods Automatisation Microbiol., 14*, 218–236.

Rollins, D. M., & Colwell, R. R. (1986). Viable but non-culturable stage of *Campylobacter jejuni* and its role in survival in the natural aquatic environment. *Appl. Environ. Microbiol., 52*, 531–538.

Rosenstraus, M., Wang, Z., Chang, S. Y., DeBonville, D., & Spadoro, J. P. (1998). An internal control for routine diagnostic PCR: design, properties, and effect on clinical performance. *J. Clin. Microbiol., 36*, 191–197.

Rossen, L., Nøskov, P., Holmstrøm, K., & Rasmussen, O. F. (1992). Inhibition of PCR by components of food samples, microbial diagnostic assays and DNA extraction solution. *Int. J. Food Microbiol., 17,* 37–45.

Rudi, K., Naterstad, K., Dromtorp, S.M., & Holo, H. (2005a). Detection of viable and dead *Listeria monocytogenes* on gouda-like cheeses by real-time PCR. *Lett. Appl. Microbiol., 40,* 301–306.

Rudi, K., Moen, B., Dromtorp, S. M., & Holck, A. L. (2005). Use of ethidium monoazide and PCR in combination for quantification of viable and dead cells in complex samples. *Appl. Environ. Microbiol., 71,* 1018–1024.

Rudi, K., Hoidal, H. K., Katla, T., Johansen, B. K., Nordal, J., & Jakobsen, K. S. (2004). Direct real-time PCR quantification of *Campylobacter jejuni* in chicken fecal and cecal samples by integrated cell concentration and DNA purification. *Appl. Environ. Microbiol., 70,* 790–797.

Rueckert, A., Ronimus, R. S., & Morgan, H. W. (2005). Rapid differentiation and enumeration of the total, viable vegetative cell and spore content of thermophilic bacilli in milk powders with reference to *Anoxybacillus flavithermus. J. Appl. Microbiol., 99,* 1246–1255.

Scheu, P. M., Berghof, K., & Stahl, U. (1998). Detection of pathogenic and spoilage microorganisms in food with the polymerase chain reaction. *Food Microbiol., 15,* 13–31.

Sekse, C., Solberg, A., Petersen, A., Rudi, K., & Wasteson, Y. (2005). Detection and quantification of Shiga toxin-encoding genes in sheep faeces by real-time PCR. *Mol. Cell. Probes, 19,* 363–370.

Seo, K. H., Valentin-Bon, I. E., & Brackett, R. E. (2006). Detection and enumeration of *Salmonella enteritidis* in homemade ice cream associated with an outbreak: comparison of conventional and real-time PCR methods. *J. Food Prot, 69,* 639–643.

Sharma, V. K. (2002). Detection and quantitation of enterohemorrhagic *Escherichia coli* O157, O111, and O26 in beef and bovine feces by real-time polymerase chain reaction. *J. Food Prot, 65,* 1371–1380.

Stewart, G. S. (1997). Challenging food microbiology from a molecular perspective. *Microbiol., 143,* 2099–2108.

Stirling, D. (2003). Quality control in PCR. *Methods Mol. Biol., 226,* 21–24.

Tholozan, J. L., Cappelier, J. M., Tissier, J. P., Delattre, G., & Federighi, M. (1999). Physiological characterization of viable-but-nonculturable *Campylobacter jejuni* cells. *Appl. Environ. Microbiol., 65,* 1110–1116.

Uyttendaele, M., Schukkink, R., Van Gemen, B., & Debevere, J. (1994). Identification of *Campylobacter jejuni, Campylobacter coli* and *Campylobacter lari* by the nucleic acid amplification system NASBA. *J. Appl. Bacteriol, 77,* 694–701.

Uyttendaele, M., Schukkink, R., Van Gemen, B., & Debevere, J. (1995a). Detection of *Campylobacter jejuni* added to foods by using a combined selective enrichment and nucleic acid sequence-based amplification (NASBA). *Appl. Environ. Microbiol., 61,* 1341-1437.

Uyttendaele, M., Schukkink, R., Van Gemen, B., & Debevere, J. (1995b). Comparison of a nucleic acid amplification system NASBA and agar isolation for detection of pathogenic campylobacters in poultry. *Med. Fac. Landbouww. Univ. Gent., 60,* 1863-1866.

Uyttendaele, M., Schukkink, R., Van Gemen, B., & Debevere, J. (1995c). Development of NASBA, a nucleic acid amplification system, for identification of *Listeria monocytogenes* and comparison to ELISA and a modified FDA method. *Int. J. Food Microbiol., 27,* 77-89.

Uyttendaele, M., Bastiaansen, A., & Debevere, J. (1997). Evaluation of the NASBA nucleic acid amplification system for assessment of the viability of *Campylobacter jejuni. Int. J. Food Microbiol., 37,* 13–20.

Uyttendaele, M., Debevere, J., & Lindqvist, R. (1999). Evaluation of buoyant density centrifugation as a sample preparation method for NASBA-ELGA detection of *Campylobacter jejuni* in foods. *Food Microbiol., 16,* 575–582.

Wallace, D. J., Van Gilder, T., Shallow, S., et al., FoodNet Working Group, . (2000). Incidence of Foodborne Illnesses Reported by the Foodborne Diseases Active Surveillance Network (FoodNet)-1997. *J. Food Prot, 63*, 807–809.

Wang, S., & Levin, R. E. (2006). Discrimination of viable *Vibrio vulnificus* cells from dead cells in real-time PCR. *J. Microbiol. Methods, 64*, 1–8.

Waring, M. J. (1965). Complex formation between ethidium bromide and nucleic acids. *J. Mol. Biol., 13*, 269–282.

Wilson, I. G. (1997). Inhibition and facilitation of nucleic acid amplification. *Appl. Environ. Microbiol., 63*, 3741–3751.

Wolffs, P., Knutsson, R., Norling, B., & Rådström, P. (2004). Rapid quantification of *Yersinia enterocolitica* in pork samples by a novel sample preparation method, flotation, prior to real-time PCR. *J. Clin. Microbiol., 42*, 1042–1047.

Wolffs, P. F., Glencross, K., Thibaudeau, R., & Griffiths, M. W. (2006). Direct quantitation and detection of salmonellae in biological samples without enrichment, using two-step filtration and real-time PCR. *Appl. Environ. Microbiol., 72*, 3896–3900.

Yang, C., Jiang, Y., Huang, K., Zhu, C., & Yin, Y. (2003). Application of real-time PCR for quantitative detection of *Campylobacter jejuni* in poultry, milk and environmental water. FEMS Immunol. *Med. Microbiol., 38*. 265-171.

Yang, C., Jiang, Y., Huang, K., Zhu, C., Yin, Y., Gong, J. H., & Yu, H. (2004). A real-time PCR assay for the detection and quantitation of Campylobacter jejuni using SYBR Green I and the LightCycler. Yale *J. Biol. Med., 77*, 125–132.

CHAPTER 15

Review of Currently Applied Methodologies used for Detection and Typing of Foodborne Viruses

Artur Rzeżutka *and* **Nigel Cook**

Contents

Abstract

In most foodborne outbreaks viruses are only indirectly implicated as a source of infection, due to difficulties in their detection. Therefore, the application of adequate detection methods allows us to recognize an accurate number of viral gastroenteritis cases in humans, as a result of consumption of contaminated food. Finding a disease-producing viral agent that contaminates food items is not an easy task because the amount of viruses present in a food can often be very low, and the analytical procedures are complicated. Presently, available virological methods (e.g. cell culture infectivity assay, electron microscope) cannot be fully used for the analysis of food samples, because they are laborious, time-consuming and have a low sensitivity. Methods used for the detection of foodborne viruses consist of two main stages: extraction and concentration of viruses from food samples and detection of viruses using nucleic acid amplification techniques. These methods are the most

Global Issues in Food Science and Technology
© 2009 Elsevier Inc.

ISBN 9780123741240
All rights reserved.

promising for detection of foodborne viruses, as they are fast, sensitive, and can allow typing of these viruses. Despite the usefulness of such methods, there are differences in the sensitivity, specificity for different food matrices, and detection limits of individual assays. This chapter deals with existing methodologies used for the detection of noroviruses, hepatitis A virus, and rotaviruses in foodstuffs, especially fresh produce, and discusses the current state of development and the requirements for their implementation in routine use. Finally, some significant needs in method development are highlighted, and recommendations to fill these needs are made.

I. INTRODUCTION

The last decade has seen an increasing amount of gastroenteritis in humans caused by consumption of food contaminated with viruses. The cases of foodborne viral infections in humans are not only associated with developing countries with poor hygienic standards, but they are also common in highly developed countries. Foodstuffs of animal origins such as shellfish, delicatessen meat, and fresh produce (vegetables and fruits) have been identified as vehicles of transmission for enteric viral disease (Pebody *et al.*, 1998; Hutin *et al.*, 1999; Lees, 2000; Schwab *et al.*, 2000; Dentinger *et al.*, 2001; Nygard *et al.*, 2001; Le Guyader *et al.*, 2004). Food is clearly an important exposure route, but it should not be considered in isolation since other direct environmental exposures via air, soil, and water can be also important. On the other hand contamination of the environment with enteric viruses can result in further contamination of food.

In many cases of gastroenteritis, food is shown to be the sole source of viruses on the basis of epidemiological investigation only, where it is not exactly confirmed as being a prime source of the disease. Therefore, without application of analytical procedures the exact number of viral gastroenteritis cases in humans as the result of consumption of contaminated food is not fully known. Another reason for low recognition of viral foodborne diseases is the inability to test the implicated food. By the time an outbreak is recognized, the food has either been consumed or discarded (Nainan *et al.*, 2006). Finally, the lack of methods that can be routinely applied for virus detection is the main cause of unsuccessful identification of the source of a foodborne disease.

This chapter deals with existing methodologies used for the detection of norovirus (NoV), hepatitis A (HAV), and rotavirus (RV) in foodstuffs, especially fresh produce. It further discusses the requirements for implementation of these methods in routine use.

II. VIRUSES TRANSMITTED BY FOOD

Several types of enteric viruses, including noroviruses, rotaviruses, astroviruses, hepatitis A and E virus can be transmitted by food causing gastroenteritis or other illnesses in humans. Among these, NoV, HAV, and RV are currently the most significant viral foodborne agents known worldwide (Klein, 2004; Atreya, 2004). NoV are considered to be a common cause of non-bacterial gastroenteritis infections due to consumption of contaminated foods. Data generated by a European surveillance network showed that more than 85% of reported cases of gastroenteritis in humans are caused by NoV (Lopman et al., 2003). Besides NoV, HAV is also considered to be a major cause of foodborne gastroenteritis. According to the World Health Organization, the annual incidence of HAV in Europe is approximately 278,000 cases (WHO, 2000). RV are often isolated from acute gastroenteritis in infants and young children. They contribute significantly to morbidity and mortality worldwide (Anderson and Weber, 2004). However, the number of all foodborne viral illness associated with RV infection is about 1% (Atreya, 2004).

III. MODES OF TRANSMISSION AND SOURCES OF INFECTION

For many years food has been considered as a source of infectious pathogens including viruses. Viruses can not grow in or on foods but they may be present as a result of fecal contamination. The main vehicles of virus transmission are foodstuffs that are eaten raw or only slightly cooked. These include seafood, especially shellfish, and fresh produce (Seymour and Appleton, 2001). Fresh produce such as soft fruits and vegetables can become contaminated on the farm during harvesting by infected persons; viruses are transferred to fresh produce from the hands or fingertips of harvest pickers contaminated with fecal material (Bidawid et al., 2000a). Predominantly, the viral contamination is associated with the surface of fruits, but some viruses could also be present inside as a result of damage to the fruit. Further contamination of food items can take place before consumption during preparation of meals or food handling by infected food handlers. At this time viruses can be easily transmitted to particular food items, making them a potential source of infection for consumers (Cook and Rzeżutka, 2006). Contamination can also occur through contact with inadequately treated sewage or contaminated water. Pollution of the water environment is the main source of contamination for shellfish, which being filter-feeding invertebrates, causes them to accumulate pathogenic microorganisms in their gut.

Animals too can be sources of virus infection. There is increasing evidence that strains of animal rotaviruses can infect humans (Cook *et al.*, 2004), and they have been isolated from several human cases of this disease (Das *et al.*, 1993; Nakagomi *et al.*, 1994; Holmes *et al.*, 1999). The excreta from infected cattle, pigs, and sheep contain large numbers of infectious rotaviruses, which are potential sources for contamination of the environment. Viruses in excreta deposited in fields could pass via run-off water into fresh waters such as rivers or lakes. In addition, excreta from cattle and pigs (including young animals, commonly infected with RV) can be stored and then spread onto land as fertilizer. Therefore it is highly likely that rotaviruses could, via manure or slurry, contaminate arable land, water sources, and possibly farm crops.

IV. GENERAL CHARACTERISTICS OF METHODS USED FOR DETECTION OF FOODBORNE VIRUSES

Routine assays for the detection of enteric viruses in clinical specimens can not be directly applied for food sample analysis. Therefore many laboratories have developed methodologies that allow them to detect viruses in contaminated food items. Current methodologies used for the detection of foodborne viruses consist of two main stages: (1) extraction and concentration of viruses from food samples and (2) detection using molecular biology techniques (Figure 15.1) (Cook and Myint, 1995). There are several basic methods employing different procedures that can be used for virus extraction and detection. Each one has advantages and drawbacks. Some methods are insensitive, while others do not have the specificity allowing identification of virus type or strain. This can produce variable efficiency and sensitivity of the method and can lead to false or negative results. Viruses are usually present in contaminated food in very low numbers, which is why virions must be extracted from the samples and concentrated before the detection system can be applied. A high efficiency of virus extraction can allow detection of viruses in food samples even when the detection method has low sensitivity. Direct virus identification in food items is difficult due to various chemical compositions of food matrices. This can result in the application of two or more methods, which, used together, can give a high detection limit.

IV.A. Extraction of viruses

Before the viruses can be detected, they have to be liberated and concentrated from the food samples. This can be achieved by using an extraction method. Nowadays, methods used for virus extraction and concentration consist of several steps, for example, the initial sample treatment of washing

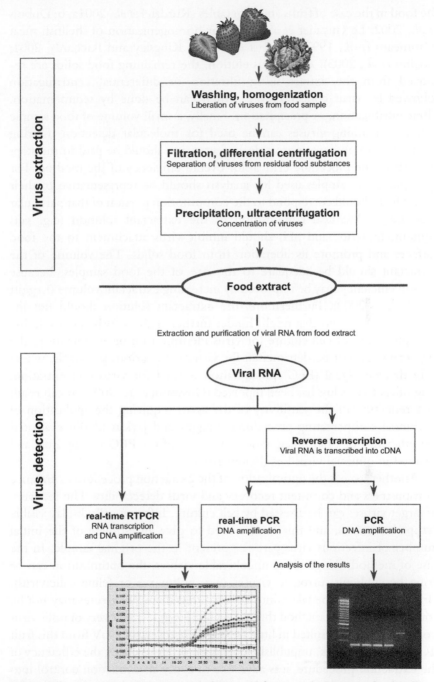

Figure 15.1   Flowchart showing procedure used for virus detection in foodstuffs.

the food in the case of fruits and vegetables (Kurdziel *et al.*, 2001a, b; Dubois *et al.*, 2002; Le Guyader *et al.*, 2004) or homogenization of shellfish meat (Cromeans *et al.*, 1997; Croci *et al.*, 1999; Kingsley and Richards, 2001; Coelho *et al.*, 2003). After virus elution, the remaining food solids are removed from the extract using filtration or differential centrifugation followed by virus concentration, which can be done by sedimentation, ultracentrifugation, or precipitation. Finally, a small volume of food sample extract containing viruses can be used for molecular detection. During methods development, a great deal of attention should be paid to the steps or parts of the procedure critical for overall efficiency of the method. For example, food samples used for analysis should be representative of their size, which should correspond to the consumption portion of that particular food item. Chemical composition of the extractant solution (e.g. salts content, molarity, and pH) should inhibit virus attachment to the food surfaces and promote its liberation from food solids. The volume of the extractant should be adequate to the size of the food samples, because elution efficiency may be improved by increasing extractant volume (Leggitt and Jaykus, 2000). Furthermore, the extractant solution should not introduce components that might hinder further analysis. When considering the application of cell culture for virus identification or enumeration, the extractant can not be detrimental for viruses. In various protocols, where polyethylene glycol (PEG) precipitation is used for virus concentration, a significant virus loss has been reported (Goswami *et al.*, 2002). It can result in a reduction of the sensitivity of the assay, requiring the application of additional compensating procedures. Leggitt and Jaykus (2000) evaluated the efficiency of PEG precipitation. The use of 6% PEG was optimal and resulted in minimal virus loss (below 10%).

Another issue is the optimization of the extraction procedure to enhance its robustness and consistent recovery and virus detectability. The recovery of target viruses can be assessed by cell culture. Infected cells display visible cytopathic effects, and this can be used to give an estimate of the initial number of infectious virus particles present in the original sample. In the case of methods designed for uncultivable viruses, the optimization can be done by applying surrogate viruses (e.g. poliovirus or feline calicivirus). However, it has to be taken into account that surrogate viruses may not be completely useful; a method that gave a high ($\sim$40 %) recovery of poliovirus from strawberries resulted in little (< 0.1 %) recovery of NoV from this fruit (Rzeżutka and Cook, unpublished data). In order to check the efficiency of the extraction procedure, it is advisable to apply an extraction control into each sample tested. Contemplating which virus could be a candidate for such a control, we have to remember that it should not naturally contaminate food samples having similar features to the target virus. A good candidate for controlling the whole process of sample analysis (extraction and detection) seems to be mengovirus (Costafreda *et al.*, 2006). This virus belongs to the *Picornaviridae* family, is not pathogenic for humans, does not occur in food, and has similar features to other picornaviruses. The necessity of extraction controls during food sample analysis was also discussed by Le Guyader *et al.* (2004), who emphasized its significance when dealing with virus detection in food.

IV.B. Detection of viruses

Previously available protocols for detection of viruses are based mainly on nucleic acid amplification by the polymerase chain reaction (PCR), which is the most popular detection technique in microbiology laboratories worldwide. Most PCR-based assays target individual viruses. They are fast and sensitive, but they do have some limitations. The main limitation is too large a volume of food sample extract that as a whole, cannot be taken for nucleic acid extraction, and a small portion of nucleic acid extract used in the RTPCR, since only extraction methods concentrating viruses in a small volume of extractant are efficient. The quality of purified RNA is also an important factor in determining the sensitivity of the assay. Nowadays available extraction and concentration procedures can not sufficiently eliminate the residual of food-related inhibitors; they themselves can use substances that may be considered as inhibitory (Bidawid *et al.*, 2000b). They can be controlled by incorporating into the same reaction tube an internal amplification control (IAC) that could serve as an internal standard (Hoorfar *et al.*, 2003). An IAC is a RNA molecule that is simultaneously amplified with the target virus sequence. Its presence provides general information on the influence of the chemical ingredients and cycling profiles of the reactions. Compared to the external standards, this control is more reliable, because it can detect inhibition in each reaction tube (Ballagi-Pordány and Belák, 1996). However, inappropriate IAC usage can result in a reduction in the efficiency of amplification for the virus genome due to template competition (Rzeżutka, unpublished data).

Another problem is the standardization of novel molecular diagnostic methods, in terms of their specificity and sensitivity. This should include efforts to refine RNA extraction methods to decrease the level of RTPCR inhibitors co-extracted from food matrices (Rossen *et al.*, 1992; Shieh *et al.*, 1999) and identification of broadly reactive primer sets (Sair *et al.*, 2002). It seems to be a very important issue, especially for NoV detection, where virus sequence diversity requires the use of generic primers or the application of several detection methods to avoid false-negative results (Le Guyader *et al.*, 2004). Multiplex RTPCR offers some of these advantages, allowing the simultaneous detection of two or more viral nucleic acids in a single test (Rosenfield and Jaykus, 1999).

Recently significant improvements in novel detection methods have been achieved. These have resulted in the appearance of a new PCR system, called real-time PCR, where PCR amplification and probe detection of target nucleic acid takes place in the same reaction vessel. One of the biggest advantages of the real-time PCR over the traditional PCR systems is the possibility of the evaluation of target amplification after each PCR cycle. Another advantage is the elimination of amplicon 'carryover' contamination. The combination of the automated nucleic acid extraction system with the real-time PCR allows performing complete analysis of food samples within a few hours (Cockerill and Uhl, 2002).

As an alternative to RTPCR, a nucleic acid sequence-based amplification (NASBA) technique has been developed. It allows for the direct and efficient amplification of viral RNA in a single step and at one temperature, compared with two steps for RTPCR and broad temperature profile (Jean *et al.*, 2002). It involves the action of three enzymes, reverse transcriptase, T7 RNA polymerase, and RNase-H, allowing the rapid amplification of target sequences at more than 10^8 fold (Jean *et al.*, 2002; Abd el-Galil *et al.*, 2005).

V. TRADITIONAL METHODS

Traditional methods such as cell culture or using an electron microscope are still commonly applied in virology studies. Electron microscopy (EM) is a valuable technique used in virology, allowing direct virus identification in the investigated sample. However, to confirm its presence, the food sample has to contain thousands of virus particles, which makes this method less sensitive. Despite the main advantage of this method, such as direct pathogen recognition, it is hardly ever used for the detection of foodborne viruses (Seymour and Appleton, 2001). The main reason for its lack of use is the low sensitivity of EM, and also its availability, since not all laboratories dealing with food sample analysis possess such expensive equipment.

There is no single universal cell culture system that will detect all human enteric viruses (Rosenfield and Jaykus, 1999). However, the cell culture allows direct virus identification by producing characteristic damage to susceptible cells, described as cytopathic effect (CPE) formation. Another advantage of this method is its sensitivity and ability to identify infective virus particles. However, this method is infrequently used as an analytical tool for food sample analysis because some viruses (e.g. noroviruses) are uncultivable or require several blind passages, making the whole analysis too long to be used in routine diagnosis (Hollinger and Emerson, 2001). Moreover, working with cell cultures requires that the laboratory maintains several mammalian or human cell culture systems suitable for propagation of different virus types. It has been shown that HAV strains were successfully adapted to the African green monkey kidney cells (Vero) and a fetal rhesus monkey kidney cell line (FRhK-4) or FRP-3. In the case of rotaviruses, they usually are grown on MA-104 and Vero cells, requiring supplementation of the medium components with proteases to enhance virus growth. Before inoculation of a suitable cell culture, viruses have to be extracted and concentrated from food samples. Despite various methods used for virus concentration, removal of food-related chemical substances from the investigated sample might not be sufficient. These substances quite often are toxic for cells, resulting in their damage or early atrophy before visible CPE is produced. So far, not very many methods based on a cell culture have been used for detection of foodborne viruses. To improve the sensitivity of the assay, especially for non-cytopathogenic virus types, some authors have combined the cell culture with RTPCR (cc-RTPCR) (De Medici et al., 2001). Dilution of the sample with cell culture media reduces the effect of RTPCR-inhibitory substances.

VI. MOLECULAR DETECTION METHODS

VI.A. Detection of noroviruses in food

Methods used for detection of NoV and HAV in shellfish meat samples have been extensively revised by other authors (Lees, 2000). There are several methodologies that combine extraction and detection methods for noroviruses in fresh produce. In most of these methods, after initial sample treatment, viruses are concentrated, then viral RNA is extracted and purified (Leggitt and Jaykus, 2000; Sair et al., 2002; Rzeżutka et al., 2005; Guévremont et al., 2006). Other methods utilize organic solvents, which first facilitate concentration of virus particles, then allow viral RNA isolation instead of whole-virus extraction (Dubois et al., 2002; Le Guyader et al., 2004). All these procedures reduce the large volume of a food sample and allow the food extract or purified RNA solution to be used for further analysis. The original methodology for NoV detection in food samples (oysters, various salads, or fresh strawberries) without the virus concentration step was devised by Tian and Mandrell (2006). They used immuno-PCR, which does not rely only on detection of the viral genome like traditional molecular methods, but combines antibody capture of antigen and DNA amplification.

Virus identification can be done by detecting the viral nucleic acid using mainly RTPCR. However, the low quantity of virus in contaminated samples requires the use of a hybridization step as a confirmation method to enhance the sensitivity and specificity of the assay (Loisy *et al.*, 2005). The constant need for the development of more sensitive methods results in new protocols utilizing non-traditional PCR nucleic acid detection techniques. For example, the method developed by Schwab *et al.* (2001) for detection of NoV and HAV in shellfish join together RTPCR and ELISA, where RTPCR-generated amplicons were detected by virus-specific oligoprobes in an enzyme-linked immunosorbent assay. The most promising methods for virus detection are real-time PCR (Jothikumar *et al.*, 2005; Butot *et al.*, 2007) or NASBA (Jean *et al.*, 2004). These methods are fast, highly specific, and have the potential to detect single virus particles.

VI.A.1. Molecular typing of noroviruses

Most human caliciviruses belong to the genus *Norovirus*, which is further divided into genogroups and genotypes. All of the isolated NoV strains were classified to one from over 15 currently recognized genotypes. The identification of each virus genotype detected in clinical or food samples is difficult, because it requires the application of advanced and expensive methods. The present classification of noroviruses, as well as their division into particular genotypes, is based on molecular differences and homology across the capsid protein or RNA dependent RNA polymerase viral genes (Vennema *et al.*, 2002; Vinje *et al.*, 2004). These genes were shown to be the most conserved among all up-to-date characterized NoV genotypes (Duizer and Koopmans, 2006). However, not all currently available PCR-based assays for NoV detection allow typing of viruses. The variability of the NoV genome causes difficulties in the development of the single generic detection test, which is able to identify the NoV genogroup or strain (Koopmans and Duizer, 2004). Therefore some methods utilize additional procedures such as probe hybridization (Rosenfield and Jaykus, 1999; Schwab *et al.*, 2000; Schwab *et al.*, 2001; Sair *et al.*, 2002; Le Guyader *et al.*, 2004) or sequence analysis of the amplified genome fragment (Schwab *et al.*, 2000; Le Guyader *et al.*, 2004) to identify the NoV genotype. Another example is multiplex NASBA for NoV detection and genogroup identification in meat samples (Jean *et al.*, 2004).

Nowadays, there is a new perspective based on DNA microarray technology, which undoubtedly will become an alternative method to sequence analysis in the near future. Both methods (sequencing and microarray) consist of several steps, where RTPCR amplification of chosen gene fragments takes place, followed by sequencing or hybridization of the specific oligonucleotide probes to the immobilized DNA. Oligonucleotide microarrays are able to detect single-point mutations in the analyzed sequence, and are useful in epidemiological or phylogenetic studies (Kostrzynska and Bachand, 2006). A distinct advantage of microarray technology is a simultaneous detection and differentiation of individual virus isolates, among thousands of specific genome sequences present in the sample. In addition, during an outbreak investigation, this method is able to provide valuable information about sources and routes of virus transmission. Previously, the available microarray method for the detection and typing of noroviruses was only applied for clinical samples (Jaaskelainen and Maunula, 2006). Indeed, the results obtained were promising, but the direct application of this method as a powerful tool for food sample analysis requires additional studies.

VI.B. Detection of HAV in food

Methods used for HAV extraction from fresh produce and shellfish consist of several steps previously described for NoV (Section VI.A). There is sample washing or homogenization followed by virus concentration and molecular detection. Brief characteristics of current developed methods for extraction and detection of HAV in food with regard to efficiency of recovery and detection limits are presented in Table 15.1. An alternative approach to the whole-virus extraction procedure was described by Goswami et al. (2002). In their method, the virus precipitation step was replaced with phenol extraction. Isolation of total RNA followed by isolation of poly(A)-containing viral RNA enriches the nucleic acid concentrate and removes RTPCR inhibitory materials. New approaches to HAV concentration based on immunomagnetic separation (IMS) from clam samples (Suñén et al., 2004) as well as fresh produce and fruit (Shan et al., 2005) have been described. In these methods magnetic beads were covered by specific anti-HAV monoclonal antibodies, which were able to bind viruses from the food sample extract. Beads could be easily separated from residual food debris, leaving behind substances that can hinder virus detection. A benefit of the IMS method is efficient virus concentration from a high volume of food extract, inhibitor-free concentrate, and simplicity of application.

Similar to detection of NoV, detection methods for HAV are based on PCR or its modifications. There are RT-semi-nested PCR (Le Guyader et al., 1994), nested-RTPCR (Croci et al., 1999; De Medici et al., 2001; Rzeżutka et al., 2006), and immunocapture-RTPCR (Cromeans et al., 1997). To increase the sensitivity of RTPCR, method application of non-isotopic cDNA (Sincero et al., 2006) or oligonucleotide (Calder et al., 2003) might be introduced. A rapid, sensitive, and specific method for detection of infectious HAV in food and environmental samples was described by Jiang et al. (2004). This method is a combined cell culture and strand-specific RTPCR, where the detection of viral replicative intermediate RNA takes place. As an alternative to RTPCR, Jean et al. (2001) developed a NASBA assay for lettuce and blueberries contaminated by HAV. This technique

Table 15.1 Characteristics of currently developed methods for extraction and detection of HAV in food with regard to efficiency of recovery and detection limits

Food	Extraction procedure	Detection method	Recovery (%)	Detection limit / sample size	Reference
Fresh lettuce, strawberries	IM[a]	IM-PCR	40	2.5–10 PFU/ml	Bidawid et al. (2000a)
Fresh lettuce, strawberries	Filtration and virus concentration by IM[a]	F-IM-PCR[b]	35	10–50 PFU/ml	Bidawid et al. (2000a)
Fresh raspberries, strawberries	Washing and virus concentration by ultracentrifugation	RT-nested-PCR	2.5–25	40–400 PCRU/60–90 g	Rzeżutka et al. (2006)
Frozen raspberries	Homogenization and virus concentration by PEG[c] precipitation	RTPCR	15–25	40 TCID$_{50}$/100 g	Dubois et al. (2002)
Green onion	Washing and virus concentration by PEG[c] precipitation	RTPCR	NE	10 TCID$_{50}$/25 g	Guevremont et al. (2006)
Green onion, strawberries	Washing and IMS[d]	real-time RTPCR	21–27	10 PFU/ml	Shan et al. (2005)
Fresh lettuce, blueberries	Washing	NASBA	80	NE	Jean et al. (2001)

Sample	Treatment	Detection		Detection limit	Reference
Ham, turkey, roast beef	Washing	RTPCR	NE	10–100 PCRU/20–30 g	Schwab et al. (2000)
Oysters, clams	Homogenization and virus concentration by PEG[c] precipitation	RTPCR	NE	0.15 PFU/3.75 g 0.015PFU/3.75 g	Kingsley and Richards (2001)
Mussels	Homogenization and virus concentration by PEG[c] precipitation	RT-nested-PCR	NE	1 TCID$_{50}$/10 g	Croci et al. (1999)
Oysters	Homogenization and virus concentration by freon precipitation	RTPCR	30–50	8–40 PFU/g	Cromeans et al. (1997)
Cockles	Homogenization and virus concentration by PEG[c] precipitation	RT-seminested-PCR	NE	20 TCID$_{50}$/g	Le Guyader et al. (1994)
Clams	Homogenization and virus concentration by PEG[c] precipitation	real-time PCR	NE	7.5 x 10^{3}/g^{e}	Costafreda et al. (2006)

NE, not estimated.
a Immunomagnetic beads.
b Positively charged virosorb filters, immunomagnetic separation and polymerase chain reaction.
c PEG, Polyethylene glycol.
d Immunomagnetic separation.
e HAV genome copies.

allows detection of specific viral RNA in samples containing high loads of extraneous microorganisms. Recently, successful application of real-time PCR for HAV detection in shellfish samples has been shown by Costafreda *et al.* (2006). All of these methods have confirmed their usefulness for food investigation, despite differences in sensitivity.

There is no need for HAV typing although isolated viruses have been classified into seven genotypes. This is due to high genetic identity between virus strains (Hollinger and Emerson, 2001).

VI.C. Detection of rotaviruses in food

Rotaviruses extracted from food samples can be easily propagated in cell culture. However, cell culture-based methods are lengthy and laborious, and therefore they are rarely used even for clinical specimens. As for other enteric viruses, PCR is also applied for detection and typing of group A of rotaviruses in shellfish (Le Guyader *et al.*, 1994) and fresh produce (Hernández *et al.*, 1997; Van Zyl *et al.*, 2006).

VII. CHALLENGES OF IMPLEMENTATION

Current trends in foodborne virology are focused on elaboration of robust, fast, sensitive, and less-expensive methodologies allowing comprehensive food matrix analysis. The ideal method used for extraction and concentration of viruses should be universal, meaning that it could be applied to different food samples regardless of the contaminating virus type. In addition, with existing assays, although they work well for individual food items, their applications for different samples are still questionable. Virus loss due to multiple steps in the procedure should be minimized towards maximizing virus recovery, which increases the sensitivity of the method. Furthermore, the method should be easy to perform in suitably equipped virological laboratories. Future work should also be focused on the development of the extraction procedure, which removes or minimizes the inhibitors' presence in food extractant. These inhibitory substances or food particulates can decrease efficiency or inhibit nucleic acid amplification. This can be controlled by applying suitable controls for both extraction and detection methods. A useful virus detection method must not only be able to amplify the very low levels of viral genetic material present in contaminated food, but lend itself to further post-PCR analysis (Goswami *et al.*, 2002). In addition, it should allow the identification of virus type or strain without applying additional procedures. Finally, the methods used for detection of foodborne viruses in fruits, vegetables, and shellfish should be suitable for monitoring and screening of food samples for food safety purposes.

Another challenge is virus infectivity. This is important because the presence of viral RNA in food samples does not necessarily indicate risk of infection for consumers (De Medici *et al.*, 2001). There is a need for standardization of methodologies used for extraction and detection of viruses. However, the lack of a culture system suitable for propagation of NoV and difficulties in propagation of wild-type HAV from environmental samples are the main obstacles in work concerning methods standardization.

REFERENCES

Abd el-Galil, K. H., el-Sokkary, M. A., Kheira, S. M., Salazar, A. M., Yates, M. V., Chen, W., & Mulchandani, A. (2005). Real-time nucleic acid sequence-based amplification assay for detection of hepatitis A virus. *Appl. Environ. Microbiol., 71*, 7113–7116.

Anderson, E. J., & Weber, S. G. (2004). Rotavirus infection in adults. *Lancet Infect. Dis., 4*, 91–99.

Atreya, C. D. (2004). Major foodborne illness causing viruses and current status of vaccines against the diseases. *Foodborne Pathog. Dis., 1*, 89–96.

Ballagi-Pordány, A., & Belák, S. (1996). The use of mimics as internal standards to avoid false negatives in diagnostic PCR. *Mol. Cell Probes, 10*, 159–164.

Bidawid, S., Farber, J.M., & Sattar, S.A. (2000a). Contamination of foods by food handlers: experiments on hepatitis A virus transferred to food and its interruption. *Appl. Environ. Microbiol., 66*, 2759–2763.

Bidawid, S., Farber, J.M., & Sattar, S.A. (2000b). Rapid concentration and detection of hepatitis A virus from lettuce and strawberries. *J. Virol. Methods, 88*, 175–185.

Butot, S., Putallaz, T., & Sanchez, G. (2007). Procedure for rapid concentration and detection of enteric viruses from berries and vegetables. *Appl. Environ. Microbiol., 73*, 186–192.

Calder, L., Simmons, G., Thornley, C., Taylor, P., Pritchard, K., Greening, G., & Bishop, J. (2003). An outbreak of hepatitis A associated with consumption of raw blueberries. *Epidemiol. Infect., 131*, 745–751.

Cockerill, F. R., & Uhl, J. R. (2002). Applications and challenges of real-time PCR for the clinical microbiology laboratory. In U. Reischl, C. Wittwer, & F. Cockerill (Eds.), *Rapid cycle real-time PCR- methods and applications, microbiology and food analysis* (pp. 3–27). Berlin, Heidelberg, New York: Springer Verlag.

Coelho, C., Heinert, A. P., Simões, C. M. O., & Barardi, C. R. M. (2003). Hepatitis A virus detection in oysters (*Crassostrea gigas*) in Santa Catarina state, Brazil, by reverse transcription-polymerase chain reaction. *J. Food Prot., 66*, 507–511.

Cook, N., Bridger, J., Kendall, K., Gómara, M. I., El-Attar, L., & Gray, J. (2004). The zoonotic potential of rotavirus. *J. Infect., 48*, 289–302.

Cook, N., & Myint, S. H. (1995). Modern methods for the detection of viruses in water and shellfish. *Rev. Med. Microbiol., 6*, 207–216.

Cook, N., & Rzeżutka, A. (2006). Hepatitis viruses. In Y. Motarjemi, & M. Adams (Eds.), *Emerging foodborne pathogens* (pp. 282–308). Cambridge: Woodhead Publishing Limited.

Costafreda, M. I., Bosch, A., & Pintó, R. M. (2006). Development, evaluation, and standardization of a real-time TaqMan reverse transcription-PCR assay for quantification of hepatitis A virus in clinical and shellfish samples. *Appl. Environ. Microbiol., 72*, 3846–3855.

Croci, L., De Medici, D., Morace, G., Fiore, A., Scalfaro, C., Beneduce, F., & Toti, L. (1999). Detection of hepatitis A virus in shellfish by nested reverse transcription-PCR. *Int. J. Food Microbiol., 48*, 67–71.

Cromeans, T. L., Nainan, O. V., & Margolis, H. S. (1997). Detection of hepatitis A virus RNA in oyster meat. *Appl. Environ. Microbiol., 63,* 2460–2463.

Das, M., Dunn, S. J., Woode, G. N., Greenberg, H. B., & Rao, C. D. (1993). Both surface proteins (VP4 and VP7) of an asymptomatic neonatal rotavirus strain (I321) have high levels of sequence identity with the homologous proteins of a serotype 10 bovine rotavirus. *Virology, 194,* 374–379.

De Medici, D., Croci, L., Di Pasquale, S., Fiore, A., & Toti, L. (2001). Detecting the presence of infectious hepatitis A virus in molluscs positive to RT-nested-PCR. *Lett. Appl. Microbiol., 33,* 362–366.

Dentinger, C. M., Bower, W. A., Nainan, O. V., Cotter, S. M., Myers, G., Dubusky, L. M., Fowler, S., Salehi, E. D., & Bell, B. P. (2001). An outbreak of hepatitis A associated with green onions. *J. Infect. Dis., 183,* 1273–1276.

Dubois, E., Agier, C., Traore, O., Hennechart, C., Merle, G., Cruciere, C., & Laveran, H. (2002). Modified concentration method for the detection of enteric viruses on fruits and vegetables by reverse transcriptase-polymerase chain reaction or cell culture. *J. Food Prot., 65,* 1962–1969.

Duizer, E., & Koopmans, M. (2006). Tracking emerging pathogens: the case of noroviruses. In Y. Motarjemi, & M. Adams (Eds.), *Emerging foodborne pathogens* (pp. 77–110). Cambridge: Woodhead Publishing Limited.

Goswami, B. B., Kulka, M., Ngo, D., Istafanos, P., & Cebula, T. A. (2002). A polymerase chain reaction-based method for the detection of hepatitis A virus in produce and shellfish. *J. Food Prot., 65,* 393–402.

Guévremont, E., Brassard, J., Houde, A., Simard, C., & Trottier, Y. L. (2006). Development of an extraction and concentration procedure and comparison of RT-PCR primer systems for the detection of hepatitis A virus and norovirus GII in green onions. *J. Virol. Methods, 134,* 130–135.

Hernández, F., Monge, R., Jiménez, C., & Taylor, L. (1997). Rotavirus and hepatitis A virus in market lettuce (*Latuca sativa*) in Costa Rica. *Int. J. Food Microbiol., 37,* 221–223.

Hollinger, F. B., & Emerson, S. U. (2001). Hepatitis A virus. In B. N. Fields, P. M. Howley, & D. E. Griffin (Eds.), *Fields Virology* (4th ed.). (pp. 799–840). New York: Lippincott Williams & Wilkins.

Holmes, J. L., Kirkwood, C. D., Gerna, G., et al. (1999). Characterization of unusual G8 rotavirus strains isolated from Egyptian children. *Arch. Virol., 144,* 1381–1396.

Hoorfar, J., Cook, N., Malorny, B., Wagner, M., De Medici, D., Abdulmawjood, A., & Fach, P. (2003). Making internal amplification control mandatory for diagnostic PCR. *J. Clin. Microbiol., 41,* 5835.

Hutin, Y. J. F., Pool, V., Cramer, E. H., et al. (1999). A multistate, foodborne outbreak of hepatitis A.N. Engl. *J. Med., 340,* 595–602.

Jaaskelainen, A. J., & Maunula, L. (2006). Applicability of microarray technique for the detection of noro- and astroviruses. *J. Virol. Methods, 136,* 210–216.

Jean, J., Blais, B., Darveau, A., & Fliss, I. (2001). Detection of hepatitis A virus by the nucleic acid sequence-based amplification technique and comparison with reverse transcription-PCR. *Appl. Environ. Microbiol., 67,* 5593–5600.

Jean, J., Blais, B., Darveau, A., & Fliss, I. (2002). Simultaneous detection and identification of hepatitis A virus and rotavirus by multiplex nucleic acid sequence-based amplification (NASBA) and microtiter plate hybridization system. *J. Virol. Methods, 105,* 123–132.

Jean, J., D'Souza, D. H., & Jaykus, L. A. (2004). Multiplex nucleic acid sequence-based amplification for simultaneous detection of several enteric viruses in model ready-to-eat foods. *Appl. Environ. Microbiol., 70,* 6603–6610.

Jiang, Y. J., Liao, G. Y., Zhao, W., Sun, M. B., Qian, Y., Bian, C. X., & Jiang, S. D. (2004). Detection of infectious hepatitis A virus by integrated cell culture/strand-specific reverse transcriptase-polymerase chain reaction. *J. Appl. Microbiol., 97,* 1105–1112.

Jothikumar, N., Lowther, J. A., Henshilwood, K., Lees, D. N., Hill, V. R., & Vinje, J. (2005). Rapid and sensitive detection of noroviruses by using TaqMan-based one-step reverse transcription-PCR assays and application to naturally contaminated shellfish samples. *Appl. Environ. Microbiol., 71*, 1870–1875.

Kingsley, D. H., & Richards, G. P. (2001). A rapid and efficient extraction method for reverse transcription-PCR detection of hepatitis A and Norwalk-like viruses in shellfish. *Appl. Environ. Microbiol., 67*, 4152–4157.

Klein, G. (2004). Spread of viruses through the food chain. *Dtsch. Tierarztl. Wochenschr, 111*, 312–314.

Koopmans, M., & Duizer, E. (2004). Foodborne viruses: an emerging problem. *Int. J. Food Microbiol., 90*, 23–41.

Kostrzynska, M., & Bachand, A. (2006). Application of DNA microarray technology for detection, identification, and characterization of food-borne pathogens. *Can. J. Microbiol., 52*, 1–8.

Kurdziel, A.S., Wilkinson, N., Gordon, S.H., & Cook, N. (2001a). Development of methods to detect foodborne viruses. In: S. A. Clark, K. C. Thompson, C. W. Keevil, & M. S. Smith (eds) Rapid Detection Assays for Food and Water, (pp. 175–177). Royal Society of Chemistry, Cambridge.

Kurdziel, A.S., Wilkinson, N., Langton, S., & Cook, N. (2001b). Survival of poliovirus on soft fruit and salad vegetables. *J. Food Prot., 64*, 706–709.

Lees, D. (2000). Viruses and bivalve shellfish. *Int. J. Food Microbiol., 59*, 81–116.

Leggitt, P. R., & Jaykus, L. A. (2000). Detection methods for human enteric viruses in representative foods. *J. Food Prot., 63*, 1738–1744.

Le Guyader, F., Dubois, E., Menard, D., & Pommepuy, M. (1994). Detection of hepatitis A virus, rotavirus and enterovirus in naturally contaminated shellfish and sediment by reverse transcription-seminested PCR. *Appl. Environ. Microbiol., 60*, 3665–3671.

Le Guyader, F., Mittelholzer, S., Haugarreau, C. L., Hedlund, K. O., Asterlund, R., Pommepuy, M., & Svensson, L. (2004). Detection of noroviruses in raspberries associated with a gastroenteritis outbreak. *Int. J. Food Microbiol., 97*, 179–186.

Loisy, F., Atmar, R. L., Guillon, P., Le Cann, P., Pommepuy, M., & Le Guyader, F. (2005). Real-time RT-PCR for norovirus screening in shellfish. *J. Virol. Methods, 123*, 1–7.

Lopman, B. A., Reacher, M. H., van Duijnhoven, Y., Hanon, F. X., Brown, D., & Koopmans, M. (2003). Viral gastroenteritis outbreaks in Europe, 1995–2000. *Emerg. Infect. Dis., 9*, 90–96.

Nainan, O. V., Xia G., Vaughan, G., & Margolis, H. S. (2006). Diagnosis of hepatitis A virus infection: a molecular approach. *Clin. Microbiol. Rev., 19*, 63–79.

Nakagomi, O., Isegawa, Y., Ward, R. L., Knowlton, D. R., Kaga, E., Nakagomi, T., & Ueda, S. (1994). Naturally occurring dual infection with human and bovine rotaviruses as suggested by the recovery of G1P8 and G1P5 rotaviruses from a single patient. *Arch. Virol., 137*, 381–388.

Nygard, K., Andersson, Y., Lindkvist, P., et al. (2001). Imported rocket salad partly responsible for increased incidence of hepatitis A cases in Sweden, 2000–2001. *Euro Surveil., 6*, 151–153.

Pebody, R. G., Leino, T., Ruutu, P., Kinnunen, L., Davidkin, I., Nohynek, H., & Leinikki, P. (1998). Foodborne outbreaks of hepatitis A in a low endemic country: an emerging problem? *Epidemiol Infect., 120*, 55–59.

Rosenfield, S. I., & Jaykus, L. A. (1999). A multiplex reverse transcription polymerase chain reaction method for the detection of foodborne viruses. *J. Food Prot., 62*, 1210–1214.

Rossen, L., Nørskov, P., Holmstrøm, K., & Rasmussen, O. F. (1992). Inhibition of PCR by components of food samples, microbial diagnostic assays and DNA-extraction solutions. *Int. J. Food Microbiol., 17*, 37–45.

Rzeżutka, A., Alotaibi, M., D'Agostino, M., & Cook, N. (2005). A centrifugation-based method for extraction of norovirus from raspberries. *J. Food Prot., 68*, 1923–1925.

Rzeżutka, A., D'Agostino, M., & Cook, N. (2006). An ultracentrifugation-based approach to the detection of hepatitis A virus in soft fruits. *Int. J. Food Microbiol., 108*, 315–320.

Sair, A. I., D'Souza, D. H., Moe, Ch. L., & Jaykus, L. A. (2002). Improved detection of human enteric viruses in foods by RT-PCR. *J. Virol. Methods, 100,* 57–69.

Schwab, K. J., Neill, F. H., Fankhauser, R. L., et al. (2000). Development of methods to detect "Norwalk-like viruses" (NLVs) and hepatitis A virus in delicatessen foods: application to a food-borne NLV outbreak. *Appl. Environ. Microbiol., 66,* 213–218.

Schwab, K. J., Neill, F. H., Le Guyader, F., Estes, M. K., & Atmar, R. L. (2001). Development of a reverse transcription-PCR-DNA enzyme immunoassay for detection of "Norwalk-like" viruses and hepatitis A virus in stool and shellfish. *Appl. Environ. Microbiol., 67,* 742–749.

Seymour, I. J., & Appleton, H. (2001). Foodborne viruses and fresh produce. *J. Appl. Microbiol., 91,* 759–773.

Shan, X. C., Wolffs, P., & Griffiths, M. W. (2005). Rapid and quantitative detection of hepatitis A virus from green onion and strawberry rinses by use of real-time reverse transcription-PCR. *Appl. Environ. Microbiol., 71,* 5624–5626.

Shieh, Y. C., Calci, K. R., & Baric, R. S. (1999). A method to detect low levels of enteric viruses in contaminated oysters. *Appl. Environ. Microbiol., 65,* 4709–4714.

Sincero, T. C., Levin, D. B., Simoes, C. M., & Barardi, C. R. (2006). Detection of hepatitis A virus (HAV) in oysters (*Crassostrea gigas*). *Water Res., 40,* 895–902.

Suñén, E., Casas, N., Moreno, B., & Zigorraga, C. (2004). Comparison of two methods for the detection of hepatitis A virus in clam samples (*Tapes* spp.) by reverse transcription-nested PCR. *Int. J. Food Microbiol., 91,* 147–154.

Tian, P., & Mandrell, R. (2006). Detection of norovirus capsid proteins in faecal and food samples by a real time immuno-PCR method. *J. Appl. Microbiol., 100,* 564–574.

Van Zyl, W. B., Page, N. A., Grabow, W. O., Steele, A. D., & Taylor, M. B. (2006). Molecular epidemiology of group A rotaviruses in water sources and selected raw vegetables in southern Africa. *Appl. Environ. Microbiol., 72,* 4554–4560.

Vennema, H., de Bruin, E., & Koopmans, M. (2002). Rational optimization of generic primers used for Norwalk-like virus detection by reverse transcriptase polymerase chain reaction. *J. Clin. Virol., 25,* 233–235.

Vinje, J., Hamidjaja, R. A., & Sobsey, M. D. (2004). Development and application of a capsid VP1 (region D) based reverse transcription PCR assay for genotyping of genogroup I and II noroviruses. *J. Virol. Methods, 116,* 109–117.

WHO (2000). WHO/CDS/CSR/EDC/2000.7, Hepatitis A. World Health Organization Department of Communicable Disease Surveillance and Response. Website: www.who. int/emc.

CHAPTER 16

Tracing Antibiotic Resistance along the Food Chain: Why and How

Reiner Helmuth, Andreas Schroeter, Burkhard Malorny, Angelika Miko, *and* **Beatriz Guerra**

Contents

Abstract

Since the introduction of antimicrobial agents into human and veterinary medicine the emergence of resistant microorganisms has been observed. In the 1960s the Swann Report emphasized the link between the agricultural and veterinary use of antimicrobials with drug resistance. This report initiated a still ongoing debate on all public health aspects of this topic. Recently progress was made applying the Codex Alimentarius risk-assessment methodologies to the question of how far the use of antimicrobials in the food chain poses a threat to animal and human therapy. Among other activities in this field, a joint FAO/OIE/ WHO expert workshop on 'Non-Human Antimicrobial Usage and Antimicrobial Resistance' (Geneva, 2003) concluded that the non-human use of antimicrobial agents contributes to resistance problems in human medicine. In order to mitigate this development a precautionary principle was recommended.

Along this line, the European Union (EU) demands in their directive 2003/ 99/EC the phenotypic monitoring of antimicrobial resistance in zoonotic agents including *Salmonella* and *Campylobacter*. For these programs micro broth dilution according to CLSI M31-A2 is one of the methods of choice. However, in order to elucidate the spread of resistant microorganisms or clones along the food chain, molecular studies are necessary. Today the identification of underlying genetic elements like

Global Issues in Food Science and Technology
© 2009 Elsevier Inc.

plasmids, transposons, integrons, gene cassettes, resistance genes, and mutations in target molecules by PCR or DNA microarrays allows the detection and characterization of resistant microorganisms from the 'stable to the table.'

I. GENERAL AND PUBLIC HEALTH ASPECTS OF ANTIBIOTIC RESISTANCE

Antimicrobial agents belong to one of the most powerful classes of pharmaceuticals and their speed and efficiency in curing human patients or animals is still impressive. In contrast, soon after the introduction of most of these agents it became quite obvious that resistant microorganisms had become more prevalent, leading to serious therapeutic problems. As a consequence, the non-human use of antimicrobial agents, especially, became a public health issue leading to numerous reports and meetings and a large body of scientific literature, as recently summarized by Aarestrup (2006).

For many decades the non-human use of antimicrobial agents has been discussed intensively. However, recent risk assessment studies based on newer scientific data resulted in a much clearer picture of this controversial field, and today it is well recognized that antimicrobial resistance should be mitigated in human medicine, agriculture, veterinary medicine, and food production (World Health Organization, 2000). In 1999 the Codex Alimentarius Commission laid down its 'Principles and guidelines for the conduct of microbiological risk assessment, CAG/GL-30,' which recommends a four-step approach to performing risk assessment studies in the food sector.

Accordingly, the process of microbiological risk assessment is broken down as: (1) hazard identification, (2) hazard characterization, (3) exposure assessment, and (4) risk characterization. Following this approach several reports reviewing the current literature and knowledge have been produced (Food and Drug Administration, 2001; Helmuth and Hensel, 2004; Helmuth, 2004; World Health Organization, 2004). Taken together, it can be stated from these data that the most important selection factors involved for resistant microorganisms and resistance genes include sub-therapeutic doses, mass medication, length of treatment, and use of antibiotics with a broad spectrum of action and their combination instead of use with a narrow spectrum of action. Furthermore, the prophylactic and meta-phylactic uses are special risk factors as well.

There are several studies confirming the causal link between the incidence of resistant bacteria in both human beings and domestic food-producing animals. Human patients infected with resistant *Salmonella* or

Campylobacter run a greater risk of dying within the next two years (Helms *et al.*, 2004; Molbak, 2004; Helms *et al.*, 2006) and/or are susceptible to bloodstream infections and hospitalizations (Varma *et al.*, 2004).

The exposure paths for humans are food, water, human beings themselves, and animals. In the case of meat products, chicken, turkey, beef, pork, and resulting products are major vehicles for transmission. During the processing of meat the bacteria from animal origins can contaminate other food items, the processing plant, or workers. On the other hand, it is possible that resistant organisms are introduced from the outside, e.g. by food handlers into the production line. However, there are already plenty of options to limit exposure to antimicrobial-sensitive organisms during food processing, e.g. the HACCP (Hazard Analysis Critical Control Point) concept, heat treatment and other measures of disinfection, and they are well suited to prevent the spread of resistant organisms as well.

Taking this into account and based on the report of the EU Microbial Threat Conference 1998 (Frimodt-Moller, 2004) and the Scientific Steering Committee 1999 (Fries, 2004), the current directive 2003/99/EC of the European Parliament and of the Council on the monitoring of zoonoses and zoonotic agents (Anonymous, 2003) states in its preamble: 'the alarming emergence of resistance to antimicrobial agents is a characteristic that should be monitored.' The annex II specifies that monitoring of antimicrobial resistance pursuant to article 7 must ensure a representative number of isolates of *Salmonella* spp., *Campylobacter jejuni*, and *C. coli* from cattle, pigs, and poultry, as well as from foods of animal origin derived from those species. Each member state is required to submit a report to the Commission every year by the end of May on the trends and sources of zoonoses, zoonotic agents, and antimicrobial resistance in their state. This is the basis used by risk assessors and managers to reduce the antimicrobial resistance in zoonotic agents in the countries of the European Union.

Monitoring antimicrobial resistance is special in that not only resistant (even multi-resistant) bacteria are passed along the food chain. In addition, there are resistance determinants that can spread and lead to the emergence of new resistant microorganisms, which can only be detected and monitored when the building blocks of resistance traits are understood on the molecular level.

II. PHENOTYPIC DETECTION OF RESISTANCE

By definition, according to the Official *Journal of the European Union and European Food Safety Authority*, antimicrobial resistance is 'the ability of microorganisms of certain species to survive or even grow in the presence of a given concentration of an antimicrobial agent, that is usually sufficient to inhibit or kill microorganisms of the same species' (Anonymous, 2003). Consequently phenotypic detection methods rely on detecting the presence or absence of visible growth. For monitoring purposes mainly two systems, agar diffusion and broth dilution, are applied. Both continue to be standardized and updated by the former National Committee for Clinical Laboratory Standards, recently renamed Clinical and Laboratory Standard Institute (CLSI) in the U.S. (CLSI, 2006b).

The frequently used agar diffusion test is a quick method with an inherent flexibility in drug selection and low cost. It is one of the most established methods and relies on the diffusion of an antimicrobial agent onto an agar plate on which a defined lawn of bacteria has been spread. In the case of susceptibility no growth occurs in the diffusion zone. For monitoring purposes the agar diffusion tests produce qualitative results only. Quantitative data are obtained by MIC (Minimal Inhibitory Concentration) determination using micro- or macro-broth dilution methods, which can be partly or fully automated (CLSI, 2006a). A concentration range from different antimicrobials can be tested in commercialized ready-to-use microtiter plates (broth microdilution method) with an automatic inoculation of the plates and a semiautomatic/automatic reading of growth after 18–24 h of incubation. The advantages are that the technique can be fully automated, quantitative results are generated, and many anaerobic and fastidious species can be tested more accurately. The disadvantage is that broth dilution methods are expensive (Aarestrup, 2004; Watts and Lindemann, 2006).

An important factor is the selection of the antimicrobial agents and microorganisms to be tested. For those microorganisms mentioned in the EU directive 2003/99/EC, which concentrates on zoonotic agents, basic principles have been laid out by the EU project on 'Resistance in Bacteria of Animal Origin' (ARBAO I) (Caprioli et al., 2000). Currently *Salmonella*, *Campylobacter*, *E. coli*, and *Enterococcus* spp. are the recommended bacterial species. The antimicrobial classes to be tested for *Salmonella* include aminoglycosides, phenicols, beta-lactams, cephalosporins, quinolones, fluoroquinolones, sulphonamides, and tetracyclines. In the case of *Campylobacter* spp. macrolides are added and for *Enterococci* glycopeptides and strepogramins as well. A comprehensive summary of the methodologies and antimicrobial agents for each microorganism has been previously described (Aarestrup, 2004; Anonymous, 2006; Watts and Lindemann, 2006).

A subsequent EU concerted action (ARBAO-II, Antimicrobial Resistance in Bacteria of Animal Origin – II) concentrates on quality assurance and data collection issues (http://www.dfvf.dk/Default.asp?ID=9753). ARBAO-II has created a network of national veterinary reference laboratories in Europe and established a surveillance system for monitoring the occurrence and emergence of antibiotic resistance among bacteria from food animals.

III. MOLECULAR BACKGROUND OF ANTIMICROBIAL ACTION AND RESISTANCE

Antibacterial activity is achieved by the inhibition of the biosynthesis pathways of essential components in the bacterial cell (mainly cell wall, proteins, nucleic acids, folic acid, etc.). Different antimicrobials act on different cell targets, and bacteria become resistant when they are able to interfere with the action of substances with antimicrobial activity. Consequently, bacteria have developed different mechanisms for resistance: (1) enzymatic drug inactivation, (2) modification or replacement of the drug target, and (3) reduced intracellular accumulation of the antimicrobial agents (decrease uptake by changes in the permeability of the cell or active transport of the agent to the outside of the cell). Other minor mechanisms (protection and overproduction of the target sites) have been described. In some cases, resistance to one antimicrobial can result from the combination of more than one resistance mechanism (Schwarz and Chaslus-Dancla, 2001; Walsh, 2003; Guardabassi and Courvalin, 2006).

Enzymatic inactivation is the main mechanism of resistance to β-lactams and amynoglycosides. The drug is modified (i.e. hydrolysis by β-lactamases, hydrolases, estearases or addition of a chemical group by acetyl-, adenyl-, and phospo- transferases) and cannot bind to its target site. Chemical modifications of the target play an important role for resistance against penicillin and glycopeptides in Gram-positive bacteria, and target replacement is the main mechanism of acquired resistance to trimethoprim and sulfonamides. Reduced uptake of certain antimicrobials (i.e. quinolones, β-lactams) is normally related to changes in the outer membrane of Gram-negative bacteria such as loss or reduced expression of porins. This mechanism is usually not mediated by resistance genes. Active efflux is an energy-dependent mechanism, which reduces the concentration of the drug in the cytoplasm. Substrate specificity of the efflux pumps is extremely variable (quinolones, tetracyclines, fenicols, β-lactams, etc.). Several resistance genes that code for a number of membrane-associated efflux proteins have been described (Schwarz and Chaslus-Dancla, 2001; Walsh, 2003; Guardabassi and Courvalin, 2006).

Within some bacterial groups such as genus and species, there are some natural structural or functional traits (i.e. inaccessibility or lack of target) that allow tolerance to different antimicrobials/antimicrobial classes by all members of the group. This type of resistance is intrinsic. On the contrary, acquired resistance is conferred by changes in the bacterial genome of one strain, including mutations in housekeeping genes or acquisition of mobile genetic elements carrying resistance genes (Schwarz and Chaslus-Dancla, 2001; Guardabassi and Courvalin, 2006).

Bacteria contain extremely efficient genetic transfer systems capable of exchanging and accumulating antimicrobial resistance genes. Resistance genes can move between chromosomal and extra-chromosomal DNA elements, and they may move between bacteria of the same or different species or to bacteria of different genera by horizontal gene transfer. The most important vehicles for transfer of resistance genes in bacteria are mobile genetic elements, such as plasmids, transposons, integrons and gene cassettes, and genomic islands (Bennett, 1995; Schwarz and Chaslus-Dancla, 2001).

Plasmids are extrachromosomal, replicable circular DNA molecules that vary in size between less than two and several hundred kb. They replicate independently of bacterial chromosomal DNA. Plasmids have been identified in most bacterial species and may have the capacity to be transferred (conjugative plasmids) or co-transferred (non-conjugative plasmids) from one bacterium to another, thus resulting in widespread dissemination of plasmid-encoded characteristics within a bacterial population. In addition to manifold traits, such as metabolic or virulence properties, plasmids may code for resistance to antimicrobial agents. These resistance plasmids carry one or more resistance genes, thus a single plasmid may code for resistance to up to 10 different antimicrobial agents simultaneously. Plasmids are known to be vectors for transposons and integrons/gene cassettes (Carattoli, 2003).

Transposons (jumping genes) are short sequences of DNA that can move between plasmids, between a plasmid and the bacterial chromosome, or between a plasmid and a bacteriophage. Unlike plasmids, transposons are not able to replicate independently; their stable maintenance necessitates integration into the chromosomal or plasmid DNA. Sizes and structures of transposons vary, but usually they harbor a transposase gene encoding their integration and excision, and in most cases one or more resistance genes. Transposons can either be conjugative or non-conjugative, and they are easily acquired by plasmids and then incorporated into bacterial DNA. Often several transposons are clustered on the same plasmid, resulting in the transfer of multiple resistance determinants with a single conjugation event (Liebert et al., 1999).

Integrons are naturally occurring gene expression elements. Among the known classes of integrons, class 1 and class 2 are most frequently found. They represent intact or defective transposons and are usually composed of a 5'- and a 3'-conserved region and an interposed variable region, which mostly contains gene cassettes for antimicrobial resistance. Gene cassettes are small elements (less than 2 kb) that include a single gene and a recombination site. They possess no replication and no transposition systems, but move by site-specific recombination, thus they are considered as mobile elements. The 5′-conserved region harbors both the integrase gene, which is responsible for the site-specific insertion of the cassettes, and the promoter, essential for the expression of the cassettes. The 3'-conserved region carries the $qacE\Delta 1$ gene encoding resistance to some disinfectants, and the sulfonamide resistance gene $sul1$ in class 1 integrons and trans-position genes in class 2 integrons. Integrons can be located either on the bacterial chromosome or more often on broad-host-range plasmids, and this association of a highly efficient gene capture and expression system, together with the capacity for vertical and horizontal transmission of resistance genes, represents a powerful weapon of bacteria to combat the assault of antimi-crobial agents (Hall, 1997; Carattoli, 2001).

Since the 1990s mobile genomic islands harboring antimicrobial resistance genes have been described. These elements are able to integrate site-specific into the bacterial chromosome. A well-studied example is the *Salmonella* genomic island 1 (SGI1) that carries a multidrug resistance (MDR) gene cluster, which is a complex class 1 integron. The 43-kb SGI1 first identified in multidrug-resistant *Salmonella enterica* serovar Typhimurium DT104 has also been found in other serovars, such as Agona, Paratyphi B, Newport, Albany and Meleagridis, suggesting its ability to spread by hori-zontal transfer. Usually the MDR region encodes resistance to ampicillin (bla_{PSE-1}), chloramphenicol/florfenicol ($floR$), streptomycin/spectinomycin ($aadA2$), tetracycline [tet(G)] and sulfonamides ($sul1$). Recently however, different combinations of resistance genes have been found in several *Salmonella enterica* serovars (Boyd *et al.*, 2001; Levings *et al.*, 2005).

Besides the vertical spread by the division of the host cell, all mobile elements can spread by horizontal transfer between bacteria. The principal mechanisms identified for inter-bacterial transfer of genetic material are transformation, transduction, and conjugation. Transformation involves the uptake and incorporation of naked (free) DNA into competent recipient bacteria. Such DNA becomes available in the environment following bacterial autolysis. Transformation plays a minor role in the transfer of resistance genes under in vivo conditions (Davison, 1999).

Transduction means the transmission of exogenous DNA carried by bacteriophages to recipient bacteria. Bacteriophages inject their DNA into host cells, thus forcing the direct production of new phage particles or the integration of the DNA as a prophage into the recipient chromosome. The bacteriophage-mediated transfer of resistance genes is no common mech-anism, since there are limits in the amount of transferable DNA as well as in host range (Kokjohn, 1989; Davison, 1999). Conjugation is the most frequently recognized mechanism for horizontal gene transfer. DNA molecules, such as conjugative plasmids, transposons, and episomes, can be transferred from a donor to a recipient cell, via a contact-dependent transmission using a special transfer apparatus. Small non-conjugative plasmids residing in the same host cell may use this apparatus under certain conditions (mobilization). Conjugation can occur between bacteria of the same species, within species of the same genera or between species of different families (Davison, 1999).

IV. GENOTYPIC DETECTION OF RESISTANCE

Today it is well known that different resistance determinants and mechanisms can be responsible for the same resistance phenotype. This observation has been confirmed in several studies carried out on the molecular epidemiology of resistant bacterial clones (Guerra et al., 2000; Miko et al., 2002; Guerra et al., 2003). The characterization of the genes and genetic determinants implicated in the resistance of a particular strain is a first step in the survey of resistant strains and their epidemiological importance. The screening of mobile genetic elements responsible for the horizontal spread of resistance such as integrons, transposons, and other genes associated with these structures, as well as plasmids, is a further prerequisite for the complete identification and characterization of multidrug resistance to estimate the risk of the dissemination capability of certain resistances within a population. The detection of resistance determinants is also useful for estimating their prevalence in bacterial populations, including human and animal pathogens, zoonotic bacteria, and bacteria from the normal flora (Soto et al., 2003; Guerra et al., 2004; Gibreel et al., 2004; Miko et al., 2005). The application of genotypic methods gives information about the diversity and distribution of resistance genes and mechanisms. The presence or absence of genes does not always imply resistance or susceptibility of the strain, but reflects the potential to express resistance once a selective pressure is applied. Thus the information provided by this analysis is important for risk analysis and control of the spread of resistance (Aarestrup, 2006).

During the last several years a large number of molecular methods have been developed and applied for the detection of resistance genes and mobile vehicles implicated in their spread (Aarts et al., 2001; Tenover and Rasheed, 2004; Aarts et al., 2006). Most of the methods are based on the application of DNA probes and PCR assays, including classic and real-time PCR procedures, which allow the detection of genes encoding resistance to most of the antimicrobials (i.e. phenicols, aminoglycosides, glycopeptides, β-lactams, sulfonamides). PCR-detection of different antimicrobial genes, integrons, and transposons, using different sets of primers, is at the moment the method of choice and is most widely used for the genotypic characterization of resistance. In combination with hybridization of the whole bacterial DNA with the resistance PCR products, the location of the genetic elements responsible for resistance can be achieved. PCR followed by sequencing can be also used for detecting point mutations, leading to resistance to some antimicrobials such as quinolones, erythromycin, etc. PCR technique is rapid and easy to perform, as well as sensitive and specific. The primers used for detection of genes should be designed at targeting nucleic acid sequences within the coding region of the genes. A great amount of literature on primers that detect most of the known genes conferring resistance, as well as integrons and transposons, has been published (Levesque *et al.*, 1995; Guerra *et al.*, 2003; Tenover and Rasheed, 2004). Despite the higher cost of PCR/sequencing it is important to characterize new unknown sequences associated with resistance. As resistance is a dynamic phenomenon, and the number of new genes and other mechanisms is continuously increasing, the number of targets can be too large for an efficient characterization by PCR, and thus the use of new molecular methods for detecting a large number of targets simultaneously (i.e. microarrays, discussed later) will be needed in future.

V. MOLECULAR TYPING METHODS FOR THE CHARACTERIZATION OF ANTIMICROBIAL-RESISTANT BACTERIAL STRAINS

During the last few decades various molecular typing methods for the analysis of zoonotic bacterial pathogens have been developed and nowadays these methods are a major part of epidemiology, disease monitoring, intervention, and food safety research. The aims of molecular typing are many. On the one hand, it enables detection of differences between bacterial isolates within a species, so that particular clones or subtypes can be pursued. This helps, for example, to differentiate pathogenic from non-pathogenic isolates, to recognize particular virulent or antimicrobial-resistant strains, or to identify subpopulations with special traits. On the other hand, it enables grouping together those isolates that belong to identical strains, clones, subtypes, or clusters. This grouping then helps in identifying epidemiological links, for instance to recognize outbreaks, to link a certain food source with disease, or to identify increased risks from animal husbandry or social behavior. Because there are some good publications in existence that review the available molecular techniques applied for the typing of pathogenic bacterial strains, including antimicrobial resistant strains (Wassenaar, 2003; Aarts *et al.*, 2006), only the most important techniques are presented here. These methods are either based on the analysis of the extrachromosomal plasmids or directed at specific loci on the chromosomal DNA, or increasingly, at various parts of the complete genome.

Plasmid profiling was the first of the molecular biology techniques to be applied to bacterial strain typing. Plasmids often carry antimicrobial resistance determinants, so that in epidemiological studies the relatedness of strains can be determined from the number and size of the plasmids and their restriction profiles. However, a few disadvantages restrict the applicability of this method. First, many bacterial species do not harbor plasmids, and secondly, plasmids are mobile genetic elements that can be lost or acquired easily; thus, in some cases, epidemiologically related strains may reveal different plasmid profiles.

Various regions of the bacterial genome are usable as targets for locus-specific analysis of chromosomal DNA. Some years ago the ribotyping technique, a RFLP–Southern blotting procedure, was successfully applied for bacterial typing by using a DNA probe, which was derived from the highly conserved 16S and 23S ribosomal RNA genes. The analysis of results was easy, but closely related strains could not be differentiated. Other techniques, for instance IS200 fingerprinting, use DNA probes from an Insertion Sequence, a mobile DNA element that can integrate within the bacterial genome. For typing, the varying positions and number of DNA elements in different strains of *Salmonella enterica* have been used. In the last few years, instead of using labor-intensive RFLP-blotting techniques, PCR-based locus-specific RFLP methods have been developed.

Since its initial description in 1983, pulsed-field-gel electrophoresis (PFGE) has emerged as the pre-eminent (i.e. 'gold standard') molecular approach to the epidemiological analysis of most bacterial pathogens. This method requires the gentle purification of bacterial DNA and its cleavage with rare cutting restriction enzymes. In this way much fewer (only 20–60) and larger (about 50–1,000 kb) restriction fragments can be produced. Fragments of these very large sizes cannot be separated by the conventional agarose gel electrophoresis techniques, but require specific electrophoretic conditions, such as an electrophoretic current 'pulsed' in different directions over a gradient of time intervals. PFGE is a well-established high-discriminatory method for typing of foodborne pathogens and has therefore been chosen as the primary molecular typing method for PulseNet (www.cdc.gov/pulsenet) and PulseNet Europe (www.pulsenet-europe.org), the molecular surveillance networks for foodborne infections in the USA and Europe covering the whole food-chain from farm to fork. These internationally unique networks include institutions from the veterinary, food, and human science fields. By harmonizing the PFGE procedures used in the partner labs it is possible to generate data that are comparable between laboratories in the same or in different countries (Figure 16.1). The main aim in

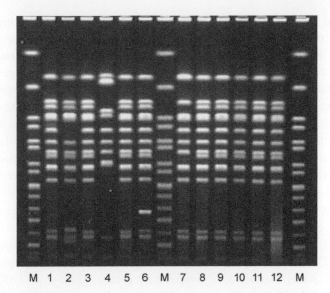

M 1 2 3 4 5 6 M 7 8 9 10 11 12 M

Figure 16.1 *Xba*I-digested genomic DNA from 12 German isolates of multi-drug-resistant *Salmonella enterica* serovar Saintpaul, analyzed by the harmonized PFGE protocol of PulseNet Europe (6 V/cm and 14°C, with a 120° included angle, with CHEF III switching from 2 to 64 s for 20 h). M; *Salmonella enterica* serovar Braenderup H9812 (PulseNet Molecular Weight Standard showing fragments from 20.5–1135 kb).

PulseNet Europe was to establish a real-time linked surveillance database system to detect infection clusters and investigate outbreaks of *Salmonella*, verocytotoxigenic *E. coli* (VTEC), and *Listeria monocytogenes*.

One molecular typing method, multilocus sequence typing (MLST), is based upon the sequencing of short internal fragments of housekeeping genes in order to define sequence types that correspond to alleles at each of the selected loci. MLST is usable for long-term epidemiology and for the identification of lineages that have an increased propensity to cause diseases.

Another method, amplified fragment length polymorphism (AFLP), combines the use of restriction enzymes to generate characteristic genomic fragments and PCR-mediated amplification of the subset of these fragments to produce complex DNA fingerprints. AFLP is a versatile method that can be adapted to the genomic constraints of any bacterial species and allows the delineation of genomic species of bacteria.

VI. DNA MICROARRAYS FOR GENOTYPIC DETECTION OF RESISTANCE

DNA microarray technology, developed within the last decade, has become a powerful molecular-based tool. Although microarrays have been used mainly for gene expression studies, they are becoming increasingly popular in diagnostic microbiology. Microarrays can be used for antibiotic resistance determinations and are a promising alternative to PCR because they enable the simultaneous screening of large sets of targets. PCR only allows the detection of a limited number of genes. The basic principle of microarray technology is similar to the Southern blot hybridization. A nucleic acid probe fixed on a solid surface matches with complementary nucleotide sequences of the DNA tested, which can be detected through reporter molecules such as fluorescence molecules. A DNA microarray consists of an orderly arrangement of spotted probes on a solid support.

During the last few years several DNA microarrays identifying antibiotic resistance genes have been published. For β-lactam antibiotic resistance families PSE, OXA, FOX, MEN, CMY, TEM, SHV, OXY, and AmpC, Lee *et al.* (2002) developed a microarray to analyze PCR amplified target DNA. Call *et al.* (2003) identified 17 different *tet* genes. Chen *et al.* (2005) used PCR products of 24 specific resistance genes including *intI1* (integrase 1) to detect them in *E. coli* and *Salmonella* isolates. Van Hoek *et al.* (2005) used 60mer oligonucleotides to characterize genes causing antibiotic resistance in *Salmonella* serovars by direct labeling of the target bacterial DNA. Antimicrobial resistance, which is caused by single point mutations, needs shorter oligonucleotides as probes. For some classes of antibiotic resistance

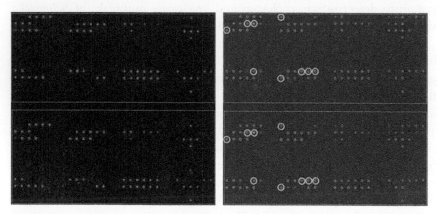

Figure 16.2 Microarray of a susceptible *Salmonella* Typhimurium strain LT2 (left) and a resistant *Salmonella* Typhimurium strain SUO5 (right). SUO5 encodes the resistance associated genes *aadA1*, *bla*$_{oxa31}$, *int1* (integase 1), *qacE*delta, *strA*, *strB*, *sul1*, *sul2* and *tet*(B). The corresponding fluorescence signals are circled. The oligonucleotide probes were printed in two replicates (white boxes), each in eight blocks (6 × 4 spots). Genomic *Salmonella* DNA was labeled with Cy5 and hybridized overnight to the probes. After a short washing protocol the glass slide was scanned and fluorescence signals detected.

such microarrays are published, showing for example, quinolone resistance (Yu *et al.*, 2004), rifampin resistance (Yue *et al.*, 2004), or identifying the single nucleotide polymorphisms (SNPs) of the TEM ß-lactamase variants (Grimm *et al.*, 2004).

Nevertheless a prerequisite of applying DNA microarrays for tracing resistance along the food chain and in diagnostic issues would be its ease of use, rapidity, and accuracy of results. At the National Salmonella Reference Laboratory at BfR (Federal Institute for Risk Assessment) a microarray is currently validated to identify in *Salmonella* isolates potentially important resistance determinants, important virulence genes located within or outside the pathogenicity islands, flagellar and somatic antigens encoding genes, prophage-associated genes, mobility elements, and genes associated with metabolic pathways. This microarray is a new tool for studying the epidemiology of *Salmonella* strains on the genotypic level and might become a powerful method for risk analysis of *Salmonella* along the food chain. The current version harbors 300 specific oligonucleotide probes (Figure 16.2). The plan after the validation phase is to transfer this new tool to other reference laboratories, and then after additional validation studies, to the end-users for routine use in tracing *Salmonella*, including resistance as an important marker group for consumer protection.

REFERENCES

Aarestrup, F. M. (2004). Monitoring of antimicrobial resistance among food animals: Principles and limitations. *J. Vet. Med. B Infect. Dis. Vet. Public Health, 51*, 380–388.

Aarestrup, F. (2006). *Antimicrobial Resistance in Bacteria of Animal Origin*. Washington, DC: ASM Press.

Aarts, H. J., Boumedine, K. S., Nesme, X., & Cloeckaert, A. (2001). Molecular tools for the characterisation of antibiotic-resistant bacteria. *Vet. Res., 32*, 363–380.

Aarts, H. J., Guerra, B., & Malorny, B. (2006). Molecular methods for detection of antibiotic resistance. In F. Aarestrup (Ed.), *Antimicrobial Resistance in Bacteria of Animal Origin* (pp. 37–48). Washington, DC: ASM.

Anonymous (2003). Directive 2003/99/EC of the European Parliament and the Council on the Monitoring of Zoonoses and Zoonotic Agents. Official Journal of the European Union L 325/31.

Anonymous (2006). Manual for Reporting on Zoonoses, Zoonotic Agents, Antimicrobial Resistance and Food-borne Outbreaks in the framework of Directive 2003/90/EC. EFSA.

Bennett, P. M. (1995). The spread of drug resistance. In S. Baumberg, J. P. W. Young, E. M. H. Wellington, & and J. R. Saunders (Eds.), *Population Genetics in Bacteria* (pp. 317–344). Cambridge: University Press.

Boyd, D., Peters, G. A., Cloeckaert, A., Boumedine, K. S., Chaslus-Dancla, E., Imberechts, E., & Mulvey, M. R. (2001). Complete nucleotide sequence of a 43-kilobase genomic island associated with the multidrug resistance region of *Salmonella enterica* serovar Typhimurium DT104 and its identification in phage type DT120 and serovar Agona. *J. Bacteriol., 183*, 5725–5732.

Call, D. R., Bakko, M. K., Krug, M. J., & Roberts, M. C. (2003). Identifying antimicrobial resistance genes with DNA microarrays. *Antimicrob. Agents Chemother., 47*, 3290–3295.

Caprioli, A., Busani, L., Martel, J. L., & Helmuth, R. (2000). Monitoring of antibiotic resistance in bacteria of animal origin: epidemiological and microbiological methodologies. *Int. J. Antimicrob. Agents, 14,* 295–301.

Carattoli, A. (2001). Importance of integrons in the diffusion of resistance. *Vet. Res., 32,* 243–259.

Carattoli, A. (2003). Plasmid-mediated antimicrobial resistance in *Salmonella enterica. Curr. Issues Mol. Biol., 5,* 113–122.

Chen, S., Zhao, S., McDermott, P. F., Schroeder, C. M., White, D. G., & Meng, J. (2005). A DNA microarray for identification of virulence and antimicrobial resistance genes in *Salmonella* serovars and *Escherichia coli. Mol. Cell Probes, 19,* 195–201.

CLSI (Clinical and Laboratory Standards Institute). (2006a). Methods for Dilution Antimicrobial Susceptibility Tests for Bacteria that Grow Aerobically. Approved Standard - Seventh Edition. M7-A7, p. 26. Clinical and Laboratory Standards Institute, Wayne, PA.

CLSI. (2006b). Performance Standards for Antimicrobial Disk and Dilution Susceptibility Tests for Bacteria Isolated from Animals, M100-S16, p. 26. Clinical and Laboratory Standards Institute, Wayne, PA.

Codex Allimentarius Commission.. (1999). *Principles and Guidelines for the Conduct of Microbiological Risk Assessment, CAG/GL-30.* Rome: Food and Agricultural Organization (FAO).

Davison, J. (1999). Genetic exchange between bacteria in the environment. *Plasmid, 42,* 73–91.

Food and Drug Administration (FDA) (2001) Risk assessment on the human health impact of fluoroquinolone resistant *Campylobacter* attributed to the consumption of chicken. Website: http://www.fda.gov/cvm/Risk_asses.htm/.

Frimodt-Moller, N. (2004). Microbial Threat - the Copenhagen Recommendations Initiative of the EU. *J. Vet. Med. B Infect. Dis. Vet. Public Health, 51,* 400–402.

Fries, R. (2004). Conclusions and recommendations of previous expert groups: The Scientific Steering Committee of the EU. *J. Vet. Med. B Infect. Dis. Vet. Public Health, 51,* 403–407.

Gibreel, A., Tracz, D. M., Nonaka, L., Ngo, T. M., Connell, S. R., & Taylor, D. E. (2004). Incidence of antibiotic resistance in *Campylobacter jejuni* isolated in Alberta, Canada, from 1999 to 2002, with special reference to tet(O)-mediated tetracycline resistance. *Antimicrob. Agents Chemother., 48,* 3442–3450.

Grimm, V., Ezaki, S., Susa, M., Knabbe, C., Schmid, R. D., & Bachmann, T. T. (2004). Use of DNA microarrays for rapid genotyping of TEM beta-lactamases that confer resistance. *J. Clin. Microbiol., 42,* 3766–3774.

Guardabassi, L., & Courvalin, P. (2006). Modes of antimicrobial action and mechanisms of bacterial resistance. In F. Aarestrup (Ed.), *Antimicrobial Resistance in Bacteria of Animal Origin* (pp. 1–18). Washington, DC: ASM Press.

Guerra, B., Laconcha, I., Soto, S. M., Gonzalez-Hevia, M. A., & Mendoza, M. C. (2000). Molecular characterisation of emergent multiresistant *Salmonella enterica* serotype [4,5,12: i:-] organisms causing human salmonellosis. *FEMS Microbiol. Lett., 190,* 341–347.

Guerra, B., Junker, E., Schroeter, A., Malorny, B., Lehmann, S., & Helmuth, R. (2003). Phenotypic and genotypic characterization of antimicrobial resistance in German *Escherichia coli* isolates from cattle, swine and poultry. *J. Antimicrob. Chemother., 52,* 489–492.

Guerra, B., Junker, E., Miko, A., Helmuth, R., & Mendoza, M. C. (2004). Characterization and localization of drug resistance determinants in multidrug-resistant, integron-carrying *Salmonella enterica* serotype Typhimurium strains. *Microb. Drug Res., 10,* 83–91.

Hall, R. M. (1997). Mobile gene cassettes and integrons: moving antibiotic resistance genes in gram-negative bacteria. *Ciba Found Symp., 207,* 192–202.

Helms, M., Vastrup, P., Gerner-Smidt, P., & Molbak, K. (2004). Mortality associated with foodborne bacterial gastrointestinal infections. *Ugeskr. Laeger, 166,* 491–493.

Helms, M., Simonsen, J., & Molbak, K. (2006). Foodborne bacterial infection and hospitalization: a registry-based study. *Clin. Infect. Dis., 42*, 498–506.

Helmuth, R., (ed.) (2004). Towards a risk analysis of antibiotic resistance. *J. Vet. Med. B Infect. Dis. Vet. Public Health, 51*, 8–9.

Helmuth, R., & Hensel, A. (2004). Towards the rational use of antibiotics: Results of the first International Symposium on the Risk Analysis of Antibiotic Resistance. *J. Vet. Med. B Infect. Dis. Vet. Public Health, 51*, 357–360.

Kokjohn, T. A. (1989). Transduction: mechanism and potential for gene transfer in the environment. In S. B. Levy, & and R. V. Miller (Eds.), *Gene Transfer in the Environment* (pp. 73–79). New York: McGraw-Hill Book Co..

Lee, Y., Lee, C. S., Kim, Y. J., Chun, S., Park, S., Kim, Y. S., & Han, B. D. (2002). Development of DNA chip for the simultaneous detection of various beta-lactam antibiotic-resistant genes. *Mol. Cells, 14*, 192–197.

Levesque, C., Piche, L., Larose, C., & Roy, P. H. (1995). PCR mapping of integrons reveals several novel combinations of resistance genes. *Antimicrob. Agents Chemother., 39*, 185–191.

Levings, R. S., Lightfoot, D., Partridge, S. R., Hall, R. M., & Djordjevic, S. P. (2005). The genomic island SGI1, containing the multiple antibiotic resistance region of *Salmonella enterica* serovar Typhimurium DT104 or variants of it, is widely distributed in other S. enterica serovars. *J. Bacteriol., 187*, 4401–4409.

Liebert, C. A., Hall, R. M., & Summers, A. O. (1999). Transposon Tn*21*, flagship of the floating genome. *Microbiol. Mol. Biol. Rev., 63*, 507–522.

Miko, A., Guerra, B., Schroeter, A., Dorn, C., & Helmuth, R. (2002). Molecular characterization of multiresistant d-tartrate-positive *Salmonella enterica* serovar Paratyphi B isolates. *J. Clin. Microbiol., 40*, 3184–3191.

Miko, A., Pries, K., Schroeter, A., & Helmuth, R. (2005). Molecular mechanisms of resistance in multidrug-resistant serovars of *Salmonella enterica* isolated from foods in Germany. *J. Antimicrob. Chemother., 56*, 1025–1033.

Molbak, K. (2004). Spread of resistant bacteria and resistance genes from animals to humans – the public health consequences. *J. Vet. Med. B Infect. Dis. Vet. Public Health, 51*, 364–369.

Schwarz, S., & Chaslus-Dancla, E. (2001). Use of antimicrobials in veterinary medicine and mechanisms of resistance. *Vet. Res., 32*, 201–225.

Soto, S. M., Lobato, M. J., & Mendoza, M. C. (2003). Class 1 integron-borne gene cassettes in multidrug-resistant *Yersinia enterocolitica* strains of different phenotypic and genetic types. *Antimicrob. Agents Chemother., 47*, 421–426.

Tenover, F., & Rasheed, K. (2004). Detection of antimicrobial resistance genes and mutations associated with antimicrobial resistance in microorganisms. In D. H. Persing, F. C. Tenover, J. Versalovic, Y. U. E. R. Tang, & and W. T. J. Relman (Eds.), *Molecular Microbiology. Diagnostic Principles and Practice* (pp. 391–406). Washington, DC: ASM Press.

Van Hoek, A. H., Scholtens, I. M., Cloeckaert, A., & Aarts, H. J. (2005). Detection of antibiotic resistance genes in different *Salmonella* serovars by oligonucleotide microarray analysis. *J. Microbiol. Methods, 62*, 13–23.

Varma, J. K., Molbak, K., Barret, J. L., et al. (2004). *Antimicrobial-resistant nontyphoidal Salmonella is associated with excess bloodstream infections and hospitalizations.* Atlanta: GA. NARMS Scientific Meeting. 3–4.

Walsh, C. (2003). *Antibiotics. Actions, origins, resistance.* Washington, DC: ASM Press.

Wassenaar, T. M. (2003). Molecular typing of pathogens. *Berl. Munch. Tierarztl. Wochenschr., 116*, 447–453.

Watts, J. L., & Lindemann, C. J. (2006). Antimicrobial susceptibility testing of bacteria of veterinary origin. In F. Aarestrup (Ed.), *Antimicrobial resistance in bacteria of animal origin.* Washington, DC: ASM Press.

World Health Organization (WHO) (2000). WHO Global Principles for the Containment of Antimicrobial Resistance in Animals Intended for Food: Report of a WHO Consultation. Geneva. WHO//CDS/CSR/APH/2000.4.

World Health Organization (2003). Joint FAO/OIE/WHO Expert Workshop on Non-Human Antimicrobial Usage and Antimicrobial Resistance: Scientific Assessment. Website: http://www.who.int/foodsafety/publications/micro/en/amr.pdf.

Yu, X., Susa, M., Knabbe, C., Schmid, R. D., & Bachmann, T. T. (2004). Development and validation of a diagnostic DNA microarray to detect quinolone-resistant *Escherichia coli* among clinical isolates. *J. Clin. Microbiol., 42,* 4083–4091.

Yue, J., Shi, W., Xie, J., Li, Y., Zeng, E., Liang, L., & Wang, H. (2004). Detection of rifampin-resistant *Mycobacterium tuberculosis* strains by using a specialized oligonucleotide microarray. *Diagn. Microbiol. Infect. Dis., 48,* 47–54.

CHAPTER 17

Lessons Learned in Development and Application of Detection Methods for Zoonotic Foodborne Protozoa on Lettuce and Fresh Fruit

H.V. Smith *and* N. Cook

Contents

Global Issues in Food Science and Technology
© 2009 Elsevier Inc.

Abstract

With the aim of producing standard protocols for detection of *Cryptosporidium* and *Giardia* in foods, a project to develop suitable methods, using lettuce and raspberries as matrices for this study, was conducted at two UK institutes – the Scottish Parasite Diagnostic Laboratory and the Central Science Laboratory. The detection was based on fluorescence microscopy preceded by immunomagnetic separation (IMS) to concentrate (oo)cysts.[1] These stages were adapted from methods developed for detection of *Cryptosporidium* and *Giardia* in water, but specific new sample treatments had to be developed for primary extraction of (oo)cysts from food samples. Importantly, these treatments were optimized by varying physico-chemical parameters to produce sample extracts, which were compatible with the IMS and microscopy materials. The optimization was monitored through the recovery efficiencies achieved through parameter variation. When the final protocols were developed, it was necessary to evaluate their robustness in a multicenter collaborative trial. Conducting a pre-collaborative trial was found to be very useful in highlighting potential ambiguities in the Standard Operating Procedures, and in pinpointing potential obstacles for the collaborative trial. This chapter describes the key features of the methods developed, the trial activities, and details the lessons learned. This information should be useful for future similar studies.

I. INTRODUCTION

Increased demand, global sourcing, and rapid transport of salad vegetables and soft fruits can enhance both the likelihood of surface contamination and survival of the transmissive stages of protozoan pathogens to humans. Some of the more important human intestinal protozoan pathogens that cause foodborne (and waterborne) outbreaks of disease are listed in Table 17.1. *Cryptosporidium parvum, C. hominis,* and *Giardia duodenalis* are the most common etiological agents of foodborne disease, and here, we focus on lessons learned in developing and applying detection methods for these zoonotic foodborne parasites.

---

[1] The word oocyst describes the transmissive stage of coccidian parasites (e.g. *Cryptosporidium, Toxoplasma*), and when used in the context of *Cryptosporidium* sp. oocysts, describes the oocysts of *Cryptosporidium*. Similarly for *Giardia*, the word cyst describes the transmissive stage of this parasite. The term (oo)cysts, invented by the author (H. Smith) many years ago, combines the terms oocyst and cyst when describing oocysts of *Cryptosporidium* and cysts of *Giardia* (which reduces the tedium of reading 'oocysts and cysts' repeatedly in the text). Over time, this term has been accepted by the majority of authors when writing about *Cryptosporidium* and *Giardia*.

Table 17.1 Human intestinal protozoan pathogens responsible for foodborne (and waterborne) outbreaks of disease

Parasite	Food (zoonotic)	Water
Giardia duodenalis	8 (2)	~140
Cryptosporidium spp.	6 (3)	~90
Toxoplasma gondii	Numerous [a]	>3
Blastocystis sp.	(unknown) [b]	2
Microsporidia	(unknown)	1

a Consumption of raw or undercooked food associated with specific cultures and practices.
b Numbers are unknown because, although there have been reports (papers, abstracts, meeting reports), there remains a question about their accuracy or completeness.

I.A. The parasites

Currently, the genus *Cryptosporidium* consists of 18 species and over 40 genotypes, with unique molecular signatures, but, as yet, unattributed species names (Xiao *et al.*, 2004; Caccio *et al.*, 2005; Smith *et al.*, 2006, 2007). *Giardia* is similarly classified into six species, with further genetically distinct assemblages within the *G. duodenalis* species complex (Caccio *et al.*, 2005). Eight described *Cryptosporidium* species (*C. hominis*, *C. parvum*, *C. meleagridis*, *C. felis*, *C. canis*, *C. suis*, *C. muris*, and *C. andersoni*) and five undescribed species of *Cryptosporidium* (cervine, monkey, skunk, rabbit and chipmunk genotypes) infect immunocompetent and immunocompromised humans (Xiao *et al.*, 2004; Smith, 2008; Smith and Evans, 2009), but *C. hominis* and *C. parvum* are the most commonly detected (Cacciò *et al.*, 2005). The prevalences of the two *Giardia duodenalis* assemblages (A and B) that infect humans vary considerably from country to country (Cacciò *et al.*, 2005). The taxonomy of both *Cryptosporidium* and *Giardia* remains only partially resolved and the species status of genetic variants of both parasites (especially in the absence of other data) will be an important and controversial future issue to resolve (Cacciò *et al.*, 2005).

I.A.1. Transmission

Transmission of *Cryptosporidium* and *Giardia* can be direct, from host to host, or indirect, following the ingestion of contaminated food or water. Transmission is via the fecal–oral route following contact with the transmissive stages (*Cryptosporidium* oocysts and *Giardia* cysts ((oo)cysts)). A multitude of transmission cycles therefore exist involving domestic animals and wildlife, which in some instances result in human infections.

Table 17.2   Possible sources of food contamination for salads and fresh fruits that receive no/minimal heat treatment[a]

• Use of oocyst-contaminated feces (night soil), farmyard manure, and slurry as fertilizer for crop cultivation
• Pasturing infected livestock near crops
• Defecation of infected feral hosts onto crops
• Direct contamination of foods following contact with oocyst-contaminated feces transmitted by coprophagous transport hosts (e.g. birds and insects)
• Use of oocyst-contaminated wastewater for irrigation
• Aerosolization of oocyst-contaminated water used for insecticide and fungicide sprays and mists
• Aerosols from slurry spraying and muck spreading
• Poor personal hygiene of food handlers
• Washing 'salad' vegetables, or those consumed raw, in oocyst-contaminated water
• Use of oocyst-contaminated water for making ice and frozen/chilled foods
• Use of oocyst-contaminated water for making products that receive minimum heat or preservative treatment

a Source: Nichols and Smith (2002).

Understanding how the cycles interact, and the frequency of transmission, requires molecular epidemiological studies in defined endemic locations (Caccio et al., 2005; Hunter and Thompson, 2005; Savioli et al., 2006). Knowledge of C. parvum and G. duodenalis transmission routes was fundamental to determining the organisms to analyze and we chose those with the potential for zoonotic transmission (Smith et al., 2007; see Table 17.2; see Section IIIA).

I.A.2. Foodborne transmission

Possible sources of food contamination for salad vegetables and fresh fruits receiving no/minimal heat treatment are presented in Table 17.2.

About 10% of all U.S. cryptosporidiosis and giardiasis cases are estimated to be foodborne, accounting for 64% of the 357,000 cases of parasitic foodborne illness estimated (~700 hospitalizations, 8 deaths; Mead et al., 1999). In England and Wales >3,700 cases of foodborne cryptosporidiosis and giardiasis (0.5% of total) were reported (44 hospital admissions, 3 deaths; Adak et al., 2002). The identification of specific food vehicles is rare, probably because of the long incubation period, supraregional distribution of foodstuffs, and global sourcing of ingredients. Small numbers of (oo)cysts can contaminate vegetable and fruit surfaces (Millar *et al.*, 2002), but their occurrence is underestimated because recovery methods are suboptimal (1–85%; Nichols and Smith, 2002). The (oo)cyst contamination of fresh vegetables and fruits is an important public health consideration, as these products are frequently consumed raw, without any thermal processing. Foodborne cryptosporidiosis outbreaks due to *C. parvum* (in apple juice) and *C. hominis* (on green onions, fruits, and other vegetables) have been described (Quiroz *et al.*, 2000; Millar *et al.*, 2002; Nichols and Smith, 2002; Smith *et al.*, 2007). Foodborne giardiasis is well documented; however no information is available on the *Giardia* assemblage(s) responsible.

II. CONSIDERATIONS PRIOR TO DEVELOPING METHODS

Given the low infectious doses (9–100 oocysts; cysts of *Cryptosporidium* and *Giardia*) (Rendtorff, 1954, 1979; Okhuysen *et al.*, 1999), surface contamination with visually undetectable amounts of viable parasites in produce, which normally receives minimal washing prior to ingestion, poses a threat to public health. Foodborne outbreaks of cryptosporidiosis caused by *C. parvum* and *C. hominis,* and giardiasis by *G. duodenalis,* have been documented in both developed and developing countries (Nichols and Smith, 2002).

Currently underdetected in the UK, such outbreaks may occur following consumption of either UK or imported surface-contaminated produce, particularly when eaten raw or with produce receiving minimal heat treatment prior to consumption. Standard methods are required for isolating, concentrating, and identifying *Cryptosporidium* spp. and *G. duodenalis* (oo)cysts, and can be used to determine their presence on salad vegetables and soft fruits, which normally receive minimal washing prior to consumption. Such methods must be robust and reliable, detecting small numbers of organisms, and should be validated prior to use in accredited analytical laboratories. The method(s) developed must be capable of detecting small numbers of organisms because currently their numbers cannot be increased reproducibly by *in vitro* culture prior to enumeration.

In general, there are no standard methods for detecting the transmissive stages of protozoan parasites on/in foods, although the potential for contamination of foodstuffs, particularly with the transmissive stages of *C. parvum*, is gaining increasing attention (Girdwood and Smith, 1999; Rose and Slifko, 1999; Nichols and Smith, 2002). Fresh produce in particular, as it is consumed with minimal preparation, is a potential vehicle of transmission, and *C. parvum* has been detected on produce in several countries (De Oliveira and Germano, 1992; Monge and Arias, 1996; Ortega *et al.*, 1997; Robertson and Gjerde, 2000, 2001b; Robertson *et al.*, 2002). Practical and reliable detection methods for monitoring foodstuffs will aid prevention of parasitic disease outbreaks associated with contaminated food (Jaykus, 1997).

Published methods for detecting the transmissive stages of protozoan parasites on foods in general are primarily modifications of previously published methods utilizing steps and reagents devised for other purposes (e.g. concentration of protozoan cysts and oocysts from water) (Robertson and Gjerde, 2000, 2001a, b). Recovery efficiencies ranging from 1–85% have been documented (Nichols and Smith, 2002), mostly on small numbers of replicates, indicating the large variability associated with these methods. Few studies have attempted to optimize (oo)cyst release from the foodstuffs tested, and failure to optimize this step leads to reduced recoveries and increased variability. Similarly, a further limitation of all previous studies is that the steps and reagents used in commercial kits for concentrating (oo)cysts by immunomagnetizable separation (IMS) have not been optimized for specific food matrices.

II.A. Specific considerations

We chose lettuce and raspberries as matrices for this study as these have been implicated in foodborne transmission of protozoan parasites (Nichols and Smith, 2002). In addition we sought to identify and challenge specific steps in our proposed method(s) that might cause inherent and/or synergistic problems in (oo)cyst extraction and identification. Ideally, from method development, validation, and user-friendliness/user-uptake perspectives, a single method for 'all analytes and matrices' would be the preferred option, however we realized that this might compromise recoveries making this option unacceptable.

A standard detection method should be straightforward, comprising as few steps as possible. Our methods are based on four steps (Figure 17.1). These are (i) extraction of (oo)cysts from the foodstuffs, (ii) concentration of extract and separation of the (oo)cysts from soluble and suspended food materials, (iii) staining of the (oo)cysts to allow their visualization, and (iv) identification of (oo)cysts by microscopy.

Microscopy is the only accepted method for identifying (oo)cysts of *Cryptosporidium* and *Giardia* in water concentrates in the UK and USA by regulators (Anonymous., 1998, 1999); we chose microscopy as the method of choice for detecting and enumerating (oo)cysts so that uptake of the validated methods might be more straightforward for those accredited food and water laboratories interested in determining the occurrence of parasites on foodstuffs. The criteria chosen for identifying *Cryptosporidium* and *Giardia* (oo)cysts are presented in Table 17.3.

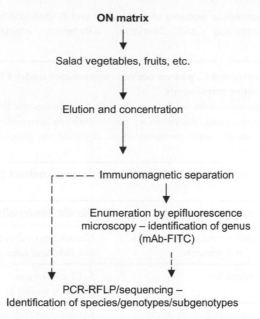

ON matrix

↓

Salad vegetables, fruits, etc.

↓

Elution and concentration

↓

┌ ─ ─ ─ ─ Immunomagnetic separation

│   ↓

│   Enumeration by epifluorescence
│   microscopy – identification of genus
│   (mAb-FITC)

│   ┆

↓   ↓

PCR-RFLP/sequencing –
Identification of species/genotypes/subgenotypes

Figure 17.1 Detection and identification of protozoan parasites on food.

Previously we had demonstrated that molecular methods could also be used to detect (oo)cysts on foods and that the microscopical and molecular methods we developed were both compatible and interchangeable (Figure 17.1; Nichols and Smith, 2004; Nichols *et al.*, 2006). The methods to detect *C. parvum* and *G. duodenalis* on lettuce and raspberries were developed to provide (i) analytical tools suitable for both routine adoption and future proposal as standards and (ii) validated standard methods that would prove useful for both surveillance and epidemiological investigations of human foodborne cryptosporidiosis and giardiasis outbreaks.

III. METHOD DEVELOPMENT

III.A. Choice of matrices and analytes

Lettuce and raspberries were chosen because of their consumption without heat treatment, large consumer demand throughout the year, and involvement in documented foodborne outbreaks. These were locally sourced and our test portions were 30 g of lettuce and 60 g of raspberries based upon previous knowledge (Anonymous, 1993). *C. parvum* and *G. duodenalis* (Assemblages A & B) were chosen because they infect humans and livestock, have a low infectious dose, and have been implicated in foodborne outbreaks of disease (Smith *et al.*, 2007).

Table 17.3 Characteristic features of *C. parvum* and *G. duodenalis* oocysts by epifluorescence microscopy and Nomarski differential interference contrast (DIC) microscopy

G. duodenalis cysts and C. parvum oocysts: appearance under FITC filters of an epifluorescence microscope
The putative organism must conform to the following fluorescent criteria: Uniform apple green fluorescence, often with an increased intensity of fluorescence on the outer perimeter of an object of the appropriate size and shape (see below).

Appearance under Nomarski differential interference contrast (DIC) microscopy	
Cryptosporidium parvum **oocysts**	*Giardia duodenalis* **cysts**
Spherical or slightly ovoid, smooth, thick walled, colorless, and refractile	Ellipsoid to oval, smooth walled, colorless, and refractile
4.5–5.5 μm in suspension	8–12 x 7–10 μm (length x width) in suspension
Sporulated oocysts contain four nuclei Four elongated, naked (i.e. not within a sporocyst(s)) sporozoites and a cytoplasmic residual body within the oocyst	Mature cysts contain four nuclei displaced to one pole of the organism Axostyle (flagellar axonemes) lying diagonally across the long axis of the cyst Two 'claw-hammer'-shaped bodies lying transversely in the mid-portion of the organism

Further considerations identified were that the proposed method must be user-friendly, consisting of as few steps as possible (thus reducing opportunities for sample cross-contamination, and reducing the possibility of a negative influence on recoveries through too many steps), and utilize equipment available in accredited analytical laboratories, thus encouraging its uptake. As we had chosen IMS as the concentration technique, following an analysis of previously published work, we decided that to optimize this step, (oo)cysts must be extracted from the food matrices into an extractant that minimizes interference with their concentration by IMS, which is performed at neutral pH in order to maximize antibody paratope[2]

---

[2] The paratope is the antigen (epitope) recognition site of an antibody that is responsible for that antibody binding to an antigenic determinant (epitope).

interaction with surface exposed (oo)cyst epitopes. The (oo)cysts concentrated from the extractant by IMS were identified and enumerated on microscope slides by epifluorescence and Nomarski differential interference contrast (DIC) microscopy to confirm the morphometry and morphology of the stained organisms (Smith *et al.*, 2002).

III.B. The need to understand surface–surface interactions

Knowledge of the surface charge of *C. parvum* and *G. duodenalis* (oo)cysts provides useful information that can be used to optimize their release from foodstuffs. Binding between (oo)cysts and foodstuffs is mediated by non-covalent interactions at the (oo)cyst–foodstuff interface. At neutral pH, (oo)cysts possess a net negative surface charge, which enhances attachment to foodstuffs. By lowering pH to make a more acidic environment, their negative surface charge decreases, and they bind less well to surfaces. Thus, decreasing surface charge by lowering pH facilitates the release of (oo)cysts from the foodstuff. Measurements of electrophoretic mobility have demonstrated zeta potentials for *C. parvum* oocysts of approximately -25 ± 6 mV at neutral pH (Ongerth and Pecoraro, 1996) and -25 ± 2.8 mV at pH 6 (Drozd and Schwartzbrod, 1996). Surface charge increases slowly with decreasing pH (-35 mV at alkaline pH, reaching zero at pH 2.5) (Drozd and Schwartzbrod, 1996) or pH 4–4.5 (Ongerth and Pecoraro, 1996). Surface charge appears reasonably stable over time, with only a slight decrease in zeta potential from 25 mV to approximately 15 mV over a one-month period (Drozd and Schwartzbrod, 1996; Smith and Ronald, 2002). Similar zeta potentials have been found for *Giardia* cysts (Engeset, 1984). Our data indicate that the zeta potential of purified, human-derived *G. duodenalis* cysts ranges from -24 mV at pH 4 to -35 mV at pH 10, with the addition of Tween 80$^{®}$ reducing the potential by only -2 mV and increasing it with sodium dodecyl sulfate. Neither detergent was found useful for destabilizing cyst suspensions (Nkweta *et al.*, 1989; unpublished observations).

IV. LESSONS LEARNED IN DEVELOPING DETECTION METHODS FOR ZOONOTIC FOODBORNE PROTOZOA ON LETTUCE AND FRESH FRUITS

In order to convey the importance of lessons learned in developing detection methods for zoonotic foodborne protozoa on lettuce and fresh fruits, we focused on the development and validation of a method for detecting *C. parvum* oocysts on lettuce. While, in our experience, the lessons learned for other parasites and matrices were similar, the learning curve in developing the method for detecting *C. parvum* oocysts on lettuce was most acute.

IV.A. Development of suitable physico-chemical extraction procedures for concentrating and identifying *C. parvum* and *G. duodenalis* (oo)cysts from lettuce and fresh fruit

IV.A.1. Influence of extractants on (oo)cyst morphology and viability

The usefulness of SOPs, developed at the SPDL to identify and enumerate oocysts of *C. parvum* and *G. intestinalis* (oo)cysts isolated from vegetables and fruits, was determined. Any adverse effects from the extractants, or lettuce or raspberry extracts, on oocyst morphology and morphometry were noted. Monitoring was performed to determine whether there was any (i) occlusion of (oo)cysts by deposits from dried extractants, (ii) deterioration in *C. parvum* oocyst morphology and viability, and (iii) deterioration in *G. duodenalis* cyst morphology.

Immersion of (oo)cysts in food eluates did not interfere with our ability to identify (oo)cysts. Only when (oo)cysts were air dried in sodium bicarbonate extractant did this extractant occlude (oo)cysts dried onto microscope slides. Similarly, no deterioration in *C. parvum* oocyst morphology or viability occurred when exposed to 1 M glycine for 3 h (control, unexposed oocysts were $92 \pm 7\%$ viable; test, eluate-exposed oocysts were $90 \pm 11\%$ viable) as determined by the fluorogenic vital dyes assay (Campbell *et al.*, 1992). *G. duodenalis* cysts exhibited no deterioration in morphology as determined by DIC microscopy. Finally, exposure to the eluates tested did not affect our ability to recognize them by epifluorescence microscopy following FITC and DAPI staining. The criteria chosen for identifying *C. parvum* and *G. duodenalis* (oo)cysts (Table 17.3) were acceptable for method development. Only when (oo)cysts were empty did acceptance of the identification criteria cause any problems due to the inability to confirm the presence of internal morphological structures by DIC.

IV.A.2. Influence of extractants on the performance of commercially available IMS kits for concentrating *C. parvum* and *G. duodenalis* (oo)cysts

Two commercially available IMS kits for concentrating *Cryptosporidium* spp. oocysts from raw and drinking water concentrates were available at the time of the study and tested. These were the Dynal Dynabeads anti-Cryptosporidium kit® IMS system (Dynal Biotech ASA, Oslo, Norway) and the ImmuCell Crypto-Scan® IMS system (ImmuCell Corporation, Portland, Maine, USA). Factors that might affect IMS recovery efficiency, including sample turbidity, availability of epitopes on (oo)cysts, contact time, and mechanical mixing were evaluated, but only sample turbidity was shown to affect kit performance. Here, the ImmuCell Crypto-Scan® IMS system outperformed the Dynal Dynabeads anti–Cryptosporidium kit® IMS system (Paton *et al.*, 2001) and was used in all further *Cryptosporidium* analyses, according to manufacturer's instructions. Only one commercial kit, the Dynal Dynabeads anti-Giardia kit® IMS system, was available for recovering G. *duodenalis* cysts from aqueous suspensions. This was used throughout the study according to the manufacturer's instructions. For both analytes, the final suspension, comprising 50 µl, was pipetted onto a well of a four-well microscope slide and air-dried. Air-dried concentrates were fixed in absolute methanol and stained with a commercially available fluorescein isothiocyanate, labeled monoclonal antibody that recognizes surface-exposed epitopes on *C. parvum* and G. *duodenalis* oocysts (FITC-mAb) according to manufacturer's instructions, and the fluorogenic DNA intercalator 4'6-diamidino-2-phenyl indole (DAPI) (Smith *et al.*, 2002). Stained organisms were evaluated and those conforming to the chosen criteria (Table 17.3) were enumerated.

IV.A.3. Optimizing *C. parvum* oocyst extraction protocols

Extraction of (oo)cysts from lettuce leaves and raspberries was identified as the most variable component of the method and care was taken to optimize the extraction buffer in order to maximize recoveries. Initial criteria for selecting extractants included minimal interference with IMS and fluorescence antibody detection of (oo)cysts. Interference with IMS was noted with some of the nine extractants tested (those containing sodium dodecyl sulfate), and of the extractants initially selected, six did not influence IMS capture of (oo)cysts compared with manufacturers' buffers on the small number of replicates tested. Both molarity and pH played an important role in the release of parasites from the matrix into suspension.

IV.A.3.a. Influence of extractants and food extracts on *C. parvum* oocyst recovery by IMS

To determine the influence of the extractants and food extracts on the *Cryptosporidium* IMS, three trials (Trials A, B, and C) were instigated. Trial A investigated whether the extractants were compatible with the *C. parvum* oocyst IMS. Trial B investigated whether soluble lettuce extracts, produced using selected extractants, were compatible with the *C. parvum* oocyst IMS. Trial C investigated whether there were any other interferents that influenced *C. parvum* oocyst recovery by IMS. The results of these trials and conclusions drawn are as follows. In Trial A, we investigated the effects of seeding extractants directly with *C. parvum* oocysts, recovering them by IMS and enumerating by immunofluorescence and DIC microscopy to determine which extractants were compatible with IMS. The results are presented in Table 17.4.

In Trial B, we investigated the effects of adding selected extractants to unseeded lettuce samples, and seeding the resulting extracts with oocysts

Table 17.4 Mean percentage recovery of C. parvum oocysts seeded directly into extractants and recovered by IMS (Trial A)

Extractant	Mean % oocyst recovery (± standard deviation) ($n = 3$)
0.1 M HEPES, pH 5.5	89.4 (± 5.1)
1 M Sodium bicarbonate, pH 6.0	88.2 (± 9.8)
1 M Glycine, pH 5.5	105.3 (± 18.4)
1 M Bicine, pH 5.6	81.1 (± 12.5)
1% Sodium dodecyl sulfate (+ Antifoam A)	90.2 (± 5.5)
PBS, 150 mM, pH 7.2	11.4 (± 5.7)
0.1 M Tricine, pH 5.4	9.8 (± 2.5)
Extraction buffer (EB [a])	5.9 (± 4.5)

a EB is the buffer recommended in the US EPA's method 1623 (Anon., 1998) for the extraction of Cryptosporidium oocysts and Giardia cysts from capsule filters. Its composition is: 0.01 M Tris pH 7.4, 0.1% sodium dodecyl sulfate, 0.005 M EDTA, 150 ppm antifoam A.

prior to IMS to determine the degree of compatibility between the extractant, foodstuff, and IMS. The results are presented in Table 17.5.

Sodium bicarbonate eluates when dried onto microscope slides left extensive debris, producing slides that were difficult to read because of oocyst occlusion. Extraction in HEPES and bicine generated lower recoveries. The trial was repeated, and tripotassium phosphate, glycine, glycine + sodium dodecyl sulfate + antifoaming agent were also tested. Extraction in HEPES, bicine, and tricine generated larger standard deviations and lower recoveries, and were deemed too expensive based on their performance. 1M glycine (pH 5.5) was chosen as the most effective extractant.

In Trial C, we investigated seeding lettuce leaves with known numbers of oocysts, extracting the oocysts with selected extractants, and determined the efficiency of recovery (Table 17.6).

This lower recovery, compared to previous results, indicated that not all oocysts were extracted from the lettuce and/or recovered by IMS. The data indicated that non-covalent interactions between oocysts and lettuce surfaces, while complex, could be minimized by judicious choice of extractant. Minimizing the oocyst surface charge reduces the attractive forces between oocysts and matrix enabling them to be released into the extractant.

Table 17.5 Mean percentage recovery of *C. parvum* oocysts seeded into lettuce contaminated extractants and recovered by IMS (trial B)

Extractant	Mean % oocyst recovery (± standard deviation) ($n = 3$)
0.1 M HEPES, pH 5.5	50.0 (± 21.8)
1 M Sodium bicarbonate, pH 6.0	Occluding debris interfered with oocyst enumeration
1 M Glycine, pH 5.5	103.1 (± 7.9)
1 M Bicine, pH 5.6	61.1 (± 23.1)
1% Sodium dodecyl sulfate + Antifoam A	Incompatible with pulsification due to excessive foaming

Table 17.6 Mean percentage recovery of *C. parvum* oocysts seeded into lettuce extractant extracted and recovered by IMS (trial C)

Extractant	Mean % oocyst recovery (± standard deviation) ($n = 4$)
1 M Glycine, pH 5.5	30.5 (± 4.9)
PBS (150 mM, pH 7.2)	11.4 (± 5.7)

IV.A.4. Influence of mechanical extraction procedures on *C. parvum* oocyst recovery from lettuce

Here, focus was placed on determining the best physical extraction method to maximize recoveries. The passive release of oocysts generated variability in the data, which could be amplified when physical extraction methods were tested. The methods tested were orbital shaking, rolling, stomaching in filtered bags, pulsifying in filtered bags, stomaching in unfiltered bags, and pulsifying in unfiltered bags. The results obtained are presented in Table 17.7.

From Table 17.7, pulsification in filtered bags and stomaching in filtered bags provided the highest recoveries. In order to verify the above result, pulsification in filtered bags and stomaching in filtered bags were compared using nine replicates each. Stomaching in filtered bags was chosen for the final stage of method development.

Table 17.7 Mean percentage recovery of C. parvum oocysts seeded onto lettuce, following physical extraction in 1 M glycine (pH 5.5) and IMS

Method	Mean % oocyst recovery (± standard deviation) ($n = 3$)
Orbital shaking	24.8 (± 5.6)
Rolling	36.4 (± 4.0)
Stomaching in filtered bags	46.6 (± 9.6)
Pulsifying in filtered bags	75.0 (± 17.0)
Stomaching in unfiltered bags	35.7 (± 14.3)
Pulsifying in unfiltered bags	34.0 (± 25.0)

IV.A.5. Enumeration of C. parvum oocyst seeds

In the studies above, while intra-trial results variation was acceptable, inter-trial results variation was large. This was due to the inherent variability in dispensing, accurately, small numbers of oocysts. Basing decisions on three replicates generated large and/or variable range and standard deviation values. Using oocyst suspensions where no microscopic evidence of clumping was apparent, adopting standardized oocyst dilution procedures for creating oocyst stocks, and increasing the number of replicates, particularly in critical method development areas, increased our ability to determine the recovery efficiencies of the methods developed.

IV.A.6. Influence of pH on C. parvum oocyst recovery

This series of experiments focused on determining whether there was an optimum pH for C. parvum oocyst extraction using 1 M glycine. This and all subsequent development experiments were performed on lettuce seeded with approximately 100 oocysts. Two peaks in oocyst recovery were observed, one between pH 3.0 and 3.5 and another between pH 5.0 and 5.5 (Figure 17.2).

We interpreted this pattern with reference to Figure 17.3, which shows the influence of adding reagents A, B, and C, provided in the Crypto-Scan® IMS kit, on the pH of 1 M glycine extractant (pH 3–7). Reagents A, B, and C are used in the Crypto-Scan® IMS procedure, to sequentially adjust the pH of an oocyst suspension to pH 7.0, the optimum for binding of bead bound mAb to C. parvum oocysts.

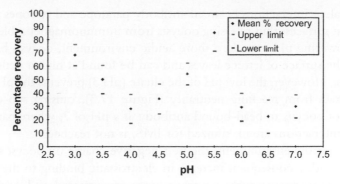

Figure 17.2   Influence of eluant pH on recovery of *Cryptosporidium* oocysts from lettuce.

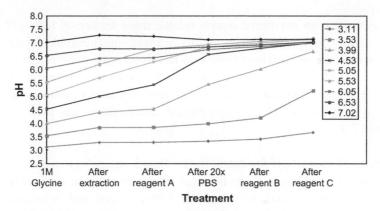

Figure 17.3   The pH of lettuce extract at various stages of extraction and IMS.

Published studies indicate that *C. parvum* oocysts and *G. duodenalis* cysts have a nett negative cell surface charge at neutral pH (Ongerth and Pecoraro, 1996). Reducing the pH reduces the oocyst surface charge. A positive linear relationship between pH and surface charge appears to exist for *C. parvum* oocysts (Drozd and Schwartzbrod, 1996; Ongerth and Pecoraro, 1996; Smith and Ronald, 2002).

In this series of experiments, processing with 1 M glycine, pH 3.0, resulted in one of the two highest recoveries of oocysts (the other occurred with 1 M glycine, pH 5.5). We suggest that when oocysts become less negatively charged under increasing acidic conditions they bind less well to the lettuce leaf surface because non-covalent interactions between oocyst and surfaces are minimized. This hypothesis is strengthened by the fact that HCl (pH 2.75) (Campbell and Smith, 1997) is used to decrease non-covalent interactions between antibody paratope and epitopes present on oocyst surfaces, thus releasing oocysts from immunomagnetizable beads.

By lowering pH to make a more acidic environment, oocysts bind less well to the surface of lettuce leaves and can be found at higher densities in the eluate. However, the low pH of the eluate (pH 3) prevents the pH of the final extract from reaching neutrality (Figure 17.3), causing sub-optimal binding of oocysts to bead-bound antibody, if a pH of 7, where paratope–epitope interactions are maximized for IMS, is not reached.

In less acidic environments the nett negative charge on oocyst surfaces increases with a consequent increase in electrostatic binding to the lettuce surfaces, thus, progressively fewer oocysts are extracted and overall recovery decreases. As pH approaches neutrality, binding to the antibody on the immunomagnetizable beads increases. At pH 7.0 binding to the beads is optimal, however more oocysts remain bound to the lettuce surface.

Attempting to maximize oocyst release from lettuce at high pH and maximizing oocyst attachment onto immunomagnetizable beads are tradeoffs. At the point where these two trends interact, recovery should be maximized. Here, sufficient oocysts should be extracted from the lettuce into the extract to produce a satisfactory recovery under optimal antibody binding conditions. These hypothetical interactions are presented in Figure 17.4.

IV.A.6.a. Choice of pH for extractant
Based on (oo)cyst surface charge data, our preferred option was to extract oocysts using an extractant with a pH between 2.5 and 4.5 (Engeset, 1984; Drozd and Schwartzbrod, 1996; Ongerth and Pecoraro, 1996). Since one of our maxima for oocyst

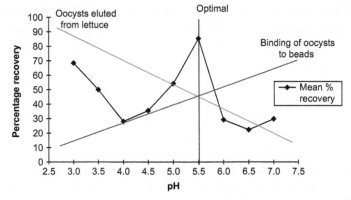

Figure 17.4 Interaction between elution of oocysts from lettuce and binding to IMS beads relative to recovery efficiencies.

extraction was pH 3, we used 1 M glycine (adjusted to pH 3) to extract oocysts followed by adjustment of pH of the released oocysts to pH 7, with 0.1 M NaOH, immediately after resuspending the pellet, in order to optimize the formation of oocyst–bead complexes in the IMS component of the method. However, when the pH was adjusted to neutrality with 0.1 M NaOH to maximize oocyst recovery by IMS, the results were inconsistent. We decided to retain the 1 M glycine extractant at pH 5.5 in the final *Cryptosporidium* and lettuce method, as the adjustment to neutrality, which was necessary for optimizing IMS, was less variable and could be accomplished using the reagent already provided in the IMS kit. Importantly, a 1 M glycine solution has a pH of 5.5, and, as such is simpler for the user to make up, as no pH adjustments have to be made to the solution prior to use. Finally, as the procedure was based on concentrating *C. parvum* and *G. duodenalis* (oo)cysts from vegetable and fruit eluates using commercially available IMS kits, issues regarding the buffering capacity of kit reagents (particularly Reagent A) were resolved when the 1 M glycine extractant was used at pH 5.5 rather than pH 3 for the (oo)cyst and lettuce methods.

IV.A.7. Optimization of a method suitable for use by routine analytical laboratories – experiences with detecting *Cryptosporidium* oocysts on lettuce

Initially, a method for detecting *Cryptosporidium* on lettuce was developed based on our data investigating non-covalent interactions between oocysts and lettuce surfaces, as well as interactions between extractants and IMS. Passive and physical removal of oocysts from lettuce was maximized using 1 M glycine extractant at pH 5.5 and mechanical agitation. Neither molarity nor pH of the extractant affected IMS recoveries. The variability of results obtained with small numbers ($n =$ 3) of replicates was not acceptable and prompted us to use larger numbers of replicates, prior to deciding on methods for pre-validation. In a series of six replicate tests, overall *C. parvum* oocyst recovery was 85 $\pm$ 14%. In a series of 30 replicate tests, overall *C. parvum* oocyst recovery was 59.0 $\pm$ 12.0%.

Given that both the release of oocysts from lettuce surfaces and extractant compatibility with IMS had been maximized, our recommended procedure was written up as a standard operating protocol (SOP). The method developed for detecting *Cryptosporidium* oocysts on lettuce was adopted as the basic method, to which modifications, where necessary, would be made for other parasites/foodstuffs. The method to detect *Cryptosporidium* on lettuce and fresh raspberries has since been published (Cook *et al.*, 2006a).

V. DETERMINING THE ROBUSTNESS OF THE SOP IN A PRE-COLLABORATIVE STUDY

Three laboratories were chosen, SPDL and CSL because of their involvement in method development, and the third laboratory because of its expertise in *Cryptosporidium* analysis. SOPs were written conjointly by SPDL and CSL. Two test materials were used, namely Webb's lettuce and fresh raspberries, which were obtained from local retail outlets; 30 g lettuce and 60 g raspberries were used per sample. Recently excreted, purified *C. parvum* oocysts were purchased commercially and G. *duodenalis* cysts were purified from fecal samples sent to the SPDL for routine examination. The (oo)cyst suspensions were prepared at SPDL and dispatched overnight to CSL, where all food samples used in the pre-collaborative trial were seeded. Each parasite was seeded onto each test material at three densities: high (100 (oo)cysts), medium (50 (oo)cysts), and low (10 (oo)cysts). Control samples of unseeded test materials were also prepared. The (oo)cyst density was confirmed by epifluorescence and DIC microscopy. In all instances, the aliquots were added at points evenly distributed on the sample as much as possible (i.e. on separate lettuce leaves or raspberries). Samples were then left at room temperature for approximately two hours until almost dry, before being packed for dispatch. On Day 0 of each round, batches of 30 samples of each density were prepared per test material. For example, in Round 1, 30 samples of 30 g lettuce seeded with 100 *C. parvum* oocysts, 30 samples with 50 oocysts, and 30 samples with 20 oocysts were prepared. Twenty control (unseeded) 30 g lettuce samples were also prepared. Each participating laboratory was sent duplicate samples of each density.

Quality assurance in the delivery of seeding doses was performed by delivering a seeding dose onto a well of a four-well multispot slide (Hendley Essex Ltd., UK) (Korich *et al.*, 2000); this was done for every fifth seeding dose, for each seeding level, and each matrix. Oocysts delivered onto welled slides were enumerated by epifluorescence microscopy. Quality assurance microscope slides and homogeneity samples (see Section V.A.4 below) were sent to SPDL for completion of analysis.

V.A. Trial protocol

All participants received copies of the SOPs and familiarized themselves with the method prior to the commencement of the trial. All FITC–mAbs and IMS test kits were purchased by SPDL and sent to each participant prior to the commencement of the trial. All commercial reagents were within their expiry dates. The trial was conducted in four segments, corresponding to the parasite/foodstuff combination. The segments were (i) *Cryptosporidium/*

lettuce, (ii) *Cryptosporidium*/raspberries, (iii) *Giardia*/lettuce, and (iv) *Giardia*/raspberries. Each segment comprised three rounds. In each round, eight blind coded samples were sent to each participant: these included two unseeded samples, and two each of samples seeded at low, medium, and high densities with the target organism. For each round, samples were prepared at CSL and sent out to each participant on Day 0.

Participants received the samples on Day 1; each was asked to analyze the samples that day and report the results back to CSL within 7 days. A report form on which to record the results was included with each batch of samples sent. Participants were also asked to send their microscope slides from each test to SPDL for cross-checking of their counts. Results were collated and analyzed by CSL. The mean number of (oo)cysts recovered from the samples was compared with the mean number of (oo)cysts in the suspension used to artificially contaminate them and the percentage recovery calculated.

V.A.1. Preparation and dispatch of samples

Lettuce samples were placed into filtered stomacher bags, which were then sealed with autoclave tape. Raspberry samples were placed in 500 ml centrifuge pots, the lids of which were sealed with tape. Cool packs were placed in all the boxes to reduce deterioration of samples during transit. Two samples from each batch were sent to each participant; all samples were coded, but no further identification was given to participants at any stage during the study. Both lettuce and raspberry samples were dispatched by standard courier.

V.A.2. Analysis of samples

Participants were told that samples should be extracted (and analyzed, if possible) within one working day (SPDL calculated that samples intended for one day's analysis could be extracted and enumerated in 6–7 h). On Day 1, each participant commenced analysis of the samples according to the SOP. Each participant was asked to keep a detailed record of sample treatment, extraction, and enumeration, noting particularly any deviation from the SOP, and to return their results to CSL within seven working days of receipt of samples.

V.A.3. Determining the (oo)cyst concentration of the seed

For each round, the (oo)cyst concentration of the seed suspensions was checked by the following procedure. For the low-density controls, 5 µl of (oo)cyst seeding suspension plus 5 µl PBS, followed by 4 × 10 µl PBS, were placed on each of 10 wells of a 10-welled slide. For the medium density, 5 × 5 µl each of PBS and (oo)cyst seeding suspension were placed on each of 10 wells of a slide. For the high-density controls, 5 × 10 µl (oo)cyst seeding suspension were placed on to each of 10 wells of a slide. The slides were then dried in an incubator at 37°C before being fixed and sent to SPDL for enumeration.

V.A.4. Homogeneity testing, trip controls, and sample dispatch

All homogeneity testing was performed at the SPDL. To confirm homogeneity, 10 samples of each seeded batch, and two samples from each unseeded batch, were selected at random. These samples were extracted at CSL to the point where the eluate had been collated into one tube, and the extracts sent to SPDL for IMS and (oo)cyst enumeration. CSL were responsible for seeding and dispatch of samples, collection, and collation of results. Two trip controls were also sent to SPDL. Coded, blind samples were packaged according to IATA regulations for dangerous pathogens for shipment to the participating laboratories, to arrive the morning of the day after seeding (Day 1).

V.B. Lessons learned from the pre-collaborative trial

Eight major lessons were learned. These were the late delivery of samples, which influenced whether the samples could be analyzed in one working day; samples arriving in poor condition due to agitation during transport; non-conformance to the SOPs; incorrect sample numbering by participating laboratories; leakage of samples during sample processing; the use of incorrect fluorescence filter blocks on microscopes; late return of results; and a lack of experience in identifying organisms, particularly with DIC microscopy. All these issues were addressed prior to and during the validation-training workshop.

The *Giardia*/raspberries results were poor for the pre-validation study, and following agreement with the Food Standards Agency, this segment was omitted from the validation study. The pre-validation study identified a serious drawback in using the same courier to transport both the lettuce and raspberry samples to the pre-validation sites. While the lettuce arrived undamaged, more often than not, the raspberries were battered or partly crushed on receipt, causing major problems in extracting (oo)cysts and concentrating them by IMS. In order to overcome this problem, we used a specialist courtier who guaranteed safe and timely delivery with minimum agitation of the samples to the two distant pre-validation sites, from round 4 of the pre-validation study, onwards. This was agreed with the Food Standards Agency.

VI. VALIDATION OF A STANDARD METHOD BY COLLABORATIVE TRIAL

Eight UK accredited laboratories with experience in analyzing environmental or clinical samples for the presence of *Cryptosporidium* and *Giardia* participated in the trial. They comprised three Health Protection Agency laboratories, three Water Company water testing laboratories, one Research Association, and one private analytical company. Each laboratory was assigned an arbitrary unique number, which remained constant throughout the trial.

VI.A. Training workshops

Two training workshops were held. The trainers comprised two SPDL staff, with particular expertise in morphological identification of (oo)cysts and IMS development, and two CSL staff who had participated in developing the food extraction procedures. All participants received copies of the SOPs prior to the workshop. The workshops were comprised of (i) a description of the SOPs and equipment, (ii) the demonstration of each method, (iii) review and demonstration of identification principles, and (iv) practice examples for each method. Discussion with the workshop participants indicated that minor points in the existing SOPs required amendment for improved clarity. Accordingly, the final versions of the SOPs were drawn up after the workshop and each participant received copies of the final versions prior to the validation study. At the training workshops, participants were told that monoclonal antibodies for detecting *Cryptosporidium* oocysts and *Giardia* cysts and IMS test kits would be sent to each participant prior to commencement of the validation trial. Participants signed for the receipt of all reagents and equipment prior to the validation trial. In addition, the SPDL sent out 'reference' FITC and DAPI stained slides of *C. parvum* and *G. duodenalis* (oo)cysts to all participants. The Olympus Optical Company (UK, Microscope Division) provided microscopes for the workshop.

VI.B. Organization of seeding, homogeneity, and trip controls

All stock suspensions of *C. parvum* and *G. duodenalis* (oo)cysts were enumerated at SPDL and dispatched overnight to CSL for seeding lettuce and raspberries. All homogeneity testing and quality analyses were performed at SPDL. CSL were responsible for seeding and dispatch of samples, collection, and collation of results. Two trip controls were also sent to SPDL. As for the pre-validation study, coded, blind samples were packaged according to IATA regulations for dangerous pathogens for shipment to the participating laboratories, to arrive the morning of the day after seeding (Day 1). Seeding of the test materials was performed at CSL on Day 0 as described previously. Each participating laboratory was sent duplicate samples of each density. All participants received copies of the SOPs prior to the workshop and commencement of the trial. Monoclonal antibodies and IMS test kits were sent to each participant prior to commencement of the trial as well.

To test the stability of the seeded samples over 24 h, 10 samples of each food type were seeded with approximately 100 *Cryptosporidium* oocysts or *Giardia* cysts on Day 0, five of which were processed that day. The remaining five samples were processed 24 h after seeding. Homogeneity testing was performed as described previously. Contractual arrangements for data analysis and evaluation resided with CSL, who calculated seeding densities, homogeneity, participant results, and evaluated the performance of each method. Repeatability and reproducibility were determined by calcu-lating the accordance and concordance values. Sensitivity and specificity values were determined by comparing true/false-positive and -negative results.

VI.C. Preparation, dispatch, and analysis of samples

Webb's lettuce and raspberries were obtained from local retail outlets. Thirty grams of lettuce and 60 g raspberries were used per sample. Samples consisted of 18 unseeded (blank) samples, 27 samples seeded at low density (approx. 10 (oo)cysts), 27 samples seeded at medium density (approx. 50 (oo)cysts) and 27 samples seeded at high density (approx. 100 (oo)cysts). Lettuce samples were sent by standard courier and raspberry samples were sent by specialist courier. Participants were told that samples should be extracted (and analyzed, if possible) within one working day. On Day 1, each participant commenced analysis of the samples according to the SOP and the results were to be reported to CSL no later than five working days after receipt of the samples. Each participant kept a detailed record of sample treatment, extraction, and enumeration, noting particularly any deviation from the SOP.

VI.D. Statistical analysis

Due to large variation in the numbers of (oo)cysts detected by these methods, from duplicate samples analyzed in one laboratory and from other samples in different laboratories, it was not possible to perform conventional statistical analysis for repeatability or reproducibility, which is applicable only to quantitative analyses. Therefore, the qualitative aspects of the methods were analyzed. However, to ensure that the methods being assessed had the same repeatability and reproducibility of quantitative methods, the statistical procedures developed by the European Commission-funded project SMT 4-CT 96-2098 were used. This project aimed to validate the main reference methods (ISO/CEN Standards) in food microbiology by means of interlaboratory trials (eight UK laboratories participated). In particular, these procedures were applied to the validation of the reference method EN ISO 11290-1 for the detection of *L. monocytogenes* in foods (Scotter *et al.*, 2001). With these statistical analyses, accordance and concordance parameters are analogous to repeatability and reproducibility values, respectively.

The accuracy of each method was determined. For seeded samples the parameter was termed 'sensitivity,' defined as the percentage of known positive test materials that were correctly identified as containing (oo)cysts. For unseeded samples the parameter was termed 'specificity,' defined as the percentage of known negative test materials that were correctly identified as negative.

Repeatability and reproducibility were determined by calculating the accordance and concordance values (Langton *et al.*, 2002). Accordance is defined as the percentage chance of finding the same result (i.e. both positive and negative, whether correctly or not) from two identical test materials analyzed in the same laboratory under standard repeatability conditions. In the present trial, the results from all three rounds were combined for this determination, 'identical' being defined as the same seeding level (e.g. 'low' for each method). The probability that two samples gave an identical result was calculated for each laboratory by the formula:

$$\{k(k-1) + (n-k)(n-k-1)\}/n(n-1) \qquad \text{(Eq. 17.1)}$$

where k = number of positive results and n = total number of results. The average of these probabilities from all laboratories was determined to give the accordance for the trial.

Concordance is defined as the percentage chance of finding the same result from two identical test materials analyzed in different laboratories, and calculated by the formula:

$$\{2r(r - nL) + nL(nL - 1) - AnL(n-1)\}/\{(n^{**}2)L(L-1)\}$$
$$\text{(Eq. 17.2)}$$

where r = total number of positives, L = number of laboratories, n = replications per laboratory, and A = accordance, expressed as a proportion.

Standard error values for accordance and concordance were calculated by the 'bootstrap' method of Davison and Hinckley (1997). The concordance odds ratio (COR): COR = accordance (100 − concordance)/ concordance (100 − accordance) was calculated in order to assess the degree of between-laboratory variation in results. This parameter reflects the relative magnitude of the accordance and concordance figures. A COR of 1.00 or less indicates that two samples sent to different laboratories will probably produce the same result as the two samples analyzed by the same laboratory. A COR of greater than 1.00 indicates that variability between laboratories is greater than intra-laboratory variation.

The percentage of *false negatives* (where no (oo)cysts were detected in seeded samples) and *false positives* (where (oo)cysts were reported in unseeded samples) was calculated for each method, for all seeding levels, individual seeding levels, and each laboratory. The latter determination gives an indication of individual laboratory performance. These determinations were performed according to the following:

% false negatives

$$= \frac{\text{number of seeded samples where no (oo)cysts were detected}}{\text{number of seeded samples}} \times 100$$

% false positives

$$= \frac{\text{number of unseeded samples where (oo)cysts were detected}}{\text{number of unseeded samples}} \times 100$$

The final total percentage of false negatives and final total percentage of false positives produced by each method were calculated using data from all rounds and all seeding levels. The percentage recoveries of each method were also calculated. The summary measures for the percentage recoveries were calculated separately for each spiking level in each round, for each spiking level combined over rounds, and for the combined data for each level and each round. Due to the skewed distribution, median and inter-quartile ranges are reported as measures of location and dispersion rather than the mean and standard deviation. However, these latter values were also calculated to provide additional information. The validated method contains a definition of repeatability and reproducibility, and the precision limits that should be expected when the method is used.

VI.E. Enhancing quality assurance criteria

All methods performed at SPDL, with the exception of those developed specifically for this study, were conducted in compliance with SOPs accredited by Clinical Pathology Accreditation (UK) Ltd. and the Drinking Water Inspectorate Regulatory *Cryptosporidium* SOPs. The density of the (oo)cyst seeds was checked for each round. The same commercially produced oocyst seeding stock was used to produce all *C. parvum* seeding samples. The stability of the seeded samples was assessed over a 24 h period. The stability of seeded lettuce samples was satisfactory, with similar recoveries achieved on Day 0 (*C. parvum*, 54.5%; *G. duodenalis*, 50.3%) and Day 1 (*C. parvum*, 64.8%; *G. duodenalis*, 42.2%). Homogeneity results were acceptable.

All participants sent microscope slides from each test to SPDL. Participant results were compared with SPDL results at CSL and deviations were reported by CSL to SPDL. Where results were grossly different from that expected (e.g. 0 counts from seeded samples, and positive counts from blank samples), SPDL examined the slides and enumerated (oo)cysts. This approach was undertaken because of the large number (>500) of samples involved. One laboratory generated consistently higher counts than those recorded at SPDL.

VI.F. Outcome of validation by collaborative trial

VI.F.1. Statistical analysis of the performance of the method developed

VI.F.1.a. **Cryptosporidium/*lettuce* method**   All data from each laboratory were used. The final total percentage of false-negative results produced by this method was 10.4%. The final total percentage of false-positive results produced by this method was 14.6%. The total percentage of false-negative results from low (8.5–14.2 oocysts per 30 g lettuce) seeded samples was 16.7%. The total percentage of false-negative results from medium (53.5–62.6 oocysts per 30 g lettuce) seeded samples was 8.3%. The total percentage of false-negative results from high (111.3–135.0 oocysts per 30 g lettuce) seeded samples was 6.3%.

The accordance and concordance values are not significantly different from each other (inferred from the 95% confidence limits of the CORs) for any seeding level in the *Cryptosporidium*/lettuce method trial. This indicates overall similar performance levels amongst the participating laboratories. The sensitivity values for the seeding levels, individually and in total, are around 90%. The final median percentage recovery of *Cryptosporidium* oocysts from lettuce (30.4%) should allow detection of at least 9–10 oocysts (approximately the lowest infectious dose of *Cryptosporidium*) in a sample of 30 g lettuce, the standard portion size in the United Kingdom (Anonymous., 1993). It is likely that with continued practice, users of the method could achieve higher sensitivities and recoveries. In the originating laboratories, a recovery of *Cryptosporidium* oocysts from sample lettuce of 59.0 g could be achieved in over 30 tests.

Several laboratories reported detection of oocysts in blank lettuce samples; however, when the slides were checked at SPDL, no oocysts were detected on slides. This may be attributed to incorrect identification of oocyst-like bodies as genuine *Cryptosporidium* oocysts; whether this occurred with the seeded samples, leading to over-counting in these laboratories, can not be determined.

VI.G. Lessons learned from validation by collaborative trial

Laboratories experienced in detecting *Cryptosporidium* (and *Giardia*) reported detecting (oo)cysts in blank samples. On re-examination of their slides, we found that validating laboratories incorrectly identified (oo)cysts.

A lack of experience in identifying cysts was manifest in the *Giardia*/lettuce trial (data not shown). The number of false-negative results was high (41.0%) whereas the number of false positives was very low (2.1%). Greater familiarization with (oo)cyst identification should minimize over-counting. Both 'in house' training and participation in accredited external quality assurance schemes should maintain standards. Finally, there remained the requirement for further encouragement for timely return of results.

VII. OVERVIEW OF LESSONS LEARNED

Although other methods for detecting protozoan parasites on produce have been published (Robertson and Gjerde, 2000, 2001a, b), none have been validated. Our experience in developing and validating similar methods indicates that this is a significant oversight as it could lead to the use of a multitude of methods with different recovery efficiencies. This could lead to confusion in understanding the significance of the data, particularly their public health significance. Apart from determining the robustness of methods at every level, validation exercises also consider the ability of external laboratories to perform methods without the inherent bias of laboratories developing methods. The methods we devised were evaluated by interlaboratory collaborative trial. The *Cryptosporidium*/lettuce (and *Cryptosporidium*/raspberry) methods had high sensitivity values, and overall performance was similar amongst the participating laboratories (Cook *et al.*, 2006a, b). High specificity values were obtained for the *Cryptosporidium*/lettuce method, but with the raspberry method some difficulty in discriminating between oocysts and oocyst-like bodies was apparent (data not shown). The *Giardia*/lettuce method did not yield high percentage recoveries, and the specificity values were low. These results should encourage their adoption as analytical tools. Their development provides an important addition to the techniques and procedures available to combat the threat that protozoan parasites pose to the food supply.

Below are the main drivers that enabled us to develop the first validated methods for detecting foodborne protozoa on lettuce and fresh fruits:

- Fully investigate physicochemical and biophysical parameters involved in the method to ensure the correct interpretation of results. Use well-researched (standardized) analytes and sufficient replicates at each stage of development.
- Food matrices contain inhibitory substances in varying quantities, decreasing detection sensitivity. Investigate what is known about each.
- Realize that detection methods work well in laboratories that develop them, but their robustness must be tested in multicenter collaborative trials.

- Investigate and discuss all aspects of collaborative trials before commencing. Ask the experts. Realize potential limitations before attempting validation (e.g. through a pre-validation phase).
- Instigate workshops covering all aspects of the analysis. Determine the competence of collaborators. Ensure that equipment is regularly serviced under contract.
- Include internal positive controls to increase quality assurance and to determine inhibitory effects. This increases the level of confidence obtained from negative results.
- Employ knowledgeable statisticians, and use the most appropriate statistical analytical procedures.
- Make contact with equipment/reagent suppliers who may be interested in being involved in the study and allow plenty of time, particularly when the equipment/reagents are in short supply or expensive.

ACKNOWLEDGMENTS

We thank the Food Standards Agency for funding this project (B09009), all participants involved in the collaborative trials, and the Olympus Optical Company (UK, Microscope Division) for providing microscopes for the workshops.

REFERENCES

Adak, G. K., Long, S. M., & O'Brien, S. J. (2002). Trends in indigenous foodborne disease and deaths, England and Wales: 1992 to 2000. *Gut, 51*, 832–841.

Anonymous.. (1993). *Ministry of Agriculture, Fisheries and Food. Food Portion Sizes*. London: HMSO.

Anonymous (1998). Method 1623, *Cryptosporidium* in water by filtration/IMS/FA. United States Environmental Protection Agency, Office of Water, Washington. Consumer confidence reports final rule. Federal Register 63, 160.

Anonymous.. (1999). *UK Statutory Instruments No. 1524. The Water Supply (Water Quality) (Amendment) Regulations 1999*. London, UK. The Stationery Office, Ltd. 5 pp.

Cacciò, S. M., Thompson, R. C. A., McLauchlin, J., & Smith, H. V. (2005). Unravelling *Cryptosporidium* and *Giardia* epidemiology. *Trends Parasitol., 21*, 430–437.

Campbell, A. T., & Smith, H. V. (1997). Immunomagnetisable separation of *Cryptosporidium parvum* oocysts from water samples. *Water Sci. Technol., 35*, 397–402.

Campbell, A. T., Robertson, L. J., & Smith, H. V. (1992). Viability of *Cryptosporidium parvum* oocysts: Correlation of in vitro excystation with inclusion or exclusion of fluorogenic vital dyes. *Appl. Environ. Microbiol., 58*, 3488–3493.

Cook, N., Paton, C. A., Wilkinson, N., Nichols, R. A. B., Barker, K., & Smith, H.V. (2006a). Towards standard methods for the detection of *Cryptosporidium parvum* on lettuce and raspberries. Part 1: Development and optimization of methods. *Int. J. Food Microbiol., 109*, 215–221.

Cook, N., Paton, C.A., Wilkinson, N., Nichols, R.A.B., Barker, K., & Smith, H.V. (2006b). Towards standard methods for the detection of *Cryptosporidium parvum* on lettuce and raspberries. Part 2: Validation. *Int. J. Food Microbiol., 109*, 222–228.

Davison, A. C., & Hinckley, D. V. (1997). *Bootstrap methods and their application*. Cambridge, UK: Cambridge University Press.

De Oliveira, C. A., & Germano, P. M. (1992). Presence of intestinal parasites in vegetables sold in the metropolitan area of Sao Paulo-SP, Brazil. II – Research on intestinal protozoans. *Rev. Saude. Publica.*, *26*, 332–335.

Drozd, C., & Schwartzbrod, J. (1996). Hydrophobic and electrostatic cell surface properties of *Cryptosporidium parvum*. *App. Environ. Microbiol.*, *62*, 1227–1232.

Engeset, J. (1984). *Optimization of drinking water treatment for G.* lamblia *cyst removal. PhD dissertation*. Seattle, USA: University of Washington.

Girdwood, R. W. A., & Smith, H. V. (1999). *Cryptosporidium*. In R. Robinson, C. Batt, & and P. Patel (Eds.), *Encyclopedia of Food Microbiology* (pp. 487–497). London and New York: Academic Press.

Hunter, P. R., & Thompson, R. C. A. (2005). The zoonotic transmission of *Giardia* and *Cryptosporidium*. *Int J. Parasitol.*, *35*, 1181–1190.

Jaykus, L-A. (1997). Epidemiology and detection as options for control of viral and parasitic foodborne disease. Emerging Infectious Diseases [serial online] 3 (4). Available from: URL: http://www.cdc.gov/ncidod/EID/eid.htm

Korich, D. G., Marshall, M. M., Smith, H. V., O'Grady, J., Bukhari, Z., Fricker, C. R., Rosen, J. P., & Clancy, J. L. (2000). Inter-laboratory comparison of the CD-1 neonatal mouse logistic dose–response model for *Cryptosporidium parvum* oocysts. *J. Euk. Microbiol.*, *47*, 294–298.

Langton, S. D., Chevennement, R., Nagelkerke, N., & Lombard, B. (2002). Analysing collaborative trials for qualitative microbiological methods: accordance and concordance. *Int. J. Food Microbiol.*, *79*, 175–181.

Mead, P. S., Slutsker, L., Dietz, V., McCaig, L. F., Bresee, J. S., Shapiro, C., Griffin, P. M., & Tauxe, R. V. (1999). Food-related illness and death in the United States. *Emerg. Infect. Dis*, *5*, 607–625.

Millar, B. C., Finn, M., Xiao, L., Lowery, C. J., Dooley, J. S. G., & Moore, J. E. (2002). *Cryptosporidium* in foodstuffs – an emerging aetiological route of human foodborne illness. *Trends Food Sci. Technol.*, *13*, 168–187.

Monge, R., & Arias, M. L. (1996). Presence of various pathogenic microorganisms in fresh vegetables in Costa Rica. *Archivo Latinoamericano de Nutrition*, *46*, 292–294.

Nichols, R. A. B., & Smith, H. V. (2002). *Cryptosporidium, Giardia* and *Cyclospora* as foodborne pathogens. In C. Blackburn, & P. McClure (Eds.), *Foodborne Pathogens: Hazards, Risk and Control, Part III, Non-bacterial and emerging foodborne pathogens, Chapter 17, pp. 453–78*. Cambridge, UK: Woodhead Publishing Limited.

Nichols, R. A. B., & Smith, H. V. (2004). Optimisation of DNA extraction and molecular detection of *Cryptosporidium parvum* oocysts in natural mineral water sources. *J. Food Protec.*, *67*, 524–532.

Nichols, R. A. B., Campbell, B., & Smith, H. V. (2006). Molecular fingerprinting of *Cryptosporidium* species oocysts isolated during water monitoring. *Appl. Environ. Microbiol.*, *72*, 5428–5435.

Nkweta, M. A., Smith, P. G., & Smith, H. V. (1989). Unpublished observations on cyst suspensions.

Ongerth, J. E., & Pecoraro, J. P. (1996). Electrophoretic mobility of *Cryptosporidium* oocysts and *Giardia* cysts. *J. Environ. Eng.*, *122*, 228–231.

Ortega, Y. R., Roxas, C. R., Gilman, R. H., Miller, N. J., Cabrera, L., Taquiri, C., & Sterling, C. R. (1997). Isolation of *Cryptosporidium parvum* and *Cyclospora cayetanensis* from vegetables collected in markets of an endemic region in Peru. *Am. J. Trop. Med. Hyg.*, *57*, 683–686.

Paton, C. A., Kelsey, D. E., Reeve, E. A., Punter, K., Crabb, J. H., & Smith, H. V. (2001). Immunomagnetisable separation for the recovery of *Cryptosporidium* sp. oocysts. In S. A. Clark, K. C. Thompson, C. W. Keevil, & and M. S. Smith (Eds.), *Rapid detection assays for food and water* (pp. 38–43). Cambridge, UK: The Royal Society of Chemistry.

Quiroz, E. S., Bern, C., MacArthur, J. R., Xiao, L., Fletcher, M., Arrowood, M. J., Shay, D. K., Levy, M. E., Glass, R. I., & Lal, A. (2000). An outbreak of cryptosporidiosis linked to a foodhandler. *J. Infect. Dis., 181,* 695–700.

Rendtorff, R. C. (1954). The experimental transmission of human intestinal protozoan parasites. II. *Giardia lamblia* cysts given in capsules. *Am. J. Hyg., 59,* 209–220.

Rendtorff, R. C. (1979). The experimental transmission of *Giardia lamblia* among volunteer subjects. In: W. Jakubowski, & J.C. Hoff (eds) Waterborne Transmission of Giardiasis, pp. 64–81. U.S. Environmental Protection Agency. Office of Research and Development, Environmental Research Centre, Cincinnati, Ohio 45268, USA, EPA-600/9-79-001.

Robertson, L. J., & Gjerde, B. (2000). Isolation and enumeration of *Giardia* cysts, *Cryptosporidium* oocysts, and *Ascaris* eggs from fruits and vegetables. *J. Food Protec., 63,* 775–778.

Robertson, L. J., & Gjerde, B. (2001a). Factors affecting recovery efficiency in isolation of *Cryptosporidium* oocysts and *Giardia* cysts from vegetables for standard method development. *J. Food Protec., 64,* 1799-1805.

Robertson, L. J., and Gjerde, B. (2001b). Occurrence of parasites on fruits and vegetables in Norway. *J. Food Protec., 64,* 1793-1798.

Robertson, L. J., Johannessen, G. S., Gjerde, B. K., & Loncarevic, S. (2002). Microbiological analysis of seed sprouts in Norway. *Int. J. Food Microbiol., 75,* 119–126.

Rose, J. B., & Slifko, T. R. (1999). *Giardia,Cryptosporidium,* and *Cyclospora* and their impact upon foods: a review. *J. Food Protec., 62,* 1059–1070.

Scotter, S. L., Langton, S., Lombard, B., Schulten, N., Nagelkerke, N., In't Veld, P. H., Rollier, P., & Lahellec, C. (2001). Validation of ISO method 11290 Part 1 – detection of *Listeria monocytogenes* in foods. *Int. J. Food Microbiol., 64,* 295–306.

Smith, H. V. (2008). Diagnostics, In: R. Fayer, & L. Xiao (eds) *Cryptosporidium* and *cryptosporidiosis.* 2nd Edition, pp. 173–208. Boca Raton, Florida, USA: Taylor and Francis.

Smith, H. V., & Ronald, A. (2002). *Cryptosporidium*: the analytical challenge. In M. Smith, & and K. Thompson (Eds.), Cryptosporidium: *the analytical challenge, Chapter 1* (pp. 1–43). Cambridge, UK: The Royal Society of Chemistry.

Smith, H. V. & Evans, R. (2009). Parasites: *Cryptosporidium, Giardia, Cyclospora, Entamoeba histolytica, Toxoplasma gondii* and pathogenic free-living amoebae (*Acanthamoeba spp.* and *Naegleria fowleri*) as foodborne pathogens. In: Blackburn, C. & McClure, P. (eds) Foodborne Pathogens: hazards, risk analysis and control. Part III Other agents of foodborne disease, 2nd Edition, Chapter 25. Cambridge CB1 6AH, UK: Woodhead Publishing Limited.

Smith, H. V., Campbell, B. M., Paton, C. A., & Nichols, R. A. B. (2002). Significance of enhanced morphological detection of *Cryptosporidium* sp. oocysts in water concentrates using DAPI and immunofluorescence microscopy. *Appl. Environ. Microbiol., 68,* 5198–5201.

Smith, H. V., Cacciò, S. M., Cook, N., Nichols, R. A. B., & Tait, A. (2007). *Cryptosporidium* and *Giardia* as foodborne zoonoses. *Vet. Parasitol., 149,* 29–40.

Xiao, L., Fayer, R., Ryan, U., & Upton, S. J. (2004). *Cryptosporidium* taxonomy: recent advances and implications for public health. *Clin. Microbiol. Rev., 17,* 72–97.

CHAPTER 18

Antimicrobial Activity of Duck Egg Lysozyme Against *Salmonella enteritidis*

Supaporn Naknukool, Shigeru Hayakawa, Takahiro Uno, *and* Masahiro Ogawa

Contents

Abstract

The objective of this study was to investigate the antimicrobial activity of native and reduced forms of duck lysozyme (dLz) against *Salmonella enteritidis*. Moreover, antimicrobial activity of dLz was compared with chicken lysozyme (cLz). Purified dLz was reduced with dithiothreitol. Free SH groups of reduced dLz were trapped with iodoacetamide. Lytic activity against *Micrococcus luteus* of reduced dLz decreased according to a period of reduction time. Whereas, antimicrobial activity of dLz against *S. enteritidis* IFO3133 and other *S. enteritidis* found in contaminated food was enhanced after reduction. The antimicrobial activity of reduced dLz depended on the time of reduction. At 1.5 h reduction, dLz showed the highest antimicrobial activity. Optimum temperature for antimicrobial activity of reduced Lz was 45–49°C. The presence of polyphosphate (0.05–0.10%) enhanced the antimicrobial effect of reduced Lz. On the other hand, addition of NaCl (0.05–0.10 M), glucose (10%), sucrose (10%) and bovine serum albumin (0.5–1.0%) decreased the antimicrobial activity

Global Issues in Food Science and Technology
© 2009 Elsevier Inc.

of reduced Lz. Meanwhile, glycine (0–0.5%) did not have any effect on reduced Lz inhibitory activity. The combination of lactoferrin (0.1 mg/ml) and reduced Lz (0.1 mg/ml) had a synergistic effect. Under all tested conditions, dLz showed higher activity than cLz in both native and reduced forms. The result of this study indicated that reduced dLz tends to be a more efficient antimicrobial agent against *S. enteritidis*.

I. INTRODUCTION

Natural food antimicrobial agents such as lysozyme, nisin, lactoferrin, and lactoperoxidase are widely used in the food industry to inhibit growth of microorganisms, including foodborne pathogens. Lysozyme (Lz) has been known as a basic protein in egg white, containing muramidase activity. It specifically hydrolyzes the 1,4-β-linkages between N-acetyl muramic acid and N-acetylglucosamine in the glycan, which stabilizes cell walls of Gram-positive bacteria. According to this effect, Lz strongly affects Gram-positive bacteria but affects Gram-negative bacteria very little because of the composition of outer membrane. Many researchers attempt to enhance the antimicrobial activity of Lz against Gram-negative bacteria by using chemical or physical treatments to disrupt the bacterial membrane (Vannini *et al.*, 2004), combining with other antimicrobial or/and chemical substances (Facon, 1996; Boland *et al.*, 2003; Branen *et al.*, 2004) and modifying the structure of Lz (Ibrahim *et al.*, 1992; Liu *et al.*, 2000; Mine *et al.*, 2004; Hunter *et al.*, 2005). Recently, reduction of Lz's disulfide bonds, exposing the hydrophobic surface of Lz, was introduced to improve the antimicrobial action of Lz against *Salmonella enteritidis* (Touch *et al.*, 2004).

Most of the reports on antimicrobial activity of lysozyme are obtained from chicken Lz (cLz). Duck Lz (dLz) shows a difference in amino acid sequence from that of cLz. It contains three isoforms (DL-1, DL-2, and DL-3), which are altered according to the displacement of Ser-37 to Gly, Gly-71 to Arg (DL-2 and DL-3), and Pro-79 to Arg (DL-3). The last two displacements induce the conformational change of dLz from β-turn to random coil. Furthermore, dLz reveals enzymatic activity that is 1.34–1.53 times higher than that of cLz (Prager *et al.*, 1971; Hermann *et al.*, 1973; Kondo *et al.*, 1982).

Salmonellosis is recognized as an important pandemic, causing major public health and economic losses worldwide. It is estimated that 1.4 million cases of salmonellosis, including about 400 fatal cases, occur annually in the United States (CDC, 2008). The total cost associated with salmonellosis is about $3 billion each year in the United States (USDA, 2006). In Japan there were 126 salmonellosis cases with 3,603 patients in 2007 (Ministry of Health, Labor and Welfare, Japan, 2007). Salmonellosis is caused by the infection of *Salmonella* spp., which are rod-shape Gram-negative bacteria. Even though there are about 2,500 serotypes known to be a cause of this disease, *S. enteritidis* has been the top serovar to cause salmonellosis in Japan since 1989 (Infectious Agents Surveillance Center, 2006). The incidence of *S. enteritidis* infection has been also found in the United States (Patrick *et al.*, 2004) and European countries (Berghold *et al.*, 2004; Gomez *et al.*, 1997; Danish Zoonosis Centre, 1998). By the year 2010, the FDA hopes to eliminate egg-related *S. enteritidis*, according to The Egg Safety Action Plan. At the same time, *Salmonella*-free carcasses are required by the European Commission. Thus many countries are concerned about the reduction of *S. enteritidis* foodborne disease.

In this chapter, the antimicrobial activity of dLz against *S. enteritidis* is investigated to create an alternative agent for food preservation. In addition, dLz was studied in various conditions to understand the effect of food components on its antimicrobial activity.

II. MATERIALS AND METHODS

Duck eggs were purchased from the Agricultural, Food and Environmental Sciences Research Centre of Osaka Prefecture (Osaka, Japan). Hen egg white lysozyme, crystallized six times, and *Micrococcus luteus* cells were supplied by Seikagaku Kogyo Co. Ltd. (Tokyo, Japan). SP-Sepharose™ Fast Flow, used for dLz purification, was purchased from Amersham Bioscience (New Jersey, USA). Dithiolthreiol (DTT) and iodoacetamide (IAM) were obtained from Nacalai Tesque Inc. (Kyoto, Japan). Unless specified, all other chemicals used were reagent grade.

II.A. Duck lysozyme purification

Duck egg white (20 ml) was mixed gently with 200 ml of sodium acetate buffer (ionic strength 0.05 mol/l, pH 5.0). Precipitate was removed by centrifugation at $6,000\,g$ for 15 min. The supernatant was applied to an SP-Sepharose column (50 ml), equilibrated with sodium acetate buffer (ionic strength 0.05 mol/l, pH 5.0), and then eluted stepwise with 50 ml of sodium acetate buffer (ionic strength 0.05 mol/l, pH 5.0), 150 ml of sodium carbonate buffer (ionic strength 0.05 mol/l, pH 9.0) (CB), 100 ml of 0.3 M NaCl in CB and 100 ml of 0.5 M NaCl in CB. Fractions containing proteins with molecular mass about 14 kDa, as determined by sodium dodecyl sulfate-polyacrylamide gel electrophoresis (SDS-PAGE) using a 15% gel (Laemmli, 1970), were collected and dialyzed using Wako size 18 seamless cellulose tubing (VA, USA) against Mili Q water at 4°C, for three days before freeze-drying for further experiments.

II.B. Reduction of lysozyme

The 1 mg/ml Lz in 10 mM Tris-HCl buffer, pH 8.0, was reduced by 2 mM DTT at 30°C for 0.5–4.0 h and treated with 5 mM IAM at 30°C for

1 h in the dark. Reduced Lz was dialyzed against Mili Q water for 3 days to remove salts and excess reagents before freeze-drying for further use.

II.C. Lytic activity assay

Lytic activity of *Micrococcus luteus* was determined according to a turbidometric method. Dried cells of *M. luteus* were suspended in 55 mM sodium phosphate buffer, pH 6.2, to the concentration of 0.4 mg/ml and stirred at 4°C overnight. The decrease of absorbance at 700 nm, 37°C, was reported after the addition of 2.88 ml of bacterial suspension and 0.12 ml of each reduced/non-reduced dLz solution (1 mg/ml) for 2 min. One unit of lytic activity is the amount of dLz producing a 0.001 decrease in absorbance per min. The activity is expressed in U/mg of protein.

II.D. Determination of antimicrobial activity

The bacteria used in this study were *S. enteritidis* IFO3313, obtained from NITE Biological Resource Centre (Chiba, Japan), and other strains from contaminated food, E991011, E990241, E990925, E990253, and E990579 obtained from National Institute of Infectious Diseases (Tokyo, Japan). They were cultured overnight at 37°C in a medium containing 1% polypeptone, 0.5% yeast extract, 0.3% glucose, 1% NaCl, 0.1% $MgSO_4$-$7H_2O$, and 1.5% agar at pH 7.0. The cells were suspended in 10 mM sodium phosphate buffer, pH 7.2, containing 0.15 mM NaCl before dilution with the same buffer to achieve the concentration of 10^5 CFU/ml as measured by the absorbance at 600 nm.

A bacterial suspension (10^5 CFU/ml) was incubated with reduced/non-reduced Lz at the final concentration of 0.1 mg/ml in 10 mM sodium phosphate buffer, pH 7.2 at 30°C for 1 h. Countable dilutions of bacterial suspensions, diluted with saline water (0.15 M NaCl), and 100 μl were streaked onto desoxycholate-hydrogen sulfide lactose agar (DHL) and incubated at 37°C for 24 h before examining for characteristic colonies. Controls were subjected to the same treatment by using Mili Q water instead of the protein solutions. The antimicrobial activity was expressed as log A_0/A_1 (A_0 = CFU/ml of control; A_1 = CFU/ml of sample). The experiment was done in triplicate.

II.E. The effect of incubation temperature

Reduced Lz 0.1 mg/ml was incubated with 10^5 CFU/ml *S. enteritidis* IFO 3313 in 10 mM sodium phosphate buffer, pH 7.2 at various incubation temperatures (30, 45, 47, and 49°C) for 1 h. Then the antimicrobial activity of reduced Lz was determined.

II.F. The effect of food components and combination with lactoferrin on antimicrobial activity of duck and chicken lysozyme

The effects of 0.1 mg/ml reduced Lz combined with different kinds of food components were investigated. The food components used in this study were 0.05 and 0.1 M NaCl, 5 and 10% glucose, 5 and 10% sucrose, 0.1 and 0.5% glycine, 0.5 and 1.0% bovine serum albumin (BSA), 0.05 and 0.10% polyphosphate (Poly), and 0.1 mg/ml lactoferrin. The mixtures of Lz and food components were incubated with 10^5 CFU/ml *S. enteritidis* IFO 3313 in 10 mM sodium phosphate buffer, pH 7.2, at 30°C, for 1 h, before determining the antimicrobial activity as described above. Moreover, the antimicrobial activity of 0.50% glycine, 0.10% polyphosphate, and 0.2 mg/ml lactoferrin against *S. enteritidis* was also determined in the same conditions as mentioned above.

III. RESULTS AND DISCUSSION

The dLz was obtained using the single-step purification method, with cation exchange chromatography. The elution pattern of proteins is shown in Figure 18.1a. Ovalbumin was eluted in region (a) and ovostatin and ovotransferrin were eluted in region (b). Proposed strong basic proteins, possessing net positive charges at pH 9, were eluted in regions (c) and (d),

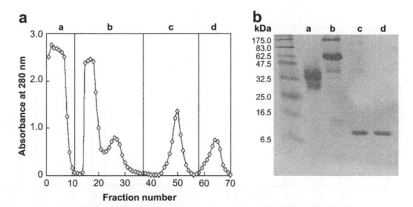

Figure 18.1 Ion exchange chromatogram (a) and SDS-PAGE pattern (b) of duck egg white eluted from SP-Sepharose column. The column was eluted with (a) sodium acetate buffer (ionic strength 0.05 mol/l, pH 5.0), (b) sodium carbonate buffer (ionic strength 0.05 mol/ml, pH 9.0), (c) 0.3 M NaCl in sodium carbonate buffer (ionic strength 0.05 mol/ml, pH 9.0) and (d) 0.5 M NaCl in sodium carbonate buffer (ionic strength 0.05 mol/l, pH 9.0). In SDS-PAGE pattern; lane a, fraction eluted by (a); lane b, fraction eluted by (b); lane c, fraction eluted by (c); lane d, fraction eluted by (d).

after applying the 0.3 M and 0.5 M NaCl in CB. Fractions of two peaks (fraction numbers 45–53 and 60–67) eluted in regions (c) and (d) contained homogeneous basic proteins with similar molecular masses, between 6.5 and 16.5 kDa, corresponding to the position on SDS-PAGE pattern in Figure 18.1b. According to the estimated molecular mass from the SDS-PAGE result, these proteins were suspected to be Lz, a major egg white basic protein having the molecular mass of 14 kDa. In order to clearly identify whether these proteins were lysozyme, the N-terminal amino acid sequence of these proteins was analyzed, resulting in similar four amino acid sequences (KVYS) that were identical to the first four amino acid sequences of dLz. Thus, the proteins that were eluted in regions (c) and (d) were dLz, as confirmed by their basicity, molecular masses, and a part of amino acid sequences. The difference between dLz exposed in region (c) and region (d) of chromatogram may be due to the multiple forms of dLz that have been mentioned in previous works (Prager *et al.*, 1971; Hermann *et al.*, 1973; Kondo *et al.*, 1982). The purified dLz from both regions were used as dLz in following results.

After being reduced by 2 mM DTT and trapped free sulfhydryl groups with IAM, the lytic activity of dLz against *M. luteus* decreased as a function of reduction time (Figure 18.2a). The reduction of SH groups unfolds the structure of native lysozyme postulated to affect the active site, which lies in two domains in its molecules, resulting in the decrease of muramidase activity of Lz (Gilquin *et al.*, 2000). On the other hand, the antimicrobial

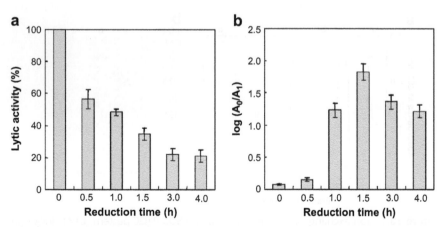

Figure 18.2 Effect of reduced lysozyme over time on (a) percentage of lytic activity of duck lysozyme against *Micrococcus luteus* and (b) antimicrobial activity of duck lysozyme against *Salmonella enteritidis* IFO3313. The experiment was done in triplicate. (A) Lysozyme 0.4 μg/ml; M. luteus 0.4 mg/ml; temperature 37°C; incubation time 2 min. (B) Lysozyme 0.1 mg/ml; *S. enteritidis* 10^5 CFU/ml; temperature 30°C; incubation time 1 h.

activity of dLz against *S. enteritidis* IFO3313 increased, depending on reduction time (Figure 18.2b). At reduction time of 1.5 h, dLz performed the highest antimicrobial activity against *S. enteritidis* IFO3313. In cLz, exposure of the hydrophobic region, which is buried in the interior of the compact lysozyme molecule, was reported after it was reduced by DTT (Touch *et al.*, 2004). Increase of the exposed hydrophobic region is an important factor in binding affinity, which may expand the spectrum of Lz antimicrobial activity against Gram-negative bacteria containing lipopolysaccharide on their outer membrane. However, after being reduced for over 1.5 h, the antimicrobial activity of the reduced Lz decreased slightly. This may be a result of the protein–protein interaction between reduced Lz molecules, occurring via the hydrophobic association that decreased the number of reduced Lz that bind to bacteria cells. As a result, the reduction time of 1.5 h was selected as the optimal condition of reduced Lz in the following experiments.

Comparing the antimicrobial activity of native and reduced Lz (Figure 18.3), the reduced Lz from duck and chicken egg white demonstrated stronger antimicrobial activity against *S. enteritidis* IFO3313 than the native counterpart. The efficiency of reduced Lz against *S. enteritidis* from contaminated food was also investigated (Figure 18.4). The results showed that reduced Lz had wide antimicrobial efficiency against food-contaminating *S. enteritidis*. Moreover, reduced dLz possessed higher antimicrobial action against all tested *S. enteritidis* than reduced cLz.

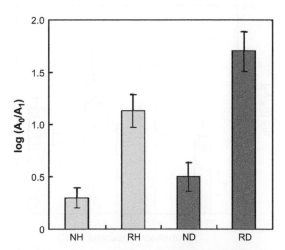

Figure 18.3 Antimicrobial activity of reduced/non-reduced lysozyme against *Salmonella enteritidis* IFO3313. The experiment conditions were reduced lysozyme 0.1 mg/ml; *S. enteritidis* IFO3313 10^5 CFU/ml; temperature 30°C; incubation time 1 h. NH, native chicken lysozyme; RH, reduced chicken lysozyme; ND, native duck lysozyme; RD, reduced duck lysozyme.

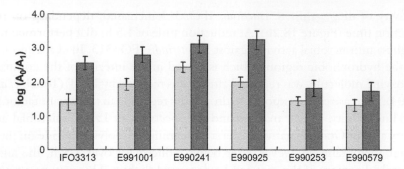

Figure 18.4 Antimicrobial activity of reduced lysozyme against several *Salmonella enteritidis* strain contaminated in food. The antimicrobial activity was measured at reduced lysozyme concentration of 0.1 mg/ml; *S. enteritidis* 10^5 CFU/ml; temperature 30°C; incubation time 1 h. Reduced chicken lysozyme (dark gray); Reduced duck lysozyme (light gray).

Since there are many factors that can affect antimicrobial activity, the effects of incubation temperature and some food components were studied. According to the experiment, incubation at 30°C resulted in the lowest antimicrobial activity of both reduced lysozyme (Figure 18.5). On the other hand, reduced lysozyme from duck and chicken showed the highest antimicrobial activity at 45°C and 47°C, respectively. Since conformational alterations in the outer membrane can take place during heating, the

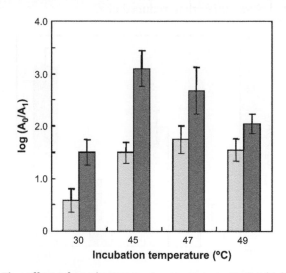

Figure 18.5 The effect of incubation temperature on antimicrobial activity of reduced lysozyme against *Salmonella enteritidis*. The antimicrobial activity was measured at reduced lysozyme concentration of 0.1 mg/ml; *S. enteritidis* IFO3313 10^5 CFU/ml; incubation time 1 h. Reduced chicken lysozyme (dark gray); Reduced duck lysozyme (light gray).

incubation temperature can alter the effect of the antimicrobial agent. Increased incubation temperature induced the bacteria physiology to be less active and the bacterial membrane to be more fluid (Li *et al.*, 2002), increased the permeability of cell membrane (Angersbach *et al.*, 1999, 2002; Russell, 2002), and increased lateral suspension movement of protein within the cell membrane (Ferris *et al.*, 1998), which can enhance the diffusion of reduced Lz into the cells. Meanwhile, not only the increase in cell surface hydrophobicity, reported in heat stress treatment (Nikodo *et al.*, 1985; Boziaris *et al.*, 2001), but also the increase of hydrophobic interaction between reduced Lz and the bacterial membrane occurs when the temperature rises. Therefore, an increase in incubation temperature can enhance the approach of reduced Lz into the bacteria cells, resulting in enhancement of antimicrobial activity. Appropriate incubation temperature was 45–47°C for reduced Lz used as an antimicrobial agent.

Salt or NaCl and sugar are widely used in the food industry as a flavor enhancer and food preservative. The addition of 0.05–0.1 M NaCl decreased the antimicrobial activity of reduced Lz against *S. enteritidis* IFO3313 (Figure 18.6a). The effect of NaCl on other antimicrobial agents was also reported. The bactericidal effect of nisin against *L. monocytogenes* is the highest at low (near 0%) or high NaCl concentrations (near 6%), while between 2 and 4% NaCl, the activity of nisin is low (Bouttefroy *et al.*, 2000). The bactericidal activity of heat–treated Lz (at 80°C for 20 min, pH 6.0), called HL80/6, against both *E. coli* and *S. aureus* is abolished by addition of 0.1% and 1% NaCl (Ibrahim *et al.*, 1996). The decrease in antimicrobial activity of reduced Lz may be due to the effect of monovalent cations (Na^+), which bind to the negatively charged head groups of the phospholipids in the cytoplasm membrane and negative charge of outer membrane as previously reported for divalent cations binding with cell membranes of Gram-positive bacteria (Abee, 1995), resulting in the decrease of reduced Lz binding area. Binding of Na^+ to the lipid ester oxygen on the outer membrane leads to compaction of the lipid (Kandasamy *et al.*, 2006), causing difficulty in diffusion of reduced Lz into bacteria cells. Moreover, Na^+ ions may stimulate the biological process at the membrane by increasing the rate of membrane synthesis to repair the damage caused by reduced Lz (Ibrahim *et al.*, 1996). The difference in the effect of NaCl may be caused by the differences between antimicrobial agents and types of bacteria.

At concentrations of 5 and 10%, respectively, glucose and sucrose slightly decreased the antimicrobial activity of reduced Lz (Figures 18.6b and 18.6c). Moreover, sucrose, it was demonstrated, had the ability to decrease the effect of HL80/6 on *E. coli* when the concentration of sucrose reached 1.5%, as also reported in a previous study (Ibrahim *et al.*, 1996). The

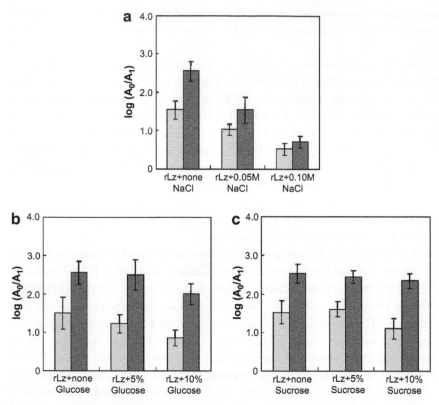

Figure 18.6 The effect of NaCl (a), glucose (b) and sucrose (c) on antimicrobial activity of reduced lysozyme against *Salmonella enteritidis*. The measurement was performed at reduced lysozyme 0.1 mg/ml; *S. enteritidis* IFO3313 10^5 CFU/ml; temperature 30°C; incubation time 1 h. rLz, reduced lysozyme; chicken lysozyme (dark gray); duck lysozyme (light gray).

addition of glucose and sucrose not only increases the foundational nutrient necessary for the growth of bacteria, but also increases the viscosity of the solution; resulting in an obstacle to the adhesion of reduced Lz to the bacterial membrane. Therefore, the presence of glucose and sucrose can decrease the antimicrobial activity of reduced Lz.

Glycine and D-alanine are known as the essential components of bacterial cells, providing an easily accessible approach into the cells, making them interested in being carrier molecules (Mishra *et al.*, 2005). Results showed that glycine alone possesses no antimicrobial activity against *S. enteritidis* (Figure 18.7a) and does not affect the antimicrobial activity of reduced Lz. Non-sufficiency of glycine at concentrations up to 0.4% in the viability of either *S. aureus* or *E. coli* was also reported (Ibrahim *et al.*, 1996). However, the effect of combining HL80/6 and glycine has a synergistic

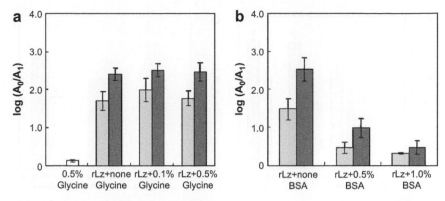

Figure 18.7 The antimicrobial activity of reduced lysozyme against *Salmonella enteritidis* in the presence of glycine (a) and BSA (b). The measurement was performed at reduced lysozyme 0.1 mg/ml; *S. enteritidis* IFO3313 10^5 CFU/ml; temperature 30°C; incubation time 1 h. rLz, reduced lysozyme; chicken lysozyme (dark gray); duck lysozyme (light gray).

effect on both bacteria. This combination was sufficient in low-salt conditions.

The addition of BSA affected the decrease in antimicrobial activity of reduced Lz as shown in Figure 18.7b. As the incubation pH is 7.2, BSA (pI = 4.6) performs as a negatively charged protein that can bind to a positive charge of Lz (pI = 10.7). This electrostatic interaction may prevent reduced Lz from binding to the bacterial membrane, resulting in the reduction of antimicrobial sufficiency of reduced Lz. Furthermore, BSA is a good source of nutrition that bacteria can use for recovery during incubation.

Even though polyphosphate alone had a small effect on *S. enteritidis* (Figure 18.8a), the combination with polyphosphate improved the antimicrobial activity of reduced Lz. Polyphosphate, which is widely used as a food additive in meat products and processing, is known as a chelating compound. It forms a metal–chelate ring by its oxygen atoms with metal ions. The chelation of structurally essential metal ions, Ca^{2+} and Mg^{2+}, in the cell walls and cytochrome system, repression of enzyme synthesis, inhibition of enzyme activity, and changes in the water activity of the media increase the antimicrobial activity by improving the penetration property of antimicrobial compounds (Cutter *et al.*, 1995; Vareltzis *et al.*, 1997; Boland *et al.*, 2003). Thus, the disruption of polyphosphate to bacterial cells can increase the activity of reduced Lz against *S. enteritidis*.

Lactoferrin, a glycoprotein, is found in milk and other exocrine secretions. It possesses an iron-binding activity and thereby sequestrates iron,

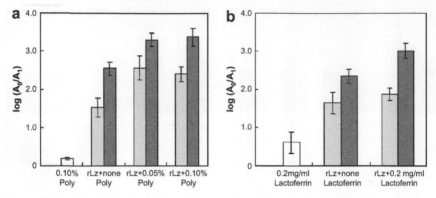

Figure 18.8 The antimicrobial activity of reduced lysozyme against *Salmonella enteritidis* in the presence of polyphosphate (poly) (a) and lactoferrin (b). The measurement was performed at reduced lysozyme 0.1 mg/ml; *S. enteritidis* IFO3313 10^5 CFU/ml; temperature 30°C; incubation time 1 h. rLz, reduced lysozyme; chicken lysozyme (dark gray); duck lysozyme (light gray).

which is an essential nutrient for the growth of bacteria (Arnold *et al.*, 1977). Not only that, the ability to release LPS from the outer membrane of Gram-negative bacteria was reported (Ellison *et al.*, 1991). The LPS-binding site is located on its N-terminal. Lactoferrin showed a synergistic effect in antimicrobial activity against *V. cholera*, *E. coli*, *S. typhimurium*, and *S. enteritidis* when combined with native cLz (Elass-Rochard *et al.*, 1995). It also showed a synergistic effect when used connately with reduced Lz as showed in this study.

As observed, the antimicrobial activity of dLz in native and reduced forms, in all conditions of this experiment, showed higher activity against *S. enteritidis* than that of cLz. It may be due to the differences in amino acid composition between these two Lz. Compared with cLz, the numbers of arginine, valine, and tyrosine residue in dLz were higher by a few residues, while those of aspartic acid, alanine, and phenylalanine residues were lower by 1–2 residues (Kondo, 1982). It was found that dLz contains 13–15 arginine residues, while only 11 residues are present in cLz (Prager *et al.*, 1971; Hermann *et al.*, 1973; Kondo *et al.*, 1982). Arginine plays an important role in the antimicrobial agent. The cationic charges of arginine provide an effective means of attracting with negatively charged surfaces such as LPS, teichoic acid, or phosphatidyl glycerol phospholipids head group. Moreover, arginine can form a complex with tryptophans via cation–π interaction, which is favorable to penetration into a lipid bilayer (Jing *et al.*, 2003). The difference in arginine content may be responsible for the high antimicrobial activity of dLz.

IV. CONCLUSION

The high antimicrobial activity of reduced Lz against *S. enteritidis* was reported in this study. It is supposed that reduced dLz could be the alternative antimicrobial agent that the food industry could supply to customers concerned about natural foods, as it proposes higher activity than reduced cLz. Also, this study investigated that food components of ingredients such as NaCl, sucrose, glucose, and BSA have antagonistic effects in antimicrobial activity of reduced Lz, while polyphosphate and lactoferrin promoted the efficiency of reduced lysozyme. In addition, glycine did not have much effect. For further applications, studies investigating the antimicrobial effect of reduced dLz against other bacteria and the effect of other food components on its antimicrobial activity are required.

REFERENCES

Abee, T. (1995). Pore-forming bacteriocins of gram-positive bacteria and self protection mechanism of produces organisms. *FEMS Microbiol. Lett., 129*, 1–10.

Angersbach, A., Heinz, V., & Knorr, D. (1999). Electrophysiological model of intact and processed plant tissues: cell disintegration criteria. *Biotechnol. Prog., 15*(4), 753–762.

Angersbach, A., Heinz, V., & Knorr, D. (2002). Evaluation of process-induced dimensional changes in the membrane structure of biological cells using impedance measurement. *Biotechnol. Prog., 18*(3), 597–603.

Arnold, R. R., Cole, R. M., & McGhee, J. R. (1977). A bactericidal effect for human lactoferrin. *Science* (Washington, DC), *197*, 263–265.

Berghold, C., Kornschober, C., Lederer, I., & Allerberger, F. (2004). Occurrence of *Salmonella enteritidis* phage type 29 in Austria: an opportunity to assess the relevance of chicken meat as source of human salmonella infections. *Euro. Surveill., 9*(10), 31–34.

Boland, J. S., Davidson, P. M., & Weiss, J. (2003). Enhanced inhibition of *Escherichia coli* 0157: H7 by lysozyme and chelators. *J. Food Prot., 66*(10), 1783–1789.

Bouttefroy, A., Mansour, M., Linder, M., & Melliere, J. (2000). Inhibitory combinations of nisin, sodium chloride, and pH on *Listeria monocytogenes* ATCC 15313 in broth by an experimental design approach. *Int. J. Food Microbiol., 54*, 109–115.

Boziaris, I. S., & Adams, M. R. (2001). Temperature shock injury and transient sensitivity to nisin in gram-negatives. *J. Appl. Microbiol., 91*, 715–724.

Branen, J. K., & Davidson, P. M. (2004). Enhancement of nisin, lysozyme and monolaurin antimicrobial activities by ethylenediaminetraacetic acid and lactoferrin. *Int. J. Food Microbiol., 90*, 63–74.

Centers for Disease Control and Prevention (CDC) (2008). Salmonellosis: Technical information. Centers for Disease Control and Prevention, GA (http://www.cdc.gov/nczved/dfbmd/disease_listing/salmonellosis_ti.html)

Cutter, C. N., & Siragusa, G. R. (1995). Population reduction of gram negative pathogens following treatments with nisin and chelators under various conditions. *J. Food Prot., 58*, 977–983.

Danish Zoonosis Centre (1998). Annual report on zoonoses in Denmark, 1997. Danish Zoonosis Centre, Copenhagen (http://zoonyt.dzc.dk/annualreport1997/ index.html)

Elass-Rochard, E., Roseanu, A., Legrand, D., Trif, M., Salmon, V., Motas, C., Montreuil, J., & Spik, G. (1995). Lactoferrin–lipopolysaccharide interaction: involvement of the 28–34 loop region of human lactoferrin in the high affinity binding to *Escherichia coli* 055B5 lipopolysaccharide. *Biochem. J., 312*, 839–845.

Ellison, R. T., & Giehl, T. J. (1991). Killing of gram-negative bacteria by lactoferrin and lysozyme. *J. Clin. Invest., 88*, 1080–1091.

Facon, M. J., & Skura, B. (1996). Antibacterial activity of lactoferricin, lysozyme and EDTA against *Salmonella enteritidis*. *Int. Dairy J., 6*, 303–313.

Food and Drug Administration (2004). Healthy people 2010 focus area data progress review, Focus area 10: food safety. (http://www.foodsafety.gov/~dms/hp2010.html)

Gilquin, B., Guilbert, C., & Perahia, D. (2000). Unfold of hen egg lysozyme by molecular dynamics simulations at 300K: insight into the role of the interdomain interface. *Proteins, 41*, 58–74.

Gomez, L., Motarjemi, Y., Miyagawa, S., Kaferstein, F. K., & Stohr, K. (1997). Foodborne salmonellosis. *World Health Stat. Q., 50*, 81–89.

Hermann, J., Jolles, J., & Jolles, P. (1973). *The disulfide bridges of duck egg-white lysozyme II.* *Arch. Biochem. Biophys., 158*, 355–358.

Hunter, H. J., Jing, W., Shibli, D., Trinh, T., Park, I. Y., Kim, S. C., & Vogel, H. J. (2005). The interactions of antimicrobial peptides derived from lysozyme with model membrane systems. *Biochim. Biophys. Acta., 1668*, 175–189.

Ibrahim, H. R., & Kato, A. (1992). Design of amphipathic lysozyme using chemical and genetic modification to achieve optimal food functionality and diverse antimicrobial action. In K. D. Schwenke, & R. Mothes (Eds.), *Food Proteins Structure and Functionality* (pp. 16–28). New York: VCH Publishers.

Ibrahim, H. R., Higashiguchi, S., Sugimoto, Y., & Aoki, T. (1996). Antimicrobial synergism of partially-denatured lysozyme with glycine: effect of sucrose and sodium chloride. *Food Res. Int., 29*(8), 771–777.

Infectious Agents Surveillance Center.. (2006). Salmonella in Japan as of June 2006. *Infectious Agents Surveillance Report, 27*(8), 191–192.

Jing, W., Demcoe, A. R., & Vogel, H. J. (2003). Conformation of a bactericidal domain of puroindoline: a structure and mechanism of action of a 13-residue antimicrobial peptide. *J. Bacteriol., 185*, 4938–4947.

Kandasamy, S. K., & Larson, R. G. (2006). Effect of salt on the interaction of antimicrobial peptides with zwitterionic lipid bilayer. *Biochim. Biophys. Acta., 1758*(9), 1274–1284.

Kondo, K., Fujio, H., & Amano, T. (1982). Chemical and immunological properties and amino acid sequences of three lysozymes from Peking-Duck egg white. *J. Biochem., 91*, 571–587.

Laemmli, U. K. (1970). Cleavage of structural proteins during the assembly of the head of bacteriophage T4. *Nature, 157*, 105–132.

Li, J., Chikindas, M. L., Ludescher, R. D., & Montcille, T. J. (2002). Temperature and surfactant induced membrane modifications that alter *Listeria monocytogenes* nisin sensitivity by different mechanisms. *Appl. Environ Microbiol., 68*, 5904–5910.

Liu, S. T., Sugimoto, T., Azakami, H., & Kato, A. (2000). Lipophilization of lysozyme by short and middle chain fatty acid. *J. Agric. Food Chem., 48*, 265–269.

Mine, Y., Ma, F., & Lauriau, S. (2004). Antimicrobial peptides released by enzymatic hydrolysis of hen egg white lysozyme. *J. Agric. Food Chem., 52*, 1088–1094.

Ministry of Health, Labour and Welfare (2007). Number of accidents and patients caused by food-borne disease in Japan, 2007 (Translated from Japanese). Ministry of Health, Labour and Welfare, Japan (http://www.mhlw.go.jp/topics/syokuchu/index.html)

Mishra, S., Narain, U., Mishra, R., & Misra, K. (2005). Design, development and synthesis of mixed bioconjugates of piperic acid-glycine, curcumin-glycine/alanine and curcumin-glycine-piperic acid and their antibacterial and antifungal properties. *Bioorgan. Med. Chem., 13*(5), 1477–1486.

Nikodo, H., & Vaara, M. (1985). Molecular basis of bacteria outer membrane permeability. *Microbiol. Rev., 49*, 1–32.

Patrick, M. E., Adcock, P. M., Gomez, T. M., Altekruse, S. F., Holland, B. H., Tauxe, R. V., & Swerdlow, D. L. (2004). *Salmonella enteritidis* infections, United States, 1985–1999. *Emerg. Infect. Dis., 10*(1), 1–7.

Prager, E. M., & Wilson, A. C. (1971). Multiple lysozymes of duck egg white. *J. Biol. Chem., 246*(2), 523–530.

Russell, N. J. (1997). Bacterial membranes: the effects of chill storage and food processing. An overview. *Int. J. Food Microbiol., 79*, 27–34.

Touch, V., Hayakawa, S., & Saitoh, K. (2004). Relationships between conformational changes and antimicrobial activity of lysozyme upon reduction of its disulfide bonds. *Food Chem., 84*, 421–428.

United States Department of Agriculture (USDA) (2006). Foodborne illness cost calculator: Salmonella. United Stated Department of Agriculture, Economic Research Service, DC (http://www.ers.usda.gov/data/foodborneillness/salm_Intro.asp)

Vannini, L., Lanciotti, R., Baldi, D., & Guerzoni, M. E. (2004). Interactions between high pressure homogenization and antimicrobial activity of lysozyme and lactoperoxidase. *Int. J. Food Microbiol., 94*, 123–135.

Vareltzis, K., Soultos, N., Koidis, P., Ambrosiadis, J., & Genigeorgis, C. (1997). Antimicrobial effects of sodium tripolyphosphate against bacteria attached to the surface of chicken carcasses. Lebensm.-Wiss. u. -*Technol., 30*, 665–669.

CHAPTER *19*

High-Pressure Homogenization for Food Sanitization

Francesco Donsì, Giovanna Ferrari, *and* **Paola Maresca**

Contents

Abstract

Homogenization has been extensively used by the dairy and food industry, especially to stabilize food emulsions and to disrupt lipid globules in liquid food. The enhancement of homogenization pressure up to 350 MPa, recently performed through the set up of a new generation of apparatus, is likely to open up new areas of application for the food industry, permitting the production of finer emulsions and the modification not only of lipid globules but also of some food constituents. Recently, great interest was also raised by the utilization of homogenization processes to inactivate pathogenic and other food spoilage microorganisms and to promote specific modifications on milk structure, suited for the

Global Issues in Food Science and Technology
© 2009 Elsevier Inc.

production of yogurt and cheese. However, the potentiality of high-pressure homogenization in the food industry is not yet fully expressed. The state of the art of high-pressure homogenization is reviewed in this chapter with special focus on food technology. Moreover, specific applications for the treatment of a wide range of microorganisms in liquid substrates are analyzed, as well as the role of the main physical and operating parameters regulating microbial inactivation. Finally, the different hypotheses about the mechanisms of inactivation are discussed.

I. INTRODUCTION

The term mechanical homogenization refers to the capability of producing a homogeneous size distribution of particles suspended in a liquid, by forcing the liquid under the effect of high pressure through a specifically designed disruption valve.

Homogenization for the stabilization of food and dairy emulsions was patented by Auguste Gaulin and presented to the public in 1900 at the Paris World's Fair. In the following years, the homogenizer became the vital unit of the dairy processing industry, for the disruption and size control of milk fat globules (Geciova *et al.*, 2002). Nowadays, pressures up to 50 MPa are used to produce food emulsions with improved texture, taste, flavor, and improved shelf life characteristics, especially for dairy products like milk, cream and ice cream, or baby food formulas, requiring a shelf life up to 6–12 months (Paquin, 1999).

High-pressure homogenization (HPH), also known as dynamic high-pressure homogenization, was recently introduced through the development of a new generation of homogenizers, capable of attaining pressures 10–15 times higher than traditional machines. HPH hence opened up new possibilities in research and product developments, as summarized in Table 19.1, ranging from the food to chemical, cosmetic, biotech, and pharmaceutical industries. Depending on the purpose of the HPH treatment, two main areas can be identified. The first area is mainly concerned with the physical changes induced in HPH-processed products, such as the reduction of size and narrowing of size distribution of particles, droplets, or micelles in suspensions or emulsions, for the preparation or stabilization of emulsions, or preparation of nanoparticles and nanosuspensions, or for attaining viscosity and texture changes.

The second area is focused on the effect of cell disruption induced by HPH, which can be either applied for recovery of intracellular material in the biotech and pharmaceutical industry, or applied to reduce the microbial load in food and pharmaceutical products (Popper and Knorr, 1990; Lanciotti *et al.*, 1994; Lanciotti *et al.*, 1996; Diels *et al.*, 2004).

Table 19.1   Fields of application of high-pressure homogenization

	Non-food	**Food**
Physical modification	Emulsions with controlled droplet size Drug nanoparticles Drug nanosuspensions	Stabilization of food emulsions Texture modification (i.e. yogurt, cheese)
Cell disruption	Recovery of intracellular products (i.e. *E. coli, S. cerevisiae, B. subtilis*)	Reduction of microorganism load in food, while preserving qualitative attributes of the fresh product (i.e. milk, fruit juices)

In this review, the state-of-the-art of HPH, intended as homogenization at pressures higher than 100 MPa, is presented, with special focus on food technology. In particular, a detailed description of the equipment used and their working principles is given. Specific applications are described for the treatment of a wide range of microorganisms in liquid substrates, with a deeper insight into the main physical and operating parameters regulating microbial inactivation. Finally, different hypotheses about the mechanisms of inactivation are presented and discussed.

II. HOMOGENIZATION TECHNIQUES

In the early 1980s a new technology was introduced for production of fine emulsions based on the availability of devices able to generate and manage very high pressures in liquids above 100 MPa and as high as 300–500 MPa, as well as on a new homogenization chamber design. Different manufacturers of high-pressure homogenizers exist, producing either prototype or industrial scale equipment, such as Microfluidics (Microfluidics™, 2006), Stansted Fluid Power (Stansted Fluid Power LTD, 2006), AVP (Invensys™ AVP), Avestin (Avestin®, 2006), and Niro Soavi (Niro Soavi S.p.A., 2006). Figure 19.1 reports the schematics of a high-pressure homogenizer of the Stansted type (Stansted Fluid Power LTD, 2006) that is equipped with two homogenizer valves to indicate pressure and temperature of the process fluid.

Any basic homogenizer design consists of a high-pressure generator, such as a positive-displacement pump coupled with a pressure intensifier that forces

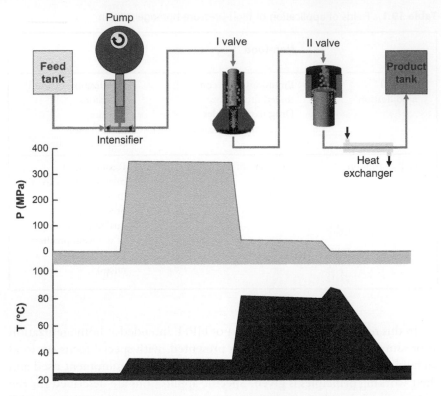

Figure 19.1 Schematic representation of a Stansted high-pressure homogenizer with a double homogenizer valve.

the process fluid through a specifically designed homogenizer valve assembly (Keshavarz-Moore *et al.*, 1990; Middelberg, 1995; Kleinig and Middelberg, 1998; Stang *et al.*, 2001; Floury *et al.*, 2004a, b), as shown in Figure 19.1.

The rapid pressurization of the process fluid (up to 350 MPa) causes a temperature increase on the order of 3°C/100 MPa (Knorr, 1999), while the instantaneous pressure drop occurring in the homogenizer valve causes more substantial heat (15–20°C/100 MPa). A secondary homogenizer valve, where a significantly lower pressure drop takes place, may be placed next to the main valve to disrupt the agglomerates eventually formed in the first stage (Briñez *et al.*, 2006a, b, c; Pereda *et al.*, 2006). Since final temperature may be high, depending on inlet temperature and pressure level of operation, a rapid cooling of the process fluid represents a good practice to preserve the thermolabile components of the product treated (Figure 19.1).

Remarkably, in dynamic high-pressure operations the fluid is exposed to high pressures for very short times (1–10 s). In this period of time, the fluid has to flow from the intensifier to the disruption valve. Hence, it is understandable that disruption is mainly determined by the passage of process fluid under high pressure through an adjustable, restricted orifice discharge valve, rather than by high-pressure exposition. In any type of homogenizer valve, the processed liquid passes under high pressure through a convergent section called the homogenization gap and then expands (Floury et al., 2004a). Pressure is controlled through an actuator, which adjusts the force on the valve (Middelberg, 1995).

The major parameters determining efficiency are operating pressure, number of passes through the valve, temperature, and homogenizer valve design (Keshavarz-Moore et al., 1990; Stang et al., 2001), while the scale of operation does not appear to influence the extent of homogenization (Siddiqi et al., 1997). It is hence understandable that the main differences between homogenizers from different manufacturers are in the homogenizer valves. Three main types of basic homogenizer valve designs can be identified, as represented in Figure 19.2.

The Manton–Gaulin APV (USA) homogenizer type valve (Figure 19.2a), which is one of the most widely used today (Middelberg, 1995; Stang et al., 2001), found application in the biopharmaceutical industry for cell breakage to release intracellular products for downstream capture and purification (Lander et al., 2000; Bury et al., 2001; Geciova et al., 2002; Miller et al., 2002), for the preparation of nanosuspensions (Muller et al., 1998; Muschwitzer et al., 2004) and solid–lipid nanoparticles, (zur Muhlen et al., 1998; Souto et al., 2005), and for citrus juice pasteurization (Clark et al., 1993).

In the Manton–Gaulin APV homogenizer, a positive displacement pump delivers the process fluid at a constant volumetric flow rate through a homogenizer valve (Shirgaonkar et al., 1998; Miller et al., 2002) consisting of a narrow gap between the valve piston and seat (Jarchau et al., 1993; Kinney et al., 1999; Jarchau et al., 2001). The fluid flows as a radial jet that impinges on an impact ring, as shown in Figure 19.2a (Miller et al., 2002; Floury et al., 2004a). Reducing the gap space increases the inlet homogenizer pressure upstream of the valve (Miller et al., 2002).

The Emulsiflex models, developed by Avestin (Canada), are fitted with a valve whose design resembles a conventional homogenization flat-bead valve (Figure 19.2a). Interestingly, by means of a micrometric adjustment of the gap, and of the use of materials highly resistant to wear such as stainless steel or ceramic, the Emulsiflex homogenizers are able to reach pressures as high as 300 MPa (Paquin, 1999; Paquin et al., 2003). The use of the Emulsiflex homogenizer is reported for disruption of cell suspensions at pressures ranging between 100 and 300 MPa (Wuytack et al., 2002; Diels et al., 2004, 2005a), for the production of nanosuspensions (Friedrich and Müller-Goymann, 2003; Hecq et al., 2005) and protein denaturation (Paquin et al., 2003).

Niro Soavi (Italy) also developed a high-pressure homogenizer of the type schematized in Figure 19.2a, which can cover a wide range of specific

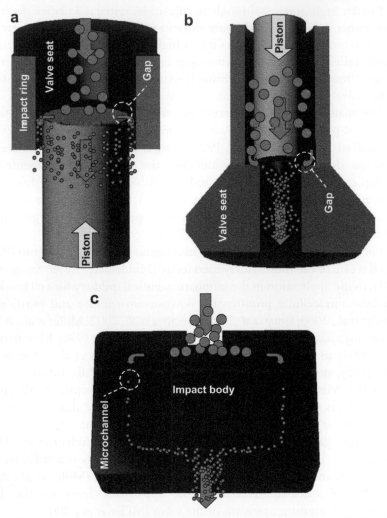

Figure 19.2 Schematics of homogenizer valves from: (a) Manton–Gaulin APV (Stevenson and Chen, 1997); (b) Stansted Fluid Power (Floury *et al.*, 2004a, 2004b; (c) Microfluidics (Stang *et al.*, 2001).

process requirements in terms of pressure and flow rate, with dependence on the specific properties of the fluids to be treated, by changing the homogenizer valve (Gandini *et al.*, 1999). The Niro Soavi equipment was used for the preparation of nanosuspensions (Trotta *et al.*, 2001) and papain extracts (Verduyn *et al.*, 1999), and for microbial inactivation (Fantin *et al.*, 1996; Vannini *et al.*, 2004).

Recently, Stansted Fluid Power Ltd introduced a significant improvement to the valve design, allowing by means of the increase of fluid pressure level up to 350 MPa, the potential use of sterilizing products in situ and modifying the texture of emulsions or biopolymers for food or pharmaceutical applications (Floury et al., 2004a). The flow through the valve in this homogenizer is reversed with respect to the classical valve design: the fluid is fed axially under high pressure along the mobile part of the valve, which is made of high-pressure-resistant ceramics and flows with high velocity through the radial narrow gap formed between the valve seat and the piston, before leaving the valve seat at atmospheric pressure (Figure 19.2b). The size of the gap and the resulting stream velocity and pressure of the liquid ahead of the valve depend on the force acting on the valve piston, which can be adjusted to regulate the homogenization intensity (Thiebaud et al., 2003; Floury et al., 2004a). The pressures (350 MPa) reached by the Stansted type homogenizer are higher than in the APV-Gaulin one (70–100 MPa), because valve gaps are much narrower (2–5 μm vs. 10–30 μm) (Floury et al., 2004a).

The Microfluidizer® (Microfluidics Corp., USA), patented in 1985 (Cook and Lagacé, 1985) and introduced in the food industry by Paquin (1999), operates on a principle different from that of the other high-pressure valve homogenizer. The process fluid is divided into two microchannels and then recombined in a reacting chamber where jets of liquid collide, dissipating the energy input almost instantaneously at the point of impact (Figure 19.2c). As the reaction chamber of the microfluidizer is static and contains no moving parts, the limiting aspect of this technology is the pressure delivered by the equipment, which is linked to the flow of the liquid and the equipment design (Middelberg, 1995; Paquin, 1999; Stang et al., 2001; Geciova et al., 2002). On the other hand, the fixed geometry of the interaction chamber ensures a high level of reproducibility.

The Microfluidizer® was used to disrupt cells at high pressure (typically 150–200 MPa), with the claimed advantage of leading to larger particle sizes after disruption with respect to the Manton–Gaulin APV homogenizer, resulting in more effective separation during the following centrifugation step (Geciova et al., 2002).

The Microfluidizer® was also used to improve the functionality of heat-denatured whey proteins (Iordache and Jelen, 2003), to reach submicron size of nanoparticles, as well as increasing the grade of monodispersity compared to previous preparation techniques (Lamprecht et al., 1999), and for the treatment of ice cream mix (Feijoo et al., 1997).

III. APPLICATIONS OF HIGH-PRESSURE HOMOGENIZATION

III.A. Background

HPH is nowadays widely applied for the physical modification of non-food and food products and for the cell disruption of yeasts or bacteria in the pharmaceutical sector and in biotechnology to recover intracellular products such as proteins and enzymes (Paquin, 1999; Floury *et al.*, 2000; Geciova *et al.*, 2002; Diels *et al.*, 2004; Sandra and Dalgleish, 2005), as summarized in Table 19.1.

Emulsions, dispersions of two or more liquid insoluble phases, are employed for detergents and cleaning agents, lubricants or agricultural protection agents and also for foods (e.g. milk and mayonnaise). In order to produce emulsions, as the two liquids are immiscible and the emulsions quickly tend to separate, the disperse phase must be distributed in finely divided form throughout the continuous phase (Paquin, 1999; Stang *et al.*, 2001). Among the numerous types of continuous emulsifying devices available, only high-pressure homogenizers with a high energy input and suitable nozzle geometry can generate sub-micron emulsions, i.e. with a droplet diameter significantly lower than 1 μm and a narrow droplet size distribution (Desrumaux and Marcand, 2002; Floury *et al.*, 2002). In the pharmaceutical field, HPH is widely applied to produce nanoparticles (50–1000 nm) as drug carrier systems as well as for the dispersions of drug nanocrystals (<1000 nm) in water, or alternatively, in mixtures of water with water-miscible liquids or non-aqueous media (Moschwitzer *et al.*, 2004). In these applications the HPH process improves the transport properties of the drug, which control absorption, metabolism, and elimination, distribution of other tissues and solubility (Mehnert and Mäder, 2001), and avoids the use of organic solvents (zur Muhlen *et al.*, 1998; Lamprecht *et al.*, 1999; Mehnert and Mäder, 2001; Friedrich and Müller-Goymann, 2003; Hecq *et al.*, 2005). HPH has been shown to be one of the most effective ways of mechanically disrupting microbial cell walls (Shamlou *et al.*, 1995; Kleinig and Middelberg, 1996; Wong *et al.*, 1997) for large-scale industrial recovery of intracellular products. Microorganisms typically employed are *Escherichia coli*, *Saccharomyces cerevisiae*, and *Bacillus subtilis* (Middelberg, 1995), even though lactic acid bacteria are also being increasingly exploited for use in production of microbial bioproducts such as enzymes for food industry applications (Geciova *et al.*, 2002). The application of HPH on food products is described in detail in the following section.

III.B. Food applications
III.B.1. Dairy products
Homogenization has been extensively used in the dairy industry since the middle of the last century, especially for the stabilization of food emulsions and the disruption of lipid globules. The recently introduced increase in the highest operating pressure made evident the applicability of this technique

not only in the production of increasingly finer emulsions and in the modification of either lipid globules or food constituents (Paquin, 1999), but also in the achievement of a partial inactivation of the endogenous microbial population by mechanical disruption of cells. In particular, HPH was proposed as a novel integrated processing milk technique, which combines many advantages of conventional homogenization and pasteurization of milk in a single process (Hayes *et al.*, 2005). It is hence understandable that many investigations focused on HPH processing of milk and milk-based products.

III.B.1.a. Milk   Intensive research was recently focused on both the size reduction of fat globules and casein micelles for preventing creaming and coalescence during long shelf-storage, as well as in texturization of milk derivatives, and in the non-thermal sanitization of milk to extend shelf life while preserving heat-labile compounds and avoiding undesirable effects, such as off-flavors, non-enzymatic browning and denaturation of certain vitamins and proteins.

Kelly and co-workers (Hayes and Kelly, 2003a, b; Hayes *et al.*, 2005) investigated size reduction of fat globules, comparing milk processed by conventional means (18 MPa, two-stage) and high-pressure homogenization (50–200 MPa). They showed that HPH yielded significantly smaller fat globules than conventional homogenization, with fat globule size decreasing at increasing HPH pressure (Hayes and Kelly, 2003a, b; Hayes *et al.*, 2005). Nevertheless, Thiebaud *et al.* (2003) showed that the use of high-pressure homogenization in milk processing led to a disadvantageous reduction of the fat globule size. In particular, they found that while increasing the pressure up to 250 MPa both the mean Sauter diameter $d_{3,2}$ and mean diameter $d_{4,3}$ decreased; for a homogenization pressure of 300 MPa an inversion in trend was observed, mainly due to the aggregation of fat globules in clusters. Similar results were observed by Guamis and co-workers (Serra *et al.*, 2007; Zamora *et al.*, 2007), who reported the broadening of the size distribution of the fat globules at high-pressure homogenization (above 300 MPa) in a similar machine, due to aggregation of small fat globules within dense proteinaceous structures. In contrast to Thiebaud *et al.* (2003), who observed this phenomenon as a consequence of single-stage homogenization at 300 MPa, Guamis and co-workers reported aggregation of fat globules only when milk was treated by double-stage homogenization (300 MPa in the first valve and 30 MPa in the second), due to the action of the second valve on the small fragments of fat globules produced in the first valve (Serra *et al.*, 2007; Zamora *et al.*, 2007). In fact, multiple steps at intermediate pressure (e.g. 200 MPa) appeared appropriate to achieve the finer particle size distribution (Thiebaud *et al.*, 2003; Zamora *et al.*, 2007).

The mechanical forces, determining the disruption of the fat globules in the homogenizer valve, act also on the microorganisms present in milk, causing their inactivation. Numerous authors hence investigated the inactivation of the microorganisms responsible for the main foodborne diseases from contaminated milk and dairy products, such as *Salmonella* spp., *Listeria monocytogenes*, *Staphylococcus aureus*, and *Escherichia coli* O157:H7.

The inactivation of the total endogenous flora present in milk strongly depended on pressure and temperature of HPH. Kelly and co-workers (Hayes *et al.*, 2005; Smiddy *et al.*, 2007) reported a reduction of total bacterial count by HPH (Stansted Fluid Power) at 250 MPa and 45°C of only ~3 log units (Hayes *et al.*, 2005), which increased to ~5 log units when increasing preheat temperature at 55 or 70°C (Smiddy *et al.*, 2007), indicating the importance of temperature in HPH-induced inactivation of bacteria. Thiebaud *et al.* (2003) reported a reduction of total bacterial count of ~1 log cycle for HPH treatment (Stansted Fluid Power) at 200 MPa and 24°C, which increased to ~2 log cycles with increasing pressure up to 300 MPa or in applying an additional homogenization step at 200 MPa. About 2 log-cycle reductions were also achieved by Kheadr *et al.* (2002) for 5 HPH steps (Emulsiflex) at 200 MPa at 28°C. A reduction between 3 and 4 log cycles was instead reported by Pereda *et al.* (2006) for homogenization (Stansted Fluid Power two-stage homogenizer) between 200 and 300 MPa and inlet temperature between 30°C and 40°C.

Psychrotrophic bacteria, which are of major concern in pasteurized milk due to storage under refrigerated conditions, are quite sensitive to HPH. For a HPH treatment (Niro Soavi), a 2 log-cycle inactivation was reported in goat milk (Guerzoni *et al.*, 1999), while between 3 and 4 log cycles of inactivation were observed by different authors (Stansted Fluid Power machines) at pressures between 200 and 300 MPa (Thiebaud *et al.*, 2003; Hayes *et al.*, 2005; Pereda *et al.*, 2006; Smiddy *et al.*, 2007).

Lactococci, lactobacilli, and enterococci are quite sensitive to HPH treatment at high pressures (200–300 MPa) and temperatures (30–40°C), exhibiting a reduction of up to 99.99% of the initial load (Pereda *et al.*, 2006), while a less effective reduction (80%) was observed under milder HPH conditions (Saboya *et al.*, 2003).

Coliform contamination was reduced to undetectable levels regardless of temperature and pressure of HPH, both in goat (Guerzoni *et al.*, 1999) and cow milk (Hayes *et al.*, 2005; Pereda *et al.*, 2006; Smiddy *et al.*, 2007).

Inactivation of *Escherichia coli* was studied both in skim (Diels *et al.*, 2005a) and in whole milk (Vachon *et al.*, 2002). Higher inactivation was detected in whole milk (~8 log units vs. ~5 log units at 300 MPa at 25°C) due to the increase of piezosensitivity of microorganisms induced by fat, as speculated by Briñez *et al.* (2006b). In particular, Briñez *et al.* (2006b) observed that the population of E. *coli* slightly decreased during 9 days of storage in refrigerated conditions (4°C).

In contrast, *Listeria innocua*, which was reduced by ~4 log units with HPH treatment in a two-stage machine (Stansted Fluid Power) at 300 MPa in the first stage and 30 MPa in the second stage, at 20°C inlet temperature, exhibited an increase of 2 log units during 9 days of storage at 4°C (Briñez *et al.*, 2006a) and 3 log units after 3 days at 37°C (Vannini *et al.*, 2004). High inactivation levels of *L. innocua* (~4 log-cycles) were reported by Kheadr *et al.* (2002) at 200 MPa and 5 homogenization steps, and of *L. monocytogenes* by Vachon *et al.* (2002) at 300 MPa and 3 cycles at 45°C.

Vachon *et al.* (2002) also investigated the inactivation of *Salmonella enterica* serotype Enteritidis in cow milk, reporting complete inactivation (~8 log units) after 5 homogenization cycles at 200 MPa. Later, *Pseudomonas fluorescens*, inoculated in milk (~10^6 CFU/mL), was reduced to undetectable levels by HPH at 200 MPa and 10°C (Hayes *et al.*, 2005). Recently, Smiddy *et al.* (2007) observed that HPH treatments (Stansted Fluid Power) at pressures between 200 and 250 MPa and temperatures between 55 and 70°C reduced the number of total bacteria, psychrotrops, pseudomonads, coliforms, *Staphylococcus aureus*, and lactobacilli to non-detectable levels. Nevertheless, high counts of total bacteria, psychrotrophs, and pseudomonads were observed after short storage periods at 5°C (~10^7 CFU/mL of storage in all samples), demonstrating therefore that HPH treatments may be scarcely effective for preservation of milk, in comparison to pasteurization, which assures a total bacterial count lower than ~10^5 CFU/mL after 14 days at 5°C (Smiddy *et al.*, 2007).

III.B.1.b. Cheese   The possibility of affecting the particle size distribution of fat globules, protein structure, and enzyme activity, in addition to significantly reducing the bacterial load in milk and its evolution during ripening, stimulated numerous studies on cheeses made from HPH-treated milk. Guerzoni and co-workers investigated the production of different kinds of cheese, such as Crescenza (Lanciotti *et al.*, 2004a), goat cheese (Guerzoni *et al.*, 1999), and Caciotta cheese (Lanciotti *et al.*, 2006) from milk treated at 100 MPa with a Niro Soavi homogenizer, highlighting the positive effect in terms of the residual microbial load (reduced with respect to untreated milk) and the chemical-physical characteristics of the products, especially in comparison with those produced from raw or heat-treated milk. Kheadr *et al.* (2002) investigated the effect of a 5-cycle HPH treatment at 200 MPa (Emulsiflex c160 homogenizer) of milk in the production of Cheddar cheese, analyzing the microbiological, textural, and microstructural properties of the material. Tunick *et al.* (2000) focused on the microstructural properties of Mozzarella cheese from milk treated with a Microfluidizer at pressures ranging from 34 to 172 MPa. Guamis and co-workers (López-Pedemonte *et al.*, 2006; Zamora *et al.*, 2007) investigated, using a Stansted Power Fluid homogenizer, both the inactivation of *Staphylococcus aureus* and its evolution during ripening in a soft-curd cheese obtained from milk treated by two-stage homogenization at 330 MPa in the first valve and 30 MPa in the second valve (López-Pedemonte *et al.*, 2006). They also investigated the cheese-making properties of milk treated at pressures varying between 100 and 330 MPa for single- and two-stage homogenization, in terms of particle size, rennet coagulation properties, yield and moisture content, whey composition, and gel microstructure (Zamora *et al.*, 2007).

Cheeses made from pressurized milk produced a yield (ratio between mass of cheese or curd obtained and milk used) that ranged from slightly (Kheadr *et al.*, 2002; Lanciotti *et al.*, 2004) to significantly higher (Guerzoni *et al.*, 1999; Lanciotti *et al.*, 2006; Zamora *et al.*, 2007) than those made from pasteurized milk, due to the increase in water-binding capacity of proteins and higher retention of the curd of whey proteins induced by the HPH treatment. Coherently, the moisture content of pressurized cheeses was remarkably higher (Kheadr *et al.*, 2002; Zamora *et al.*, 2007), since the rapid pressurization–depressurization sequence is believed to enhance the water-binding capacity of the caseins during milk treatment (Lanciotti *et al.*, 2004; Vannini *et al.*, 2004).

In addition, the HPH treatment was able to produce cheeses with different rheological and microstructural properties in comparison with those obtained from unprocessed raw materials. Guerzoni and co-workers reported that cheeses obtained from milk treated by HPH at 100 MPa exhibited firmness and textural qualities not observed in conventional dairy products (Guerzoni *et al.*, 1999; Lanciotti *et al.*, 2006), which can be attributed to the reduced fat particle sizes, below 0.8 μm (Kent *et al.*, 2003). The textural analysis of Cheddar cheese produced from milk treated with five homogenization cycles at 200 MPa confirmed that the cheese was more firm, elastic, and cohesive than cheese made from pasteurized milk, due to the more compact protein matrix, in which large numbers of very small fat globules were emulsified (Kheadr *et al.*, 2002).

Scanning electron microscopy and confocal laser scanning microscopy confirmed that HPH cheeses exhibited a more homogeneous structure (Guerzoni *et al.*, 1999), with smaller fat globules and casein micelles at increasing pressure (Tunick *et al.*, 2000; Kheadr *et al.*, 2002) and differences in the interaction of the proteinaceous matrix with the smaller fat globules (Zamora *et al.*, 2007).

Other effects of the HPH treatment were identified in the lipolytic and proteolytic activities during ripening (Guerzoni *et al.*, 1999), with consequences in the evolution of the aroma profile (Lanciotti *et al.*, 2006), in the rennet coagulation properties, and in whey composition (Zamora *et al.*, 2007). The sensory assessment of the organoleptic properties reckoned a higher overall grade to HPH cheese (Lanciotti *et al.*, 2006), suggesting that dynamic high pressure can be used to develop a variety of dairy products with different functional properties (Kheadr *et al.*, 2002; Lanciotti *et al.*, 2004).

HPH can also be used as an alternative or in combination with thermal treatments to reduce the microbial load of milk and during cheese ripening. López-Pedemonte *et al.* (2006) reported a reduction of *Staphylococcus aureus* inoculated in milk (7.3 log CFU/ml) of 2 and 4 orders of magnitude depending on treatment temperature (respectively, 6 and 20°C). In addition, during ripening the *S. aureus* population was decreased in HPH cheese, and was further reduced when the cheese received an additional high-hydrostatic-pressure treatment (400 MPa). *Listeria innocua*, inoculated at 10^5–10^6 CFU/ml was reduced 3 to 4 log cycles upon HPH treatment at 200 MPa, while the endogenous flora (total microbial count) was reduced 4 log cycles for skim milk and 2 log cycles for whole milk, due to the protective role exerted by the fat globules; the total bacterial count in cheeses made with fresh and 3-month-old pasteurized and pressurized milk cheeses did not differ markedly (Kheadr *et al.*, 2002). Also Guerzoni and co-workers reported a substantial reduction of the endogenous flora in cheese, with a 1 log-cycle reduction for yeast and lactobacilli (Lanciotti *et al.*, 2006) and 2 log-cycle reductions for psychrotropic aerobic cells (Guerzoni *et al.*, 1999).

III.B.1.c. Yogurt   Lanciotti *et al.* (2004b) investigated the effect of HPH treatment of milk at moderate pressures (30–75 MPa, using a Niro Soavi homogenizer) for yogurt production, under microbiological and rheological profiles. HPH yogurts exhibited a greater variety of textures associated with a high microbiological quality, showing an increase of yogurt viscosity with increasing pressure, which is possibly related to the modification of the fat globule and casein micelle size and their functional properties without detrimental effects on shelf life and safety (Lanciotti *et al.*, 2004b).

Milk treatment at higher pressures (200–300 MPa) with a Stansted Fluid Power homogenizer according to Serra *et al.* (2007) is suitable for yogurt production, yielding reduced syneresis and increased firmness, which is due to the modifications induced by HPH, including protein denaturation, protein–protein interactions, and fat–protein interactions. This enhanced the interactions between particles and the formation of a stable gel network, which strongly retained water, contributing to increased firmness and reduced deformability.

III.B.1.d. Milk derivatives Reconstituted skim milk powder was homogenized at different pressures (40–190 MPa) in up to six HPH cycles; revealing that the higher the pressure and number of passes employed, the smaller the average diameter of casein micelles. As a consequence, the use of processing conditions with a more pronounced impact on the product and subsequent effect on the amount of non-sedimentable caseins in the serum was discussed. Since HPH was able to modify the structural properties of casein micelles, the sequence of treatments was an important factor in determining the nature of the modifications (Sandra and Dalgleish, 2005). Fermented milk used as a food additive produced (with homogenization up to 80 MPa) better chemical-physical properties such as viscosity, particle size, total °Brix value, and amino acid content with respect to ordinary fermented milk, and also had a beneficial effect on human absorption (Gu and Zhang, 2005). The HPH technique was favorably tested in the preparation of a homogeneous whole soybean milk type product and whole soybean curd, producing better texture and higher yield than conventional techniques (Kyun and Hoon, 2005), and was further beneficial in the production of fermented products and desserts from milk-based emulsions (Schorsch, 2004).

HPH treatment could also be useful in superfine micronization of ice cream mixes for obtaining a reduced particle size in the product (Hongyan, 2005); while increasing the palatability of the ice cream compared to low pressure homogenization (below 15 MPa), which is effective in reducing ice crystal size (Ranyan and Baer, 2005), the HPH technique also contributes significantly to microbial inactivation. Feijoo *et al.* (1997) reported up to 70% inactivation of *Bacillus licheniformis* spores in an ice cream mix with the combination of mild preheating up to 50°C and homogenization up to 200 MPa in a Microfluidizer[®].

III.B.2. Fruit juices

Fruit juices represent an interesting perspective on the application of HPH for achieving the reduction of microbial load and preserving the quality attributes of fresh products, as well as affecting the physical properties, such as viscosity. Homogenization of orange juice to reduce the pulp mean particle size and the viscosity of orange juice concentrate is a well known technique (Stipp and Tsai, 1988; Grant, 1989), consisting of exposing the orange juice to high shear in a high-pressure homogenizer, at pressures up to 50 MPa. The high-pressure differential between the inlet of the homogenizer valve and the outlet results in high shear and cavitation, which may alter the size and/or the properties of the suspended/insoluble pulp in the juice.

Nevertheless, the most interesting applications are in the field of microbial inactivation. In 1993 a patent was issued for processing of fresh squeezed juices by HPH at pressure levels around 100 MPa. The homogenized juice exhibited increased shelf life and decreased microbiological activity compared to juice homogenized under conventional pressures, despite the fact that no pasteurization step was conducted. In addition, juice processed by HPH maintained good flavor and palatability for 40 days at a storage temperature of 4°C, while untreated juice deteriorated in 20 days (Clark et al., 1993).

Fliss and co-workers (Tahiri et al., 2006; Lacroix et al., 2005) focused their research on orange juice, evaluating the potential of high-pressure homogenization technology to inactivate pathogenic and spoilage microflora (Tahiri et al., 2006), and studying the effect of HPH alone or in combination with pre-warming on pectin methylesterase (PME) activity, which is known to cause the loss of opalescence of orange juice during storage (Lacroix et al., 2005). Results revealed that upon a multi-cycle treatment (5 cycles) of HPH (Emulsiflex C5) at 200 MPa and 25°C, *Lactobacillus plantarum* decreased by 2.3 log units, *Leuconostoc mesenteroides* by 1.6 log units, *Saccharomyces cerevisiae* by 2.5 log units, *Penicillium ssp.* by 4 log units, and *Escherichia coli* by 6 log units (5 log units after 3 cycles of HPH) (Tahiri et al., 2006).

HPH treatment of orange juice at 170 MPa for five cycles (Emulsiflex C50) decreased pectin methylesterase activity only slightly; however, HPH-treated juices were significantly more stable, probably due to modifications of the structure of pectin, making the substrate less available to PME, and also due to the reduction of pulp particle size. In addition, with respect to pasteurization, HPH contributes to preservation of the freshness and texture attributes of the juice (Lacroix et al., 2005).

Maresca and co-workers investigated the effect of multi-cycle homogenization (Stansted Fluid Power), up to 250 MPa on orange, apple and pineapple juices, evaluating the microbial inactivation and quality loss in treated products. Interestingly, HPH appears effective in extending the shelf life of fruit juices, keeping their organoleptic attributes over 28 days of storage at 4°C (Maresca et al., 2006; Donsì et al., 2006). Three cycles of homogenization at 150 MPa and ambient temperature represented a good balance between inactivation efficiency and preservation of the organoleptic properties. The total microbial count of unprocessed juice exhibited a significant microbial growth after 7 days of storage, in contrast to HPH treated juice, which exhibited a lower increase (always lower than 10^2 CFU/ml) after 14 days, while the changes of pH, Brix%, Vitamin C values, and color attributes were negligible during storage (Donsì et al., 2006).

Briñez et al. (2006a and 2006c) investigated the inactivation after a two-stage HPH treatment (300 MPa in first valve, 30 MPa in second valve; inlet temperature ranging between 6 and 20°C) of inoculated strains of *Escherichia coli* and *Listeria innocua* in orange juice and the evolution of survival over up to 33 days of storage at 4°C. The treatment reduced *E. coli* from 7 log CFU/ml to ~4 log CFU/ml and this level remained stable for 18 days and then decreased by ~3 log units in the remaining 15 days (Briñez *et al.*, 2006c). Analogously, *L. innocua*, which was initially reduced by ~3 log units, constantly decreased during the 21 days of storage, by 2 additional log units (Briñez *et al.*, 2006a).

III.B.3. Nanosized food additives

In food emulsions, stability is usually achieved by adding proteins as the main stabilizer. Proteins and low-molecular-weight emulsifiers lower the surface tension between the interfaces formed during the emulsification process, and form a macromolecular layer surrounding the dispersed particles, reducing the rate of coalescence (Stang *et al.*, 2001; Perrier-Cornet *et al.*, 2005). HPH treatments have the potential of inducing molecular changes in the agents with emulsifying and stabilizing properties, as shown by Corredig and Wicker (2001) for pectins, which are widely used in the industry as texture enhancers and stabilizers, and by Floury *et al.* (2002) for methylcellulose. In particular, the functional properties of whey proteins were improved by HPH treatment up to 300 MPa, by means of the dissociation of large proteins, without affecting protein solubility, which determined better foaming and stabilizing properties (Bouaouina *et al.*, 2006; Paquin *et al.*, 2003). For instance, HPH treatment of xanthan gum, which is a hydrocolloid widely used in food industries for its thickening and stabilizing properties, affected the functionality of xanthan solutions through the decrease of the molecular weight, in terms of hydration rate, with a decrease in water uptake, and in terms of rheological properties, with a reduction of the consistency index (Lagoueyete and Paquin, 1998).

The improvement of stability and the reduction of coalescence during storage of additives and ingredients for beverages and other types of foods were achieved through nanoemulsification and nanodispersion by means of HPH, both for vitamin E additives for beverage fortification, microencapsulated with a food-approval starch at a size of around 100 nm (Chen and Wagner, 2004), and for β-carotene nanodispersions, as active ingredient for food formulations (Tan *et al.*, 2005).

The size reduction resulting from HPH treatment was reported to significantly improve solubility of dried whey products (Iordache and Jelen, 2003) and that of microemulsions of spices, and also to promote the dispersibility, permeability and flavoring ability, and to change the compatibility of spices with food (Yipeng and Binghua, 2005).

IV. EFFECT OF HIGH-PRESSURE HOMOGENIZATION ON MICROORGANISMS

Several parameters have been shown to affect microbial inactivation in HPH processes. Table 19.2 summarizes and classifies the main investigations reported in the literature in terms of microbial strain, type of equipment employed, medium composition, pressure and temperature of homogenization, and corresponding level of inactivation (with precision reported by the authors), specifying eventual multiple homogenization steps and addition of antimicrobial agents.

The large variability of microbial inactivation, which can be observed from Table 19.2 not only among different microorganisms but also within the same class of microorganisms, is determined by the combined effects of the parameters, which will be discussed in the following sections.

IV.A. Microbial strain

The capability of high-pressure homogenization to inactivate microorganisms has been already reckoned by some authors, more than 15 years ago (Popper and Knorr, 1990; Lanciotti et al., 1996; Guerzoni et al., 1999, 2002). Since then, numerous microbial strains have been exposed to high-pressure homogenization, in order to evaluate either the sensitivity of the microorganism to treatment, or its efficiency in sanitization processes.

Several studies indicated that Gram-negative bacteria are more sensitive to high-pressure homogenization than Gram-positive bacteria (Vachon et al., 2002; Wuytack et al., 2002). This suggests a correlation between cell wall structure and high-pressure resistance, which indicates that high-pressure homogenization kills vegetative bacteria mainly through mechanical destruction of the cell integrity; this is caused by the spatial pressure and velocity gradients, turbulence, impingement (Engler and Robinson, 1981; Keshavarz-Moore et al., 1990), and/or cavitation (Save et al., 1994; Save et al., 1997; Shirgaonkar et al., 1998), which occur in liquids during high-pressure homogenization. Indeed, Gram-negative bacteria, characterized by a thinner cell wall membrane, are more sensitive to high-pressure homogenization than Gram-positive bacteria (Vachon et al., 2002; Wuytack et al., 2002). The wall structure of yeast and fungi is thicker than in Gram-positive bacteria, but due to the larger size and different cell wall structure, with glucans, mannans, and proteins as basic structural components (Geciova et al., 2002), the exhibited resistance is intermediate between Gram-negative and Gram-positive bacteria (Lanciotti et al., 2006; Maresca et al., 2006; Tahiri et al., 2006), as can be observed in Table 19.2.

Bacterial spores appear to be only moderately affected by HPH, with reduction lower than 1 log unit at 200 MPa, and in combination with

Table 19.2 Microbial strains exposed to high-pressure homogenization

Microbial strain	Apparatus	Medium	Pressure (MPa)	Inlet temp. (°C)	log(N_0/N)	Ref.
Gram-positive bacteria						
Bacillus subtilis	APV Gaulin Micron Lab 40	Buffer	100	37	3 3 (chitosan)	Popper and Knorr, 1990
	Niro Soavi Panda	Skim cow milk	130	n.a.	2.9 3.8 (lysozyme)	Vannini et al., 2004
Enterococcus faecalis	Niro Soavi Panda	Goat milk	100	n.a.	3	Guerzoni et al., 1999
	Emulsiflex C5	PBS	300	25	1	Wuytack et al., 2002
	Stansted DRG FPG7400H350	Whole cow milk	300	30–40	5	Pereda et al., 2006
Lactobacillus	Stansted FPG7400H	Whole cow milk	200 300	24	1 2	Thiebaud et al., 2003
	Niro Soavi Panda	Cow milk (Caciotta cheese)	100	n.a.	1	Lanciotti et al., 2006
	Stansted DRG FPG7400H350	Whole cow milk	300	30–40	4	Pereda et al., 2006
	Stansted nm-GEN 7400H	Whole cow milk	250	55–70	3.8	Smiddy et al., 2007
Lactobacillus delbrueckii	Mini-Lab Rannie, Rannie Lab 2000, APV	MRS broth	200	8	n.a.	Bury et al., 2001
	Stansted FPG 7420A.275	Water	150	2	2 (5 passes)	Donsì et al., 2006

Organism	Equipment	Medium	Pressure (MPa)	Temp (°C)	Log reduction	Reference
Lactobacillus helveticus	Stansted Fluid Power	Water	200	15	2	Saboya et al., 2003
Lactobacillus plantarum	Emulsiflex C5	PBS	300	25	0.5	Wuytack et al., 2002
	Emulsiflex C5	PBS	200	25	3.5	Tahiri et al., 2006
		Orange juice			6 (5 passes)	
					0.2	
					2.3 (5 passes)	
Lactococcus lactis	APV Gaulin Micron Lab 40	Buffer	100	37	1	Popper and Knorr, 1990
	Stansted Fluid Power	Water	200	15	1	Saboya et al., 2003
Leuconostoc dextranicum	Emulsiflex C5	PBS	300	25	2	Wuytack et al., 2002
Leuconostoc mesenteroides	Emulsiflex C5	PBS	200	25	1	Tahiri et al., 2006
		Orange juice			6 (5 passes)	
					0.2	
					1.6 (5 passes)	
Listeria innocua	Emulsiflex C160	Whole cow milk (Cheddar cheese)	200	28	3.2	Kheadr et al., 2002
		Skim cow milk (Cheddar cheese)			3.2	
	Emulsiflex C5	PBS	300	25	1	Wuytack et al., 2002
	Stansted DRG FPG7400H350	Whole milk	300 (1st valve)	6	3.2	Briñez et al., 2006a
			30 (2nd valve)	20	4.3	
		Orange juice		6	3.1	
				20	3.6	

(continued)

Table 19.2 Microbial strains exposed to high-pressure homogenization—Cont'd

Microbial strain	Apparatus	Medium	Pressure (MPa)	Inlet temp. (°C)	$\log(N_0/N)$	Ref.
Listeria monocytogenes	Emulsiflex C160	PBS Whole cow milk	300	25	8.3 (3 passes) 4 (3 passes) 5.8 (5 passes)	Vachon et al., 2002
	Niro Soavi Panda	Skim cow milk	130	n.a.	1.0 0.8 (lysozyme) 3.0 (lacto-peroxidase)	Vannini et al., 2004
Propionibacterium freudenreichii	Stansted Fluid Power	Water	200	15	0.7	Saboya et al., 2003
Staphylococcus aureus	Emulsiflex C5	PBS	300	25	0.5	Wuytack et al., 2002
	Emulsiflex C5	PBS	300	5–50	3	Diels et al., 2003
	Niro Soavi Panda	Skim cow milk	130	n.a.	1.9 2.0 (lysozyme)	Vannini et al., 2004
	Stansted DRG FPG7400H350	Whole cow milk	300 (1st valve) 30 (2nd valve)	6°C, 20°C	2.1	López-Pedemonte et al., 2006
	Stansted nm-GEN 7400H	Whole cow milk	250	55–70	3.6 3	Smiddy et al., 2007
Streptococcus lactis	APV Gaulin Micron Lab 40	Buffer	100	37	1.5 1.5 (lysozyme)	Popper and Knorr, 1990

Gram-negative bacteria

Coliforms	Stansted DRG FPG7400H350	Whole cow milk	300	30–40	5	Pereda et al., 2006
	Stansted nm-GEN 7400H	Whole cow milk	250	55–70	4.1	Smiddy et al., 2007
Escherichia coli	Microfluidizer M110T	Buffer	95	4	1	Sauer et al., 1989
	APV Gaulin Micron Lab 40	Buffer	100	37	2 (5 passes)	Popper and Knorr, 1990
	15MR APV Gaulin	Buffer	55	10	3	Wong et al., 1997
	Emulsiflex C160	PBS	300	25	6	Vachon et al., 2002
		Whole cow milk			8.3 (3 passes)	
					8.3	
	Emulsiflex C5	PBS	300	25	4.5	Wuytack et al., 2002
	Emulsiflex C5	PBS	300	5–50	7	Diels et al., 2004
	Niro Soavi Panda	Skim cow milk	130	n.a.	0.3	Vannini et al., 2004
					0.3 (lysozyme)	
					4.0 (lacto-peroxidase)	
	Emulsiflex C5	PBS	300	25	5	Diels et al., 2005a
		PBS with PEG (5 times higher viscosity)			2	
		Skim cow milk			3.5	

(continued)

Table 19.2 Microbial strains exposed to high-pressure homogenization—Cont'd

Microbial strain	Apparatus	Medium	Pressure (MPa)	Inlet temp. (°C)	log(N_0/N)	Ref.
	Emulsiflex C160	Soy milk			3.0	
		Milk drink			2.0	
		PBS	300	25	4.5	Diels et al., 2005b
					6.5 (lysozyme)	
					5.5 (nisin)	
	Stansted DRG FPG7400H350	Whole cow milk	300 (1st valve) 30 (2nd valve)	6	3.6	Briñez et al., 2006b
				20	4.3	
		Skim cow milk		6	3.0	
				20	3.5	
	Stansted DRG FPG7400H350	Orange juice	300 (1st valve) 30 (2nd valve)	6	3.4	Briñez et al., 2006c
				20	3.9	
	Stansted FPG 7420A.275	Water	250	2	7	Donsì et al., 2006
	Emulsiflex C5	PBS	200	25	2.2	Tahiri et al., 2006
					6.0 (5 passes)	
		Orange juice			2.0	
					6.0 (5 passes)	
Proteus vulgaris	Niro Soavi Panda	Skim cow milk	130	n.a.	1.8	Vannini et al., 2004
					2.3 (lysozyme)	
Pseudomonas fluorescens	Emulsiflex C5	PBS	300	25	6.5	Wuytack et al., 2002

Microorganism	Apparatus	Medium	Pressure	Temperature	Log reduction	Reference
Pseudomonads	Stansted nm-GEN 7400H	Whole cow milk	250	55–70	4.9	Smiddy et al., 2007
Pseudomonas putida	Constant Cell Disruption Systems, I.K.S. BV	Buffer	200	10	0.5	Van Hee et al., 2004
	Niro Soavi Panda	Skim cow milk	130	n.a.	2.4 / 2.7 (lysozyme)	Vannini et al., 2004
Salmonella enterica Thyphimurium	Emulsiflex C5	PBS	300	25	3	Wuytack et al., 2002
Salmonella enteridis	Niro Soavi 'PA 'NS	BHI	140	n.a.	2.4	Guerzoni et al., 2002
	Emulsiflex C160	PBS	200	25	4 / 8 (3 passes)	Vachon et al., 2002
	Niro Soavi Panda	Skim cow milk	130	n.a.	1.8 / 2.6 (lysozyme)	Vannini et al., 2004
Shigella flexneri	Emulsiflex C5	PBS	300	25	5.5	Wuytack et al., 2002
Yersinia enterocolitica	Niro Soavi Panda	PBS	130	–	3	Lanciotti et al., 1994
	Emulsiflex C5	PBS	300	25	2.5	Wuytack et al., 2002
	Emulsiflex C5	PBS	300	5–50	6	Diels et al., 2003
Spores						
Bacillus licheniformis	Microfluidizer	Ice cream mix	200	33–50	<1	Feijoo et al., 1997

(continued)

Table 19.2 Microbial strains exposed to high-pressure homogenization—Cont'd

Microbial strain	Apparatus	Medium	Pressure (MPa)	Inlet temp. (°C)	log(N_0/N)	Ref.
Yeasts						
Saccharomyces cerevisiae	Manton Gaulin Micon LAB40, APV	n.a.	50	n.a.	1	Shamlou et al., 1995
	Manton LAB40, APV Manton Gaulin LAB60, Manton Gaulin K3	NaOH buffer	160	n.a.	2	Siddiqi et al., 1997
	Niro Soavi Panda	Whole cow milk (Caciotta cheese)	100	n.a.	1	Lanciotti et al., 2006
	Stansted FPG 7420A.275	Water	200	25	5	Maresca et al., 2006
		Orange, apple, pineapple juices			5	
	Emulsiflex C5	PBS	200	25	0.5	Tahiri et al., 2006
					2.2 (5 passes)	
		Orange juice			0.9	
					2.5 (5 passes)	
Yarrowia lipolitica	Niro Soavi Panda	PBS	130	n.a.	6	Lanciotti et al., 1994

Fungi

Penicillium ssp.	Emulsiflex C5	PBS	200	25	0.2	Tahiri *et al.*, 2006
		Orange juice			2.2	
					2.0	
					4.0	

Total microbial load

Endogenous flora	Niro Soavi Panda	Goat milk	100	n.a.	2–3	Guerzoni *et al.*, 1999
	APV Gaulin lab 60–10 TBS	Bore water	70	35	1	Jyoti *et al.*, 2001
	Emulsiflex C160	Whole cow milk (Cheddar cheese)	200	28	2	Kheadr *et al.*, 2002
		Skim cow milk (Cheddar cheese)			3.5	
	Stansted FPG7400H	Whole cow milk	300	4	0.4 0.8 (2 passes)	Thiebaud *et al.*, 2003
				24	1.2 2.2 (2 passes)	
	Niro Soavi Panda	Whole cow milk (Crescenza cheese)	100	n.a.	–	Lanciotti *et al.*, 2004a
	Niro Soavi Panda	Whole cow milk (yogurt)	75	n.a.	–	Lanciotti *et al.*, 2004b
	Stansted DRG FPG7400H350	Whole cow milk	300	30–40	4	Pereda *et al.*, 2006
	Stansted nm-GEN 7400H	Raw whole milk	250	55–70	5.3	Smiddy *et al.*, 2007

(continued)

Table 19.2 Microbial strains exposed to high-pressure homogenization—Cont'd

Microbial strain	Apparatus	Medium	Pressure (MPa)	Inlet temp. (°C)	log(N₀/N)	Ref.
Psychrotropic aerobic cells	Niro Soavi Panda	Goat milk	100	n.a.	2	Guerzoni et al., 1999
	Stansted FPG7400H	Whole raw milk	300	4	1.3	Thiebaud et al., 2003
	Stansted DRG FPG7400H350	Whole cow milk	300	24	3.1	Pereda et al., 2006
				30–40	4	
	Stansted nm–GEN 7400H	Raw whole milk	250	55–70	4.6	Smiddy et al., 2007
Viruses						
Lactococcal bacteriophages	Emulsiflex C5	PBS	200	25	2	Moroni et al., 2002
					5 (5 passes)	
		Whey permeate			1	
					3 (5 passes)	
		Whole cow milk			1	
					3.5 (5 passes)	

frictional heating at high temperature derived from the passage through the HPH valve (Feijoo *et al.*, 1997). Moroni *et al.* (2002) investigated the inactivation of bacterial viruses, such as lactococcal bacteriophages, reporting a significant reduction at 200 MPa and multiple homogenization steps in PBS (phosphate-buffered saline), milk, and whey permeate.

IV.B. Pressure of homogenization

Among the process parameters, pressure and temperature are the main factors influencing the effectiveness of homogenization in microbial inactivation. As in high hydrostatic pressure processing, the level of microbial inactivation in high-pressure homogenization increases with the pressure level. Diels and Michiels (2006) reviewed the relationship proposed to correlate the level of inactivation with processing pressure. If the level of inactivation is measured as the percentage of disrupted cells, a sigmoidal shape of the curves is detected, suggesting that the mortality of the cells occurs above a certain level of pressure. If, instead, the level of inactivation is measured as logarithm of the number of surviving cells, a linear relationship is found. The modified Gompertz equation is suitable for modeling an asymptotic behavior of the number of cells, in terms of log (CFU/ml), as a function of the pressure level.

Although high hydrostatic pressure treatment shares with HPH the use of extremely high pressures (>100 MPa), the mechanism by which microbial cell death is achieved is totally different. In both processes, high pressures act on microorganisms; however, in high-pressure homogenization this is only for a very short time, on the order of a second or less, while in high-hydrostatic-pressure treatments the exposure time is on the order of minutes or more. Hydrostatic pressure is believed to act mainly by causing protein denaturation and structural and functional damage to the cellular membrane. In high-pressure homogenization microbial cells are exposed to cavitation, impingement against static surfaces, high turbulence, and fluid shear; this causes the disruption of the cells in very small fragments. The comparison of resistance of different bacteria to high-pressure homogenization (100–300 MPa) and high hydrostatic pressure (200–400 MPa) showed large differences between the two techniques (Wuytack *et al.*, 2002).

IV.C. Initial cell concentration

Interestingly, there is no consensus among results reported in the literature on the role of initial cell concentration on the inactivation level attained by HPH. Vannini *et al.* (2004) claimed that, according to the viability data, the relationship between survival level and pressure applied is linear in the pressure range investigated. In addition, Thiebaud *et al.* (2003) observed a linear dependence of cell inactivation on pressure in the treatment of endogenous flora of raw bovine milk, with a second homogenization cycle causing an additional inactivation ratio roughly equal to that achieved during the first cycle. Also, Wuytack *et al.* (2002) investigated the effect of cyclic high-pressure homogenization treatment on *Y. enterocolitica* and *S. aureus*, finding that the different rounds have an additive effect. Wong *et al.* (1997) previously claimed that it can be assumed the disruption of *E. coli* is a first-order process.

In contrast, evidence of strong dependence on initial bacterial concentration was obtained by some authors. Moroni *et al.* (2002) reported that the effectiveness of HPH in inactivating lactococcal bacteriophages in PBS was affected by the initial concentration; with greater initial load, the treatment became less effective. Tahiri *et al.* (2006) also demonstrated that the effectiveness of the HPH increases at low initial bacterial concentration of different microbial strains (*L. plantarum*, *Penicillium* ssp., and *S. cerevisiae*). Analogously, Vachon *et al.* (2002) reported that the effectiveness of treatment of *Listeria monocytogenes*, *Escherichia coli*, and *Salmonella enterica* in PBS by HPH was dependent on the initial cell concentration, with a decreasing efficacy at increasing concentration.

Recently our research group undertook a detailed investigation of HPH effects on *Saccharomyces cerevisiae* (Maresca *et al.*, 2006) and *E. coli* (Donsì *et al.*, 2006) in a Stansted homogenizer. Results showed that the extent of inactivation significantly depended on the initial cell concentration; the higher the inactivation the lower the concentration. Nevertheless, upon multiple steps of homogenization, an opposite effect was observed, showing reduced efficiency with increased number of passes. A possible explanation was attributed to the protective effect exerted by high cell concentration on one side, and the inefficiency of homogenization below a certain concentration level (Donsì *et al.*, 2006; Maresca *et al.*, 2006).

A general conclusion can not be drawn, as the results reported in the literature on the subject are contradictory and cannot be applied to the different valve geometries of the homogenizers, since a different effect due to initial cell load was reported among homogenizers with the same class of valve geometry, and similitudes can be found among homogenizers with different disruption valves.

IV.D. Medium properties influencing microbial inactivation

At a macroscopical level, the parameters that mostly affect microbial inactivation by HPH are fluid temperature, viscosity, water activity, and medium composition.

IV.D.1. Temperature effect

Temperature effects must be taken into account with HPH treatment, since upon homogenization, an important rise of temperature is observed in the fluid downstream of the valve. This can be attributed to viscous stress caused by high velocity of the fluid flow, which is then impinged on the ceramic valve, leading to dissipation of a significant fraction of mechanical energy as heat in the fluid (Floury et al., 2000).

As an example, Floury et al. (2000) reported a linear increase of water temperature with homogenization pressure of $16°C/100$ MPa at the valve exit of a Stansted homogenizer, notwithstanding the water-jacketed ($5°C$) homogenization chamber. A similar linear increase was measured also for milk and whey-protein isolate solution (Bouaouina et al., 2006). Popper and Knorr (1990) reported an increase of $20°C/100$ MPa, which is somewhat lower than the calculated values from thermodynamic equations of approximately $25°C/100$ MPa for water-like media.

According to Diels et al. (2004), heating of milk between 0 and $45°C$ is not directly responsible for E. coli inactivation, however temperature appears to play a significant though indirect role on microbial inactivation with HPH, but only through its influence on some of the inactivation mechanisms, such as cavitation and turbulence. In particular, temperature has a contrasting effect on cavitation: at low temperature, liquids have a lower vapor pressure and thus cavity formation is reduced (Save et al., 1994; Save et al., 1997), while the severity of cavitation increases at low temperature as a result of a more violent collapse when vapor pressure is low. On the other side, high temperature increases the turbulence of the fluid due to the reduction of fluid viscosity, whereupon increased turbulence will in turn increase cavitation and hence microbial inactivation (Diels et al., 2004).

HPH processing of solutions inoculated with E. coli at different temperatures, but with the same viscosity, through adjustments using different concentrations of polyethylene glycol, showed that the effect on bacterial inactivation of temperature up to $35-45°C$ can be entirely explained in terms of its indirect effect on fluid viscosity (Diels et al., 2004, 2005a). Only at outlet temperatures above $65°C$ was the additive effect of temperature, on top of the mechanical factors, observed in microbial inactivation (Diels et al., 2004).

In partial contrast, Vachon et al. (2002) reported that temperature may play a direct role on HPH treatments, by affecting the membrane lipid composition and physical state, with crystallization of phospholipids and increased rigidness of cell membranes at low temperatures (2 to $10°C$), and by weakening the hydrogen and hydrophobic bonds at high temperatures (40 to $50°C$).

IV.D.2. Characteristics of process medium

IV.D.2.a. Viscosity
Upon HPH application, viscosity in liquid foodstuff may change, as a consequence of a variation in temperature, the creation of physical changes such as emulsification or disruption of fat globules, and the

disruption of cells. For instance, some authors claim that the disruption of cells can lead to the decrease of viscosity of the process fluid (Geciova et al., 2002), hence partially explaining the dependence of homogenization effectiveness on initial microbial concentration. Nevertheless, up to now the influence of fluid viscosity on bacterial inactivation by HPH has not been systematically studied.

It is known that viscosity influences the flow patterns in high-pressure homogenizer valves (Stevenson and Chen, 1997) and has an effect on cavitation and fluid turbulence (Diels et al., 2005a), hence it can influence cell disruption. For instance, by means of CFD data obtained for standard valve geometry (Figure 19.2a), varying the gap spacing and fluid viscosities, Miller et al. (2002) showed that fluid viscosity (1–5 cP) significantly affected the cell breakage mechanisms. Diels et al. (2004) significantly contributed, shedding light on the question; they reported clear evidence that viscosity is inversely related to microbial inactivation, explaining its effect in terms of reduced cavitation, turbulence, and impact pressure in viscous fluids. The authors further showed that bacterial inactivation by HPH is strongly affected by viscosity, compared to water activity or product composition, and also that the observed effects of temperature, up to a threshold value of about 40°C, can be entirely explained by its indirect effect on fluid viscosity (Diels et al., 2004; Diels et al., 2005a).

IV.D.2.b. Water activity   Depending on the nature of the solutes present, liquid foods or other aqueous liquids of high viscosity can have a reduced water activity, which has been shown to affect microbial inactivation by HPH. Reduced water activity does indeed dramatically increase the microbial resistance against physical treatments such as heat treatment and high hydrostatic pressure treatment (Donsì et al., 2004, 2005; Diels et al., 2005a; Van Opstal et al., 2005). The effect of the variation of water activity in the range 0.953–1.000 on inactivation by HPH, tested at constant viscosity by adjusting the solution with polyethylene glycol of different molecular weight, was negligible; in contrast, water activity in the same range strongly influences inactivation by high hydrostatic pressure treatment (Diels et al., 2005a).

IV.D.2.c. Medium composition   Medium composition influences the efficacy of the HPH treatment mainly due to the interaction of the medium constituents with the microorganisms. For instance, the higher viability loss of *Salmonella enteritidis* observed in egg-based systems than that in model brain-heart infusion systems was not only explained in terms of differences in variables, such as pH and NaCl concentration, but also postulated an implication of the antimicrobial enzymes naturally occurring in eggs (Guerzoni *et al.*, 2002).

Some authors ascribe the reported reduced levels of *L. monocytogenes* inactivation in milk compared to buffer (Vachon *et al.*, 2002), and of the endogenous flora in full fat compared to skim milk (Kheadr *et al.*, 2002), as shown in Table 19.2, to a supposed protective effect of milk fat, in analogy to the well-established protective effect of fat on inactivation by heat (MacDonald and Sutherland, 1993) and hydrostatic pressure (Garcia-Graells *et al.*, 1999). Nevertheless, contrasting effects are reported for the inactivation of *E. coli* by Vachon *et al.* (2002), with a lower inactivation in buffer than in whole milk, and by Briñez *et al.* (2006b), with a lower inactivation in full fat compared to skim milk (300 MPa at the primary homogenizing valve and 30 MPa at the secondary valve).

Differences in inactivation between buffer and orange juice were observed by Tahiri *et al.* (2006), who reported greater tolerance of *L. plantarum* and *L. mesenteroides* to HPH in juice (Table 19.2). Conversely, *Penicillium* was inactivated to a greater degree in orange juice than in PBS, while for *S. cerevisiae* and *E. coli* the efficiency of HPH treatment exhibited no significant dependence on medium composition (Table 19.2). The authors explained this behavior in terms of the pH difference of the mediums investigated. In addition, Briñez *et al.* (2006a) reported a higher inactivation of *L. innocua* in milk than in orange juice (Table 19.2).

In contrast, Diels *et al.* (2005a), by evaluating the inactivation attained in different food matrices (skim milk, soy milk, and a milk drink) with buffer solutions adjusted with polyethylene glycol to achieve viscosities similar to the food matrices, showed that the extent of inactivation is regulated only by fluid viscosity, hence, indicating evidence against a specific protective role of fat.

IV.E. Antimicrobial additives

Popper and Knorr (1990) first investigated the synergy between HPH and antimicrobial agents, such as lysozyme, a lytic enzyme, and chitosan, a biopolymer with antimicrobial activity. The authors observed, notwithstanding the reduction in initial total count of *S. lactis* up to 2 log cycles, by increasing the enzyme concentration, that no synergistic effect could be detected. Similarly, the 7 log-cycle inactivation of *S. lactis* caused by the addition of chitosan (300–1000 ppm) before homogenization was not the result of a synergistic effect, but the mere superimposition of HPH and chitosan inactivations (Popper and Knorr, 1990). The effect of the addition of enzymes to the process fluids, before or after homogenization step, was investigated mainly for the purpose of determining the eventual physical damage to treated cells resulting from HPH treatment.

Instead, Vannini *et al.* (2004) observed an enhancing effect of HPH on enzyme activities, which was either attributed to increased exposure of the microbial cells (*Pseudomonas putida, Proteus vulgaris, Escherichia coli, Staphylococcus aureus, Listeria monocytogenes*) to the enzymes or to the conformational changes of the antimicrobial enzymes exposed at high pressures. In particular, they postulated that a direct effect of pressure on the integrity of the walls or outer membranes of the microorganisms existed, with a consequent increased penetration of the enzymes through the damaged walls and membranes, and an indirect effect of the process through steric changes on the enzyme molecules, whose activity is ruled by their three-dimensional configuration.

Evidence of the synergistic effect of HPH on enzymatic activity was also reported by Guerzoni *et al.* (2002) in egg-based systems, where a higher viability loss of *Salmonella entiridis* was observed than in model BHI systems, due to the enhancement of naturally occurring antimicrobial enzymes in eggs and to outer membrane damage induced by homogenization pressure. Later, Diels *et al.* (2005b) confirmed the synergistic effect of HPH with enzymes, such as lysozyme and nisin, on *E. coli*, but contradicted the reported existence of a synergistic effect with the lactoperoxidase system, attributing the discrepancy to either the different medium type (PBS versus skim milk) used or to the different high-pressure homogenizers. HPH up to 300 MPa did not sensitize *E. coli* to lactoperoxidase, ·probably because sensitization to the lactoperoxidase system requires metabolic injury to the cells and HPH does not cause metabolic sublethal injury (Wuytack *et al.*, 2002; Diels *et al.*, 2003; Briñez *et al.*, 2006a, b, c).

IV.F. A case study: *Escherichia coli*

To underline the dependence of microbial mortality on the main factors influencing the level of inactivation, a single case study is next analyzed in detail. *E. coli*, one of the microbial strains most investigated in the literature, was selected in particular. Literature data are compared in Figure 19.3 in terms of decimal reduction pressure D_p, which is defined as the pressure of homogenization required for a 1 log-cycle decrease in the initial microbial population. D_p is calculated from the regression of the inactivation data reported in the literature, using the hypothesis of linear dependence of log cycles of inactivation on the homogenization pressure. As clearly shown in Figure 19.3, a wide range of observed pressure resistance of *E. coli* is reported in literature. D_p values mainly depend on the valve geometry, medium composition, inlet temperature, and the eventual addition of antimicrobial enzymes.

In particular, flow through the homogenizer valve, and the homogenization technique applied, play an important role (Sauer *et al.*, 1989; Popper

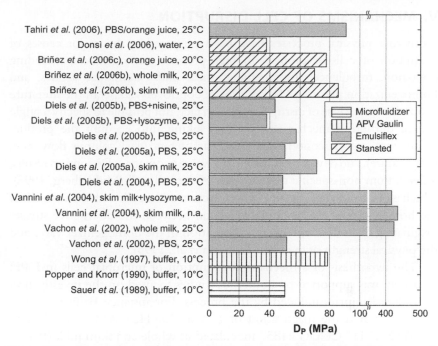

Figure 19.3 Comparison of literature data on *E. coli* inactivation by HPH in terms of decimal reduction pressure D_p.

and Knorr, 1990; Diels *et al.*, 2004) in determining significant differences between D_p values attained in the Stansted Fluid Power (Briñez *et al.*, 2006b, c; Donsì *et al.*, 2006), Emulsiflex (Vachon *et al.*, 2002; Wuytack *et al.*, 2002; Diels *et al.*, 2004, 2005a, b; Tahiri *et al.*, 2006), APV Gaulin (Popper and Knorr, 1990; Wong *et al.*, 1997), and Microfluidizer® (Sauer *et al.*, 1989) machines (Figure 19.3). Interestingly, some differences also exist within the same class of homogenizers, as highlighted in D_p values calculated from data reported by Tahiri *et al.* (2006) and Diels *et al.* (2004, 2005a, b) for HPH in PBS at 25°C (~90 vs. ~50 MPa), and Popper and Knorr (1990) and Sauer *et al.* (1997) in buffer at 10°C (~35 vs. ~80 MPa); further, the simple replacement of the homogenizer valve within the same device can cause distinct variations in disruption efficiency (Diels *et al.*, 2005b).

The resistance of *E. coli* cells to homogenization effects is clearly influenced by medium composition. The cells are more sensitive to homogenization in buffer solution than in real food, showing higher values of D_p for skim milk (Vannini *et al.*, 2004; Diels *et al.*, 2005a; Briñez *et al.*, 2006b) than whole milk (Vachon *et al.*, 2002; Briñez *et al.*, 2006b) and contrasting effects for orange juice (Briñez *et al.*, 2006c; Tahiri *et al.*, 2006), or in the case of addition of antimicrobial enzymes (Vannini *et al.*, 2004; Diels *et al.*, 2005b).

V. MECHANISMS OF CELL DISRUPTION

Numerous physical processes have been proposed as the major causes of disruption of cells or droplets in high-pressure homogenizers, including cavitation, turbulence, impingement, shear and extensional stresses, and homogenization pressure. Hence, a lot of controversy exists in the literature about the exact cause of disruption. It is, in fact, not possible to specify a single overall disruption mechanism, without taking into account the product parameters (e.g. viscosity), operating parameters (e.g. volume flow rate, temperature), and device parameters (e.g. valve geometry). Disruption results, indeed, from non-specific tearing apart of the cell wall (Middelberg, 1995), which is determined by the physical interaction of the cells with the valve slit of the homogenizer, in a co-operative action between the destructive stresses originating from the fluid dynamic condition in the homogenizer valve and the physical strength of the cells (Shamlou et al., 1995).

The hypothesis of a mechanical disruption of microbial cells in HPH treatment was supported by numerous studies focused on the investigation of sublethal injuries induced by the process. For instance Briñez and co-workers studied the inactivation of E. coli O58:H21 ATCC 10536 and E. coli O157:H7 CCUG 44857 inoculated in whole and skim milk (Briñez et al., 2006a) and orange juice (Briñez et al., 2006b), and Listeria innocua ATCC 33090 inoculated in milk and orange juice (Briñez et al., 2006c). To assess the level of injuries caused by HPH process, microbial samples were plated in duplicate in TSAYE and TSAYE supplemented with NaCl, which allows the selective growth of non-injured cells. The authors observed no significant difference in lethality values between non-selective and selective media, indicating that HPH treatment caused no sublethal injuries. Results in support of the absence of sublethal injuries were reported by Wuytack et al. (2002) for Y. enterocolitica and S. aureus, plated on PCA modified with different selective factors, such as low pH, NaCl or SDS; Diels et al. (2005b) also observed that above 150 MPa, E. coli became sensitive to lysozyme and nisin due to transient permeabilization of the outer membrane induced by the short exposure at high pressure, while neither sensitization was observed to lactoperoxidase enzyme, whose efficacy requires metabolic injury of the cells, nor could evidence of cytoplasmic membrane damage be found in cells surviving HPH treatment using SYTO 9/propidium iodide staining.

All these observations suggest that high-pressure homogenization causes an 'all or nothing' impact on microbial cells: if microorganisms are disrupted by the process they are dead, otherwise, if not disrupted, they are in-distinguishable from untreated cells. Moreover, the high pressure developed during HPH is not to any major extent responsible for inactivation of microorganisms due to the very short processing time. The mechanism of inactivation, in fact, significantly differs from that of high hydrostatic pressure treatment, which can cause an accumulation of sublethal injury, ultimately leading to cell death (Van Opstal et al., 2003; Yuste et al., 2004). In this case, cell death occurs only after inactivation of various cellular structures or functions, such as cytoplasmic membrane, ribosomes, and specific enzymes. The inability of high-pressure homogenization to cause sublethal injuries may limit applications of this technique in hurdle technology. If the cells surviving high-pressure homogenization treatment are not injured, it can be anticipated they will not be sensitive to other treatments or to unfavorable conditions.

V.A. Shear stress

Recent studies provide some evidence supporting viscous shear as one of the primary mechanisms of cell or particle disruption during HPH (Middelberg, 1995; Kleinig and Middelberg, 1998). Disruption is indeed attained in the narrow valve gap when deformation beyond a critical level is induced by the intense shear forces and elongational flow caused by the restriction between the piston and the seat of the valve. In laminar flow, which predominates in high-pressure homogenizer valves, only shear forces can bring deformation (Stang et al., 2001; Floury et al., 2004a, b). According to Kelly and Muske (2004), shear stress becomes the dominant mechanism of disruption in the case of high viscosity (5 cp in the case of E. coli), when the channel inlet pressure gradient, post-channel turbulence, and ring impingement forces are reduced below the threshold required for disruption.

V.B. Turbulence

Turbulence is instead indicated as one of the main mechanisms of disruption with process medium of low viscosity (Middelberg, 1995; Shirgaonkar et al., 1998; Floury et al., 2004b; Kelly and Muske, 2004). In particular, the inertial forces generated in the highly turbulent region, located just at the exit of the gap, may be an important mechanism of disruption, either with the standard valve geometry (Figure 19.2a) (Kleinig and Middelberg, 1998), or in the case of Microfluidizer® geometry (Figure 19.2c) (Stang et al., 2001).

The homogenization effect is due to the rapid release of the high energy concentrated within the liquid by pressurization. The passage of fluid within the cross-sections of the valve, under conditions of high pressure and controlled flow regime, exposes the liquid to elevated turbulence and mechanical solicitation of strain and impact, creating an extremely effective process that reduces the size of particles and globules.

Concerning microbial disruption, the observation that viscosity is inversely related to inactivation of E. coli, in conjunction with the consideration that at lower viscosity turbulence is increased, indicates that turbulence, whether through inducing cavitation, or oscillatory motions in the cell fluid, covers an important role in cell disruption. However extensional stress, which instead increases with viscosity, apparently does not provide an important contribution (Diels *et al.*, 2005a).

V.C. Cavitation

Many authors have provided experimental and numerical evidence supporting cavitation as one of the primary mechanisms of cell disruption in high-pressure homogenizers (Middelberg, 1995; Kleinig and Middelberg, 1998; Shirgaonkar *et al.*, 1998; Paquin, 1999; Diels *et al.*, 2004; Lanciotti *et al.*, 2004), especially in consideration that cavitation and microbial inactivation both increase at reduced fluid viscosity (Diels *et al.*, 2005a). Cavitation is the formation of cavities due to the local vaporization of the liquid under conditions of pressure lower than its vapor pressure. When cavities flowing within the fluid through the system encounter a region of higher pressure, they collapse violently, causing vibrations and noise with a disruptive effect. Increasing the fluid viscosity reduces cavitation by causing a shift from a turbulent to a laminar flow pattern, and thus dampens highly localized differences in fluid velocity and fluid pressure (Diels *et al.*, 2005a).

The onset of cavitating conditions can be predicted in hydrodynamic devices by values of the cavitation inception number $C_v \leq 1.0$, where C_v is defined as $C_v = \frac{P_2 - P_v}{1/2 \rho \, v_{gap}^2}$, P_2 is the downstream pressure, P_v is the vapor pressure of the liquid, ρ is the density of the liquid, and v_{gap} is the maximum liquid velocity through the constriction (Shirgaonkar *et al.*, 1998). Nevertheless, in addition to difficulties with accurate estimation of v_{gap}, $C_v \leq 1.0$ indicates only a possibility of cavitation occurring (Shirgaonkar *et al.*, 1998). The experimental identification of cavitation by means of detection of iodine liberated by decomposition of aqueous KI solution, which occurs only under cavitating conditions, when strongly oxidizing OH radicals are generated, revealed in a high-pressure homogenizer that cavitating conditions are already generated above 35 MPa (Shirgaonkar *et al.*, 1998).

V.D. Impingement

Impingement and impact of cells against the walls of the homogenizer valve are also reckoned as important causes of disruption in HPH (Middelberg, 1995; Kleinig and Middelberg, 1998; Shirgaonkar *et al.*, 1998; Geciova *et al.*, 2002; Diels *et al.*, 2004). Kleinig and Middelberg (1998) developed a correlation of cell disruption with the homogenizer valve pressure gradient, determined by means of a computational fluid dynamics model, and concluded that impingement was important for cell disruption by HPH, even though it was more critical for yeast than *E. coli*.

On the contrary, Shamlou *et al.* (1995) highlighted that direct collision between the cells and the walls of the valve involves a maximum impaction force that, for typical homogenization conditions, was estimated to be lower than required to cause any serious physical damage to the walls of yeast cells. Hence, impingement becomes a relevant disruption mechanism in comparison to pressure gradients, turbulence, and shear stresses only under conditions of low pressure and low viscosity (Kelly and Muske, 2004).

V.E. Extensional stress

In high-stress zones close to the surface of the impingement wall, cells or droplets can experience an extensional stress that exceeds the mechanical resistance of the droplet or cell wall (Shamlou *et al.*, 1995). However, it is not clear whether droplet disruption can be attributed to elongational flow alone; more likely, the droplets are deformed in the elongational flow regime and then disrupted due to a perturbation in a zone following the elongational flow (Stang *et al.*, 2001). According to Floury *et al.* (2004b), the disruption of yeast cells is caused by elongational stresses occurring in regions of the homogenizer valve where the flow field can be approximated to that of a plane hyperbolic flow. Experimental results showed that very strong elongational flow at the entrance of the homogenizer valve and the resulting frictional forces induced the irreversible disruption of large protein aggregates (Floury *et al.*, 2002).

VI. CONCLUSIONS AND PERSPECTIVES

HPH technology is likely to support many different applications in the dairy and food industry, enhancing food safety and quality, and/or delivering new products to the market. In spite of the high potential of this technology, not all potential applications have yet been exploited, due to the lack of well-established knowledge about the interactions between HPH and food materials. This technology, which in fact differs from simple homogenization, may also induce the physical modification of aggregation profiles or even of the molecular structure of biopolymers such as milk proteins, polysaccharides or complexes, and can produce a definite effect on microbial inactivation. This could open the doors to production of different types of functional ingredients, ranging from mimetic fat substitutes to textural agents, as well as to delivery of non-thermally stabilized liquid/paste foods.

To support the engineering developments required to set up industrial production of such types of foods, further fundamental research work is needed in the food engineering fields, to assess the stability of dairy emulsions and the structural modifications made to foods during flow through homogenization nozzles under very critical conditions. On the contrary, some basic achievements are well established and could be the starting point for process modeling, among which the effect of microbial factors, such as strain, growth phase, and growth conditions on the kinetics of inactivation are deeply investigated in the literature.

The effect of product characteristics (viscosity, water activity, and composition of processing medium) on the HPH process performance is still a matter of concern. Until now the influence of fluid viscosity on bacterial inactivation by HPH has not been systematically studied. From a more fundamental point of view, viscosity influences the flow patterns in high-pressure homogenizer valves, and has an effect on cavitation and fluid turbulence and hence can influence cell disruption. Whereas a deeper knowledge of the effect of water activity on microbial inactivation by HPH would be useful in developing applications of HPH as a microbicidal process, but it could also provide additional insights on the specific mechanism of microbial kill by HPH as compared to hydrostatic pressure treatment.

In conclusion, up-to-date, quantitative data to describe and model inactivation phenomenon by high-pressure homogenization are scarce. Evidently, the physical mechanisms responsible for rupturing the walls of microbial cells in high-pressure homogenizers remain uncertain. This is partly because in most previous studies only the overall release of intracellular material has been measured as a function of homogenizer type and its operating condition. Moreover, the extent of cell wall disruption increases with increasing pressure drop across the valve slit and the two parameters are often related by empirical equations. Experimental data relating the whole cell diameter directly to homogenizer operating parameters are needed in order to establish the physical mechanisms by which cell disruption occurs in the homogenizer valve.

Several questions arise from the investigation of bacterial inactivation by high-pressure homogenization regarding the role of cell shape in homogenization, the occurrence of sublethal injuries, and consequent limited application in hurdle technology, and finally the role of peptidoglycan layers in microbial inactivation.

To improve the efficiency of homogenization processes, as required by the envisaged applications of new ultra-high-pressure homogenizers, a deeper understanding of the mechanism of homogenization is needed. This could help the designer to improve the geometry of the homogenizer valve, which seems to be the critical design element with reference to the quality of the finished product. A proper valve design is also the key to minimizing operative pressure, which in turn makes for longer equipment life without undergoing mechanical failure.

REFERENCES

Avestin® (2006). Avestin Inc., Ottawa, Canada. Website: http://www.avestin.com/index. html.

Bouaouina, H., Desrumaux, A., Loisel, C., & Legrand, J. (2006). Functional properties of whey proteins as affected by dynamic high-pressure treatment. *Int. Dairy J., 16,* 275–284.

Briñez, W. J., Roig-Saugués, A. X., Hernández Herrero, M. M., & Guamis López, B. (2006a). Inactivation of *Listeria innocua* in milk and orange juice by ultra high pressure homogenization. *J. Food Protect., 69*(1), 86–92.

Briñez, W. J., Roig-Saugués, A. X., Hernández Herrero, M. M., & Guamis López, B. (2006b). Inactivation of two strains of *Escherichia* coli inoculated into whole and skim milk by ultrahigh-pressure homogenization. *Lait, 86,* 241–249.

Briñez, W. J., Roig-Saugués, A. X., Hernández Herrero, M. M., & Guamis López, B. (2006c). Inactivation by ultrahigh-pressure homogenization of *Escherichia coli* inoculated into orange juice. *J. Food Protect., 69*(5): 984–989.

Bury, D., Jelen, P., & Kaláb, M. (2001). Disruption of *Lactobacillus delbrueckii* ssp. bulgaricus 11842 cells for lactose hydrolysis in dairy products: a comparison of sonication, high-pressure homogenisation and bead milling. *Innovative Food Science & Emerging Technologies, 2,* 23–29.

Chen, C. C., & Wagner, G. (2004). Vitamin E Nanoparticle for Beverage Applications. Trans IChemE, Part A, November 2004. *Chemical Engineering Research and Design, 82*(A11), 1432–1437.

Clark, A. V., Rejimbal, T. R. J., & Gomez, C. M. (1993). Ultra-high pressure homogenization of unpasteurized juice. US Patent No. 5,232,726. The Coca-Cola Company.

Cook, E.J., and Lagacé, A.P. (1985). Apparatus for forming emulsions. US Patent No. 4,533,254.

Corredig, M., & Wicker, L. (2001). Changes in the molecular weight distribution of three commercial pectins after valve homogenisation. *Food Hydrocolloid, 15,* 17–23.

Desrumaux, A., & Marcand, J. (2002). Formation of sunflower oil emulsions stabilized by whey proteins with high-pressure homogenization (up to 350 MPa): effect of pressure on emulsion characteristics. *Int. J. Food Sci. Tech., 37,* 263–269.

Diels, A. M. J., Wuytack, E. Y., & Michiels, C. W. (2003). Modelling inactivation of *Staphylococcus aureus* and *Yersinia enterocolitica* by high-pressure homogenisation at different temperatures. *Int. J. Food Microbiol., 87,* 55–62.

Diels, A. M. J., Callewaert, L., Wuytack, E. Y., Masschalck, B., & Michiels, C. W. (2004). Moderate temperatures affect *Escherichia coli* inactivation by high-pressure homogenization only through fluid viscosity. *Biotechnol. Progr., 20,* 1512–1517.

Diels, A. M. J., Callewaert, L., Wuytack, E. Y., Masschalck, B., & Michiels, C. W. (2005a). Inactivation of *Escherichia coli* by high-pressure homogenisation is influenced by fluid viscosity but not by water activity and product composition. *Int. J. Food Microbiol., 101,* 281–291.

Diels, A. M. J., De Taeye, J., & Michiels, C.W. (2005b). Sensitisation of *Escherichia coli* to antibacterial peptides and enzymes by high-pressure homogenisation. *Int. J. Food Microbiol., 105*: 165-175.

Diels, A. M. J., & Michiels, C. W. (2006). High-pressure homogenization as a non-thermal technique for the inactivation of microorganisms. *Crit. Rev. Microbiol., 32*(4), 201–216.

Donsì, G., Ferrari, G., & Maresca, P. (2004). The effect of osmotic agents modifying water activity on thermo and baroresistance of yeasts. Proceedings of the International Congress on Engineering and Food ICEF9. *Montpellier 7–11 March,* 2004, 367.

Donsì, G., Ferrari, G., & Maresca, P. (2005). An analysis of the effect of operating parameters on the process of food sanitization by high hydrostatic pressure. Proceedings of 7th World Congress of Chemical Engineering. *Glasgow (Scotland) 10–14 July 2005,* 62343

Donsì, F., Ferrari, G., & Maresca, P. (2006). High-Pressure Homogenisation for Food Sanitisation. Proceedings of the 13th World Congress of Food Science and Technology 'Food is Life' *Nantes 17–21 September 2006,* 1851–1862, doi: 10.1051/IUFoST: 20060497.

Engler, C. R., & Robinson, C. W. (1981). Disruption of Candida utilis cells in high pressure flow devices. *Biotechnol. Bioeng., 23,* 765–780.

Fantin, G., Fogagnolo, M., Guerzoni, M. E., Lanciotti, R., Medici, A., Pedrini, P., & Rossi, D. (1996). Effect of high hydrostatic pressure and high pressure homogenisation on the enantioselectivity of microbial reductions. *Tetrahedron: Asymmetry,* 7(10), 2879–2887.

Feijoo, S. C., Hayes, W. W., Watson, C. E., & Martin, J. H. (1997). Effects of Microfluidizer[®] Technology on *Bacillus licheniformis* spores in ice cream mix. *J. Dairy Sci., 80,* 2184–2187.

Floury, J., Desrumaux, A., & Lardières, J. (2000). Effect of high-pressure homogenisation on droplet size distributions and rheological properties of model oil-in-water emulsions. *Innovative Food Science & Emerging Technologies, 1,* 127–134.

Floury, J., Desrumaux, A., Axelos, M. A. V., & Legrand, J. (2002). Degradation of methylcellulose during ultra-high pressure homogenisation. *Food Hydrocolloid, 16,* 47–52.

Floury, J., Bellettre, J., Legrand, J., & Desrumaux, A. (2004a). Analysis of a new type of high pressure homogeniser. A study of the flow pattern. *Chem. Eng. Sci., 59,* 843–853.

Floury, J., Legrand, J., & Desrumaux, A. (2004b). Analysis of a new type of high pressure homogeniser. Part B. study of droplet break-up and recoalescence phenomena. *Chem. Eng. Sci., 59,* 1285–1294.

Friedrich, I., & Müller-Goymann, C. C. (2003). Characterization of solidified reverse micellar solutions (SRMS) and production development of SRMS-based nano-suspensions. *Eur. J. Pharm. Biopharm., 56,* 111–119.

Gandini, M., Volpi, A., & Grasselli, S. (1999). Homogenizing valve. US Patent No. 5,887,971.

Garcia-Graells, C., Masschalck, B., & Michiels, C. W. (1999). Inactivation of *Escherichia coli* in milk by high-hydrostatic pressure treatment in combination with antimicrobial peptides. *J. Food Protect., 62,* 1248–1254.

Geciova, J., Bury, D., & Jelen, P. (2002). Methods for disruption of microbial cells for potential use in the dairy industry – a review. *Int. Dairy J., 12,* 541–553.

Grant, P. M. (1989). Citrus juice concentrate processor. US Patent No. 4,886,574. APV Gaulin.

Gu, B., & Zhang, B. (2005). A fermented milk and its preparation method. CN Patent No. CN1579177.

Guerzoni, M. E., Vannini, L., Chaves Lopez, C., Lanciotti, R., Suzzi, G., & Gianotti, A. (1999). Effect of high pressure homogenization on microbial and chemico-physical characteristics of goat cheeses. *J. Dairy Sci., 82,* 851–862.

Guerzoni, M. E., Vannini, L., Lanciotti, R., & Gardini, F. (2002). Optimisation of the formulation and of the technological process of egg-based products for the prevention of *Salmonella enteritidis* survival and growth. *Int. J. Food Microbiol., 73,* 367–374.

Hayes, M. G., & Kelly, A. L. (2003a). High pressure homogenization of raw whole bovine milk (a) effects of fat globule size and other properties. *J. Dairy Res., 70,* 297–305.

Hayes, M. G., & Kelly, A. L. (2003b). High pressure homogenization of raw whole bovine milk (b) effects on indigenous enzymatic activity. *J. Dairy Res., 70,* 307–313.

Hayes, M. G., Fox, P. F., & Kelly, A. L. (2005). Potential applications of high pressure homogenisation in processing of liquid milk. *J. Dairy Res., 72*(1), 25–33.

Hecq, J., Deleers, M., Fanara, D., Vranckx, H., & Amighi, K. (2005) . Preparation and characterization of nanocrystals for solubility and dissolution rate enhancement of nifedipine. *Int. J. Pharm.*, *299*, 167–177.

Hongyan, S. (2005). Method for preparing ice cream in fruity flavor. CN patent, CN1625961.

Invensys™ AVP (2006). http://www.apv.com/us/eng/APV+Home.htm.

Iordache, M., & Jelen, P. (2003). High pressure microfluidization treatment of heat denatured whey proteins for improved functionality. *Innovative Food Science & Emerging Technologies*, *4*, 367–376.

Jarchau, M., Priebe, R., & Lindemann, K. (1993). Homogenizing system having improved fluid flow path. US Patent No. 5,273,407.

Jarchau, M. (2001). Homogenization valve with outside high pressure volume. US Patent No. 6,238,080.

Jyoti, K. K., & Pandit, A. B. (2001). Water disinfection by acoustic and hydrodynamic cavitation. *Biochem. Eng. J.*, *7*, 201–212.

Kelly, W. J., & Muske, K. R. (2004). Optimal operation of high-pressure homogenization for intracellular product recovery. *Bioproc. Biosyst. Eng.*, *27*, 25–37.

Kent, C., Loh, J. P., & Eibel, H. (2003). Dairy products with reduced average fat particle size. *Eur. Pat. Appl.*, EP 1364583.

Keshavarz-Moore, E., Hoare, M., & Dunnill, P. (1990). Disruption of baker's yeast in a high-pressure homogenizer: new evidence on mechanism. *Enzyme Microb. Tech.*, *12*, 764–770.

Kheadr, E. E., Vachon, J. F., Paquin, P., & Fliss, I. (2002). Effect of dynamic high pressure on microbiological, rheological and microstructural quality of Cheddar cheese. *Int. Dairy J.*, *12*, 435–446.

Kinney, R. R., Pandolfe, W. D., & Ferguson, R. D. (1999). Homogenization valve. US Patent No. 5,899,564.

Kleinig, A. R., & Middelberg, A. P. J. (1996). The correlation of cell disruption with homogeniser valve pressure gradient determined by computational fluid dynamics. *Chem. Eng. Sci.*, *51*(23), 5103–5110.

Kleinig, A. R., & Middelberg, A. P. J. (1998). On the mechanism of microbial cell disruption in high-pressure homogenisation. *Chem. Eng. Sci.*, *53*(5), 891–898.

Knorr, D. (1999). Novel approaches in food-processing technology: new technologies for preserving foods and modifying function. *Curr. Opin. Biotechnol.*, *10*(5), 485–491.

Kyun, C. S., & Hoon, O. S. (2005). Process for the preparation of whole soybean milk and curd comprising multiple steps of ultra high-pressure homogenization of soybean. International Patent No. WO2005020714.

Lacroix, N., Fliss, I., & Makhlouf, J. (2005). Inactivation of pectin methylesterase and stabilization of opalescence in orange juice by dynamic high pressure. *Food Res. Int.*, *38*, 569–576.

Lagoueyete, N., & Paquin, P. (1998). Effects of microfluidization on the functional properties of xanthan gum. *Food Hydrocolloid*, *12*(3), 365–371.

Lamprecht, A., Ubrich, N., Hombreiro Pérez, M., Lehr, C.-M., Hoffman, M., & Maincent, P. (1999). Biodegradable monodispersed nanoparticles prepared by pressure homogenization-emulsification. *Int. J. Pharm.*, *184*, 97–105.

Lanciotti, R., Sinigaglia, M., Angelini, P., & Guerzoni, M. E. (1994). Effects of homogenization pressure on the survival and growth of some food spoilage and pathogenic microorganisms. *Lett. Appl. Microbiol.*, *18*, 319–322.

Lanciotti, R., Gardini, F., Sinigaglia, M., & Guerzoni, M. E. (1996). Effects of growth conditions on the resistance of some pathogenic and spoilage species to high pressure homogenization. *Lett. Appl. Microbiol.*, *22*, 165–168.

Lanciotti, R., Chaves-López, C., Patrignani, F., Paparella, A., Guerzoni, M. E., Serio, A., & Suzzi, G. (2004a). Effects of milk treatment with dynamic high pressure on microbial populations, and lipolytic and proteolytic profiles of Crescenza cheese. *Int. J. Dairy Technol.* 57(1): 19–25.

Lanciotti, R., Vannini, L., Pittia, P., Guerzoni, M. E. (2004b). Suitability of high-dynamic-pressure-treated milk for the production of yoghurt. *Food Microbiol., 21,* 753–760.

Lanciotti, R., Vannini, L., Patrignani, F., Iucci, L., Vallicelli, M., Ndagijimana, M., & Guerzoni, M. E. (2006). Effect of high pressure homogenisation of milk on cheese yield and microbiology, lipolysis and proteolysis during ripening of Caciotta cheese. *J. Dairy Res, 73*(2), 216–226.

Lander, R., Manger, W., Scouloudis, M., Ku, A., Davis, C., & Lee, A. (2000). Gaulin Homogenisation: A Mechanical Study. *Biotechnol. Progr., 16,* 80–85.

López-Pedemonte, T., Briñez, W. J., Roig-Saugués, E. X., & Guamis, B. (2006). Fate of *Staphylococcus aureus* in cheese made from inoculated milk treated by ultra high pressure homogenization with or without further high hydrostatic pressure treatment. *J. Dairy Sci., 89*(11), 1–9.

MacDonald, F., & Sutherland, A. D. (1993). Effect of heat treatment on Listeria monocytogenes and gram-negative bacteria in sheep, cow and goat milks. *J. Appl. Bacteriol., 75,* 336–343.

Maresca, P., Ferrari, G., & Donsì, G. High pressure homogenisation of fruit juices. 2009 (in preparation).

Mehnert, W., & Mäder, K. (2001). Solid lipid nanoparticles. Production, characterization and applications. *Adv. Drug Deliver. Rev., 47,* 165–196.

Microfluidics™ (2006). http://www.microfluidicscorp.com/cell_disruption/cell_disruption.html.

Middelberg, A. P. J. (1995). Process-Scale Disruption of Microorganisms. *Biotechnol. Adv., 13*(3), 491–551.

Miller, J., Rogowski, M., & Kelly, W. (2002). Using a CFD model to understand the fluid dynamics promoting *E. coli* breakage in a high-pressure homogenizer. *Biotechnol. Progr., 18,* 1060–1067.

Moroni, O., Jean, J., Autret, J., & Fliss, I. (2002). Inactivation of lactococcal bacteriophages in liquid media using dynamic high pressure. *Int. Dairy J., 12,* 907–913.

Möschwitzer, J., Achleitner, G., Pomper, H., & Müller, R. H. (2004). Development of an intravenously injectable chemically stable aqueous omeprazole formulation using nanosuspension technology. *Eur. J. Pharm. Biopharm., 58,* 615–619.

Niro Soavi S.p.A. (2006). Parma, Italy. http://www.niro-soavi.it/ndk_website/SOAVIIT/CMSDoc.nsf/WebDoc/anhe5pzbmc.

Paquin, P. (1999). Technological properties of high pressure homogenisers: the effect of fat globules, milk proteins, and polysaccharides. *Int. Dairy J., 9,* 329–335.

Paquin, P., Lacasse, J., Subirade, M., & Turgeon, S. (2003). Continuous process of dynamic high-pressure homogenization for the denaturation of proteins. US Patent No. 6,511,695 B1. Universite Laval.

Pereda, J., Ferragut, V., Guamis, B., & Trujillo, A. J. (2006). Effect of ultra high pressure homogenisation on natural occurring micro-organisms in bovine milk. *Milchwissenchaft, 61*(3), 246–248.

Perrier-Cornet, J. M., Marie, P., & Gervais, P. (2005). Comparison of emulsification efficiency of protein-stabilized oil-in-water emulsions using jet, high pressure and colloid mill homogenization. *J. Food Eng., 66,* 211–217.

Popper, L., & Knorr, D. (1990). Applications of high-pressure homogenisation for food preservation. *Food Technol., 44,* 84–89.

Ranjan, S., & Baer, R. J. (2005). Effects of milk fat and homogenization on the texture of ice cream. *Milchwissenschaft, 60*(2), 189–192.

Saboya, L. V., Maillard, M.-B., & Lortal, S. (2003). Efficient mechanical disruption of *Lactobacillus helveticus, Lactococcus lactis* and *Propionibacterium freudenreichii* by a new high-pressure homogenizer and recovery of intracellular aminotransferase activity. *J. Ind. Microbiol. Biot., 30*, 1–5.

Sandra, S., & Dalgleish, D. G. (2005). Effects of ultra-high-pressure homogenization and heating on structural properties of casein micelles in reconstituted skim milk powder. *Int. Dairy J., 15*, 1095–1104.

Sauer, T., Robinson, C. W., & Glick, B. R. (1989). Disruption of native and recombinant *Escherichia coli* in a high-pressure homogenizer. *Biotechnol. Bioeng., 33*, 1330–1342.

Save, S. S., Pandit, A. B., & Joshi, J. B. (1994). Microbial cell disruption-role of cavitation. *Chem. Eng. J., 55*, B67–B72.

Save, S. S., Pandit, A. B., & Joshi, J. B. (1997). Use of hydrodynamic cavitation for large scale microbial cell disruption. *Trans. Inst. Chem. Eng., 75*(C), 41–49.

Schorsch, C. (2004). Process for high pressure homogenization of a milk emulsion. *Eur. Pat. Appl.* EP1464230.

Serra, M., Trujillo, A. J., Quevedo, J. M., Guamis, B., & Ferragut, V. (2007). Acid coagulation properties and suitability for yogurt production of cows' milk treated by high-pressure homogenisation. *Int. Dairy J., 17*(7), 782–790.

Shamlou, P. A., Siddiqi, S. F., & Titchener-Hooker, N. J. (1995). A physical model of high-pressure disruption of baker's yeast cells. *Chem. Eng. Sci., 50*(9), 1383–1391.

Shirgaonkar, I. Z., Lothe, R. R., & Pandit, A. B. (1998). Comments on the mechanism of microbial cell disruption in high-pressure and high-speed devices. *Biotechnol. Progr., 14*, 657–660.

Siddiqi, S. F., Titchener-Hooker, N. J., & Shamlou, P. A. (1997). High pressure disruption of yeast cells: The use of scale down operations for the prediction of protein release and cell debris size distribution. *Biotechnol. Bioeng., 55*(4), 642–649.

Small, L.E., Mehansho, H., Woodly, S.A., Nunes, R.V., & Krummen, R.W. (2003). Compositions comprising protein and fatty acid and processes of their preparation. US *Pat. Appl. Publ.*, US 2003203005.

Smiddy, M. A., Martin, J.-E., Huppertz, T., & Kelly, A. L. (2007). Microbial shelf-life of high-pressure-homogenised milk. *Int. Dairy J., 17*, 29–32.

Souto, E. B., Anselmi, C., Centini, M., & Müller, R. H. (2005). Preparation and characterization of n-dodecyl-ferulate-loaded solid lipid nanoparticles (SLN®). *Int. J. Pharm., 295*, 261–268.

Stang, M., Schuchmann, H., & Schubert, H. (2001). Emulsification in high-pressure homogenisers. *Eng. Life Sci., 4*, 151–157.

Stansted Fluid Power LTD (2006). Essex, United Kingdom. http://www.sfp-4-hp.demon.co.uk/Prodcuts.htm.

Stevenson, M. J., & Chen, X. D. (1997). Visualization of the flow patterns in a high-pressure homogenizing valve using a CFD package. *J. Food Eng., 33*, 151–165.

Stipp, G. K., & Tsai, C. H. (1988). Low viscosity orange juice concentrates useful for high brix products having lower pseudoplasticity and greater dispersibility. US Patent No. 4,946,702. The Procter & Gamble Company.

Tahiri, I., Makhlouf, J., Paquin, P., & Fliss, I. (2006). Inactivation of food spoilage bacteria and *Escherichia coli* O157:H7 in phosphate buffer and orange juice using dynamic high pressure. *Food Res. Int., 39*, 98–105.

Tan, C. P., & Nakajima, M. (2005). β-Carotene nanodispersions: preparation, characterization and stability evaluation. *Food Chem., 92*, 661–671.

Thiebaud, M., Dumay, E., Picart, L., Guiraud, J. P., & Cheftel, J. C. (2003). High-pressure homogenisation of raw bovine milk. Effects on fat globule size distribution and microbial inactivation. *Int. Dairy J., 13*, 427–439.

Trotta, M., Gallarate, M., Pattarino, F., & Morel, S. (2001). Emulsions containing partially water-miscible solvents for the preparation of drug nanosuspensions. *J. Control. Release, 76*, 119–128.

Tunick, M. H., Van Hekken, D. L., Cooke, P. H., Smith, P. W., & Malin, E. L. (2000). Effect of high pressure microfluidization on microstructure of mozzarella cheese. Lebensm.-Wiss. u.-Technol, 33, 538–544.

Vachon, J. F., Kheadr, E. E., Giasson, J., Paquin, P., & Fliss, I. (2002). Inactivation of some food pathogens in milk using dynamic high pressure. J. Food Protect., 65, 345–352.

Van Hee, P., Middelberg, A. P. J., van der Lans, R. G. J. M., & van der Wielen, L. A. M. (2004). Relation between cell disruption conditions, cell debris particle size, and inclusion body release. Biotechnol. Bioeng., 88(1), 100–110.

Van Opstal, I., Vanmuysen, S. C. M., & Michiels, C. W. (2003). High sucrose concentration protects E. coli against high pressure inactivation but not against high pressure sensitization to the lactoperoxidase system. Int. J. Food Microbiol., 88(1), 1–9.

Van Opstal, I., Vanmuysen, S. C. M., Wuytack, E. Y., Masschalck, B., & Michiels, C. W. (2005). Inactivation of Escherichia coli by high hydrostatic pressure at different temperatures in buffer and carrot juice. Int. J. Food Microbiol., 98(2), 179–191.

Vannini, L., Lanciotti, R., Baldi, D., & Guerzoni, M. E. (2004). Interactions between high pressure homogenization and antimicrobial activity of lysozyme and lactoperoxidase. Int. J. Food Microbiol., 94, 123–135.

Verduyn, C., Suksomcheep, A., & Suphantharika, M. (1999). Effect of high pressure homogenization and papain on the preparation of autolysed yeast extract. World J. Microb. Biot., 15, 57–63.

Wong, H. H., O'Neill, B. K., & Middelberg, A. P. J. (1997). A mathematical model for Escherichia coli debris size reduction during high pressure homogenisation based on grinding theory. Chem. Eng. Sci., 52(17), 2883–2890.

Wuytack, E. Y., Diels, A. M. J., & Michiels, C. W. (2002). Bacterial inactivation by high-pressure homogenisation and high hydrostatic pressure. Int. J. Food Microbiol., 77, 205–212.

Yipeng, M., & Binghua, M. (2005). A nanometer microemulsion of spices and its preparation. CN Patent, CN 1561815.

Yuste, J., Capellas, M., Fung, D. Y. C., & Mor-Mur, M. (2004). Inactivation and sublethal injury of foodborne pathogens by high pressure processing: Evaluation with conventional media and thin agar layer method. Food Res. Int., 37(9), 861–866.

Zamora, A., Ferragut, V., Jaramillo, P. D., Guamis, B., & Trujillo, A. J. (2007). Effects of ultra-high pressure homogenization on the cheese-making properties of milk. J. Dairy Sci., 90, 13–23.

zur Mühlen, A., Schwarz, C., & Mehnert, W. (1998). Solid lipid nanoparticles (SLN) for controlled drug delivery – Drug release and release mechanism. Eur. J. Pharm. Biopharm., 45, 149–155.

CHAPTER 20

Key Issues and Open Questions in GMO Controls

John Davison *and* **Yves Bertheau**

Contents

Abstract

Developments in biotechnology over recent decades have resulted in the cultivation of a range of GMO plants. Many European citizens express a strong reluctance to GMOs and derived products, although a discrepancy between expressed opinions and attitudes of consumption can be observed. The EU, as well as Japan, South Korea, and Russia, implemented a set of regulations, for enabling the consumers' freedom of choice, based on clear and reliable mandatory labeling of food- and feedstuffs above the regulatory thresholds of the fortuitous presence of GMOs.

The current EU legislation established (i) a Community Reference Laboratory (CRL) supported by the European Network of European Laboratories (ENGL), (ii) mandatory traceability of food- and feedstuffs, and (iii) obligation to GMO notifiers to provide the CRL with specific and quantitative detection methods that need to meet specific quality criteria, as well as appropriate reference materials (control samples). This new legislation has taken into account several of the major problems revealed by EU research programs. However, a number of serious challenges are not covered by the current European legislative framework, e.g. (i) lack of methods for non-EU approved GMOs, (ii) cost-effectiveness of sampling,

Global Issues in Food Science and Technology
© 2009 Elsevier Inc.

(iii) rapid detection methods for GMOs, GM taxa and controls (CaMV, *Agrobacterium* spp., *Bacillus* spp., etc.), and (iv) differentiation of stacked genes from simple mixtures of GMOs.

The cost of analyses impacts the cost of end-products and is also a concern for enforcement laboratories, whose budgets are not unlimited, while the number of new detection areas (allergens, mycotoxins, etc.) is still growing. Mandatory traceability should thus lead to a reduced need for analytical controls. Availability of appropriate DSS (Decision Support Systems) that would assist the stakeholders In selectIng the most appropriate analytical strategy in a certain environment could also be a major step forward. The DSS may also be used for interpreting data and taking appropriate decisions in front of particular results, e.g. when suspecting the presence of unknown GMOs.

Most of the issues and open questions discussed here are common to several detection areas and should encourage discussions between method developers, analysts, and stakeholders.

I. INTRODUCTION

Genetically modified plants for human consumption or animal feed are mainly grown in the USA and Canada, with increasing production in Brazil, Argentina, and China. Europe cultivates only a small amount of GM crops and these are mainly GM maize grown in Spain. While GM food is readily accepted in the USA (Fernandez-Cornejo and Caswell, 2006), European consumers have shown considerable reluctance. Part of the reason is because a number of food safety scares – mad-cow disease and the related Creuzfeld-Jacob syndrome (UK), foot and mouth disease in livestock (UK), *Listeria* in refrigerated products (Europe), *Salmonella* in eggs, dioxin in chickens, milk, and meat (Belgium, France), radioactivity from Chernobyl (1986), and more recently, bird-flu – have left the general public distrustful of the food safety records of European governments, and thus they are easy prey for non-governmental, anti-GMO (genetically modified organism) organizations (Eurobarometer, 2005).

As a result, the European Community (EC) has developed a series of regulations to ensure GMO safety, detection and traceability, and consumer freedom of choice in GM products:

- **Regulation (EC) 258/97** provides for labeling of new food and includes, but not specifically, GMOs.
- **Directive 2001/18** covers the deliberate release of GMOs in the environment in the absence of specific containment measures (field trials and cultivation). It also regulates commercialization (i.e. importing, processing, and transforming) of GMOs into industrial products.

- **Regulation (EC) 178/2002** resulted in the creation of the European Food Safety Authority (EFSA) and its general obligation to trace GMO food at least one step forward and one step backwards in the food chain.
- **Regulation (EC) 1946/2003** is concerned with the trans-boundary movement and accompanying documentation for GM food and feed, which, under the terms of the Carthagena Protocol on Biosafety, must be furnished by exporters of GMOs to Europe.
- **Regulation (EC) 1829/2003** was designed to facilitate GMO detection by obligating the providers of GMOs to disclose methods of their detection. These methods are then verified and certified by the Central Reference Laboratory (CRL) of the EC Joint Research Centre (JRC), before being made public. A recent EC regulation, EC 1981/2006, provides for a fee to be paid by the applicant to the CRL for this service. The EC Regulation also provides a temporary measure, active only until 2007, that allows a 0.5% GMO threshold for the 'adventitious' presence of GMOs that have received a positive assessment by EFSA but that have not yet been formally authorized. Such GMOs, when present above the 0.5% threshold, cannot be marketed, even when labeled.
- **Regulation (EC) 1830/2003** imposes a specific traceability requirement on GMOs, over and above that of the general traceability in EC Regulation 178/2002. This labeling requirement in EC Regulation is a response to a strong European demand (which is also reflected in some other countries such as Korea and Japan). The EC Regulation imposes labeling for authorized GMOs above a threshold of 0.9%. Labeling is not required for food and feed containing the 'adventitious' or 'technically unavoidable' presence of authorized GMOs at levels less than 0.9%. Neither 'adventitious' nor 'technically unavoidable' is defined in the legal text (Craddock, 2004) though documentary proof must be submitted that suitable steps have been taken to avoid adventitious presence. The adventitious presence of unauthorized GMOs, such as the recent cases of BT10 maize and LLrice601, is not permitted.

EC guidelines have recently been developed for the co-existence of GMOs and non-GMOs in the food chain (COM2006 104), and guidelines for the fortuitous presence of GMOs in seeds for planting will also be developed. These guidelines will provide a basis for national regulations by relevant, competent authorities.

These GMO EC Regulations pose a considerable challenge for all stakeholders including GM cultivators, transporters, shippers, packers,

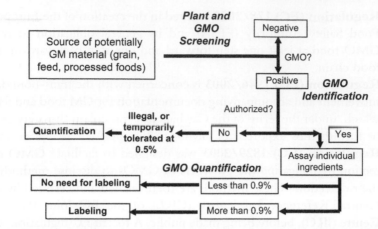

Figure 20.1 Decision tree. Reliable decision tree applicable to various potentially GMO matrices.

processors, and distributors. Their implementation raises difficulties for government regulators, food-controlling agencies, and laboratories involved in GMO detection. At the analytical level, with the support of EC research projects (e.g. QPCRGMOFOOD, GMOCHIPS, Co-Extra) in addition to efforts of the ENGL (European Network of GMO Laboratories) network and Joint Research Centre, it has been necessary to develop new methods. A typical decision tree to help implement GMO EC Regulations in routine analysis laboratories is shown in Figure 20.1.

II. OPEN QUESTIONS IN GMO CONTROLS

II.A. Interpretation of GMO percentage in adventitious presence

The threshold level of adventitious GMO presence above which food must be labeled is given by EC Regulation 1830/2003 as 0.9%. This figure represents a mass/mass ratio, which is the unit generally used by farmers, transporters, and distributors. However, this raises the problem that the current, most sensitive methods of GMO detection do not measure the mass/mass ratio but instead measure the DNA/DNA ratio, which is then converted to mass/mass using standards called Certified Reference Material (CRM), which are mixtures of homogenized ground powders containing certified mass/mass percentages of GMO material. While the DNA-based measurements can be accurately determined, for a variety of different biological reasons, the conversion from DNA to mass has numerous difficulties in interpretation. The EC recommendation, 2004/787/2000, proposes that GM copy numbers should be expressed as the percentage in relation to taxon-specific gene target DNA copy numbers calculated in terms of haploid genomes. However, this recommendation gives rise to new difficulties that may have non-negligible effects on estimating the percentage of GMO in a given sample. These problems have been discussed in depth (ENGL, 2005; Holst-Jensen et al., 2006; Weighardt, 2006) and are based on the non-evident relationship between DNA content and plant mass. For example, there may be a non-uniform nuclear DNA content between different lines of the same plant. Similarly, the same plant line grown under different environmental and agricultural conditions may have different DNA/mass ratios. Different parts of a plant may have different ploidy (e.g. the endosperm of maize is triploid). Finally, transgenic plants that are respectively homozygous diploids, heterozygous diploids, and tetraploids, with one modification per genome, have DNA percentages of 100, 50, and 25%. However, under EC Regulations all are 100% GMO.

Another difficulty involves the simultaneous presence of different GMOs in the same food sample. If these are of the same species (same ingredient), then the threshold corresponds to the sum of the individual GMOs. Thus a sample containing three maize GMOs having individual levels of 0.2, 0.5, and 0.1% (total = 0.8%) would still be below the threshold.

EC Regulations also require the labeling of highly processed GMO-derived food products (e.g. refined sugar or maize oil) where no GMO-derived analyte may be detectable in the final product. The difficulties of enforcing such regulations, where no applicable test is available for differentiating GM and non-GM products, have been discussed (Craddock, 2004).

Recommendations to improve the coherence of the legislation with real-life practical methods of GMO detection have recently been proposed (Holst-Jensen et al., 2006). The International Seed Testing Association (ISTA) and the European Association for Bioindustries (Europabio) have requested a more flexible approach with performance-based methods.

II.B. Reducing the cost of GMO detection

GMO quantification is most accurately done by using the quantitative real-time polymerase chain reaction (QRT-PCR), which is an analytical technique that requires sophisticated state-of-the-art equipment, dedicated laboratories, and highly trained personnel. The cost of these analyses could result in increased prices for both GMO and non-GMO food and feed. Cost reduction is thus a major imperative. One solution is in the choice of reliable and cost-effective control plans.

Another question is whether the analyses really require accurate determination by qualitative RT-PCR measurements, as compared to analyses using multiple control plans with attributes that give a robust statistical position relative to the designated threshold for the GMOs (Kobilinsky and Bertheau, 2002; Kobilinski et al., 2002; Laffont et al., 2005).

Most statistical sampling procedures assume a homogeneous spatial distribution of the adventitious GMO in a cargo or transport container. However, for very large trans-Atlantic cargoes the distribution of GMOs in a shipment are unlikely to be evenly distributed; in addition the shipment may contain more than one type of GMO. The KeLDA (Kernel Lot Distribution Assessment) project demonstrated experimentally that the spatial distribution in shipments is heterogeneous and therefore sampling plans must be devised to take this into account. A software sampling-package called KeSTE (Kernel Sampling Technique Evaluation) has been elaborated by the Joint Research Centre (JRC) to assist in the best choice of sampling plans (Paoletti et al., 2006; see also JRC, KeLDA, and KeSTE websites).

II.C. Certified reference materials

Certified reference materials (CRM) are produced by the Institute of Reference Materials and Measurements (IRMM) of the JRC and derived from homogenates of plant material (e.g. maize kernels). CRMs are certified to contain specific mass/mass mixtures of non-GM and GM material. The host reference genes are simultaneously amplified in the QRT-PCR reaction and serve to quantify the DNA ratios of the transgenes present in the unknown samples to give the mass/mass ratios, using the calibration curves generated with certified CRM m/m. Ideally, this CRM should be both standardized in quality and homogeneity and be readily available, but in reality there may be lapses in availability, particularly for old GMOs no longer grown on a commercial scale and for new GMOs that are recently, and often asynchronously, approved. Another difficulty is that no CRM for GM material exists that is authorized or commercialized, such as BT10 maize or LL601 rice. There may also be problems with irreproducible quality and stability. Finally, CRMs at the IRMM are costly to produce, thus adding to the cost of GMO analysis.

In principle, it would be possible to provide DNA-based CRM either from plant genomic DNA or from specially constructed bacterial plasmids containing multiple transgenes and reference genes. This would have the advantage of being rapidly available, low-cost, and having guaranteed continuity and stability. Attempts have been made to use DNA-based reference material in Europe and Japan (Kuribara et al., 2002; Mattarucchi et al., 2005), but while these have been successful, there remains the problem of recognized commutability of the results to the commonly used mass/mass ration. There is a potential problem of cross-contamination of test samples by CRM plasmid material that may be present on laboratory surfaces and equipment, though in principle this may be resolved by good laboratory practice involving cleaning and physical separation of parts of the laboratory performing different functions (ISO/DIS 24276). There is also the potential problem of construction, and maintaining plasmids containing multiple transgenes and reference genes, given that these are in constant change as new GM plants are introduced and others abandoned.

III. DNA DETECTION AND QUANTIFICATION METHODS

III.A. Types of PCR machines and chemistries

While there are several possible DNA amplification methods, using several different chemistries, on several different commercially available machines, the recognized world standard for GMO quantification is the Applied Biosystems machine, using Taqman® chemistry. Studies are presently under way, as part of the EC-sponsored Co-Extra program, to extend the portability of PCR protocols to other machines and chemistries.

Primers for the QRT-PCR reaction may target any one of the genetic elements in the cassette (promoter, terminator, or one of the transgenes). Alternatively, they may target the junctions between two genetic elements (construct specific primers). However, in most cases, the primers will target the edge- (or border-) fragment (specific insertion site) since this region clearly identifies a specific GMO (Figure 20.2). In addition, appropriate negative controls from donor organisms (e.g. *Agrobacterium*, CaMV virus) are necessary to ensure that a positive result is not simply due to contamination by these microorganisms. Finally, for quantification, primers specific for the taxon-specific reference gene (e.g. ADH coding for alcohol dehydrogenase, in the case of maize) are necessary. To economize time and cost, it may be useful to screen unknown samples using qualitative PCR or other methods, before quantitative determination by the QRTPCR reaction.

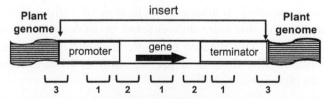

Figure 20.2 Design of PCR Primers. Primers may target: (1) the genetic control element (promoter, terminator), (2) the junction between the gene and genetic control elements (construct specific event), and (3) the junction between the genome and construct, known as the edge, or border fragment (insertion-specific event). Other necessary primers include those for the plant-specific reference gene and the donor microorganisms used in the construction.

Cost may further be decreased, and measurement uncertainty reduced, by using the modular approach (Holst-Jensen and Berdal, 2004), whereby each step from the sampling procedure through the analyte extraction to the quantitative analysis is independently validated. This system has the further advantage of being highly adaptable since it is able to integrate new modules. The modular approach was designed for GMO detection and validation but may be equally applied to other fields such as microorganisms or toxins.

Statistical analyses are necessary to reduce measurement uncertainty, which may be due to multiple causes, including matrix extraction, matrix quality, sampling, calibration, manipulator error, local and environmental conditions, and reference material (Bertheau *et al.*, 2002). Statistical programs can thus be made available by Co-Extra (http://www.coextra.eu) to minimize measurement uncertainty, to eliminate outlying points, and to validate robust methods.

III.B. Multiplex PCR

Successful attempts have been made to perform multiple simultaneous detections by using multiplex qualitative PCR. Successful quadriplex and nonaplex qualitative PCR reactions have been reported (Hernandez *et al.*, 2005; Onishi *et al.*, 2006). It is important to demonstrate that such laboratory protocols are transferable between different GMO laboratories (Hernandez *et al.*, 2005). The use of multifactorial plans, to simultaneously take into account the numerous variables of a multiplex qualitative PCR reaction, may provide optimal resolution in a single PCR run. Such qualitative PCR reactions do not give a quantitative result and are thus less used for regulatory purposes, though they serve a much-needed role for screening unknown samples or seed certification.

Attempts to apply multiplex QRT-PCR, for multiple targets higher than duplex, are generally less successful due to the overlap of the fluorescence spectrum of the different available fluorophores. Recently, a multiplex PCR system using a combination of three different fluorescent dyes and capillary electrophoresis, enabled simultaneous detection of eight targets plus an endogenous control, that were differentiated by molecular size and color (Nadal *et al.*, 2006). This method has high throughput potential and the necessary sensitivity (LOQ, limit of quantification, of 0.1%) to comply with the European directives.

III.C. The matrix approach using DNA chips

Considerable cost reduction could be achieved if test samples were simultaneously screened for the presence of a large number of transgenic constructs. Indeed, such an approach is becoming increasingly necessary as the number of possible transgenic inserts increases due to a wide variety of desired new traits (e.g. augmented nutritional value, salinity tolerance, drought tolerance, fruit-ripening characteristics, virus resistance, nematode resistance, pharmaceutical production, bioremediation). Multiple PCR amplicons from multiplex PCR reactions are hybridized to a microarray containing multiple immobilized DNA fragments (DNA chips) corresponding to the transgenes under investigation. These methods were developed under the EC 5th Framework GMOChips project (Leimanis *et al.*, 2006). From the patterns of this hybridization, the types of transgenes present in the original sample, and sometimes the exact identity of the transgenic plant, may be deduced (Figure 20.3). Discrepancies in the patterns of hybridization may indicate an unknown GMO; even so, the matrix approach may provide considerable information on this unknown GMO. Sophisticated DNA-chip analysis software, as well as a decision tree, is necessary for the interpretation of the hybridization results of the DNA chip. The advantage of the matrix approach is that it provides a multiple screening method, even for complex mixtures of GMOs. The same matrix approach strategy can be used with several methodologies such as multiplex PCR, multiplex PCR and micro-arrays hybridization, direct hybridization after whole-genome amplification, and high-throughput (or not) SNPlex (see below). The disadvantage is that it is qualitative and does not yield information on the quantities of GMOs present but can be used to position the GMO content versus a threshold using multiple control plans by attributes. To become generally acceptable the matrix approach will require standardized decision trees and a reliable publicly available database.

One approach under experimental investigation at INRA Versailles/ Evry involves the ligase chain reaction (LCR). This method, derived from the SNPlex method, was devised to detect single nucleotide polymorphisms (De la Vega *et al.*, 2005), whereby the DNA of an unknown sample is subjected to LCR in the presence of multiple probes. Only those probes finding a correct partner on the test DNA (i.e. hybridized side by side with

	GMO-1	GMO-2	GMO-3
P35S	+	+	+
Tnos	–	+	+
T35S	+	+	–
CryI(a)b	+	+	+
Bar	–	+	–
NtpII	–	–	–

Figure 20.3 The matrix approach. A possible hybridization pattern obtained using the matrix approach. Numerous small sequences detected by the GMO chip hybridizations may be common to several GMOs. Both primers are located within the same genetic element, e.g. P35S, Tnos, T35S, CryI(a)b, or may be constructed specifically.

exactly juxtaposed 5′ and 3′ ends) will prime the LCR; these ligated fragments are subsequently amplified by a standard PCR reaction using bor-
dering universal primers. In order to determine which of the multiple
probes in the LCR were in fact successful, the samples may be analyzed (by
e.g. capillary electrophoresis), and the products identified according to
molecular size. Alternatively, the SNPlex amplification products may also be
recognized by DNA hybridization to DNA chips containing an array of
known transgenic and non-transgenic samples, as described above.

IV. UNSOLVED PROBLEMS IN PCR DETECTION

IV.A. Known, unknown, and unauthorized GMOs

PCR reactions have a serious limitation, as they can only detect sequences
known (or reasonably predicted) to be present. From this it follows that a PCR
reaction cannot detect a completely unknown transgene or GMO, since prior
knowledge of the DNA sequence is necessary for the design of the primers used
to initiate the reaction. A first step towards the detection of unknown GMOs
would be to use computer modeling for the optimal design of DNA chips
(Nesvold *et al.*, 2005). Another possibility would be to use anchored PCR
primed from commonly found sequences such as promoters and terminators.

A statistical approach, a quantitative differential PCR being developed in
Versailles under the EC Co-Extra program, seeks to relate the copy numbers
of various common transgenic inserts to that of another common insert (e.g.
P35 promoter). An excess of one insert compared to another may indicate
the presence of an unknown GMO or stacked genes (next section).

EC Regulation 1830/2003 fixes a threshold of 0.9% of authorized GMO
above which labeling is required, and EC Regulation 1829/2003 fixes
a threshold of 0.5% for GMOs that have received a favorable EFSA assessment
but still await formal authorization. In contrast, there is no permissible level for
unauthorized and non-EFSA-assessed GMOs (recent examples include Bt10
maize or LL601 rice). Thus, such contaminated products are not permitted to
enter the European food and feed chains, and must be detected and rejected
beforehand. Determination of very low percentages of unauthorized GMOs
poses a problem with the limit of detection using conventional techniques,
since there is always a machine background, below which low levels cannot be
detected. Useful concepts for regulatory control are the limit of detection
(LOD) and the limit of quantification (LOQ). These limits vary according to
the system of analysis, the GMO to be detected, and the source of DNA
(practical LOD and LOQ), but the limit of detection is typically about 10
copies per reaction whereas the limit of quantification corresponds to about
100 copies per reaction.

IV.B. Stacked genes

A particular problem arises in the case of plant varieties containing 'stacked' transgenes. Such varieties typically arise from genetic crosses between two different transgenic lines containing different transgenes, so that two or more transgenic constructs may be present in the same cell (Halpin, 2005). Stacked genes may also be constructed by multiple transformation events of the same plant. The PCR reaction will be able to detect the two transgenes but will be unable to distinguish the 'stacked' gene scenario from the quite different scenario, whereby the sample contains a simple mixture of the two or more simple transgenic lines (Figure 20.4). For the moment the only solution to this problem has been to perform multiple PCR reactions on individual maize kernels, a method that is expensive, labor-intensive, and time-consuming (Akiyama *et al.*, 2005). This problem of detecting stacked GMO genes is similar to that of detecting multiresistant bacteria by direct PCR (i.e. without colony isolation) on clinical samples.

V. CONCLUSION

The issues in GMO detection and traceability resemble many issues in related fields of food contamination by microbes and toxins, and similar DNA-based techniques may be used. Due to ambiguous text in the EC Regulations, there is a major unresolved problem with the definition of

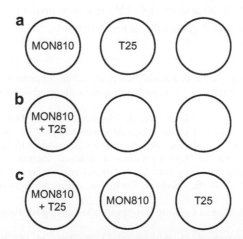

Figure 20.4 Stacked genes. The figure illustrates three real-life situations that cannot easily be distinguished: (a) shows a mixture of non-GMO seeds and two different simple GMOs, (b) shows a mixture of non-GMO kernels and a GMO carrying two stacked genes, (c) shows a mixture of two simple GMOs and a GMO carrying stacked genes.

percentage of GMO in a sample. There is also the problem relating to a practical definition, since in the farmers' fields, food distribution, food-processing, and food regulation the functional unit of the percentage of GMO is mass/mass. The conversion between these functional units is a source of difficulty and misunderstanding, and for this reason the use of CRM is preferred since this gives a direct mass/mass conversion of the DNA units. Thus, the use of CRM poses other problems and may not be sustainable in the long term, so that CRM may eventually need to be replaced by DNA-based plasmid or genomic reference material.

In ideal situations the determination of the percentage of adventitious GMO presence in a sample may be a simple and reproducible matter. In other circumstances, however, where the GMO is unknown or un-authorized, or present in mixtures or stacked versions, it may be very difficult. There is a general need for improvement of DNA detection techniques, such as quantitative differential PCR, the matrix approach, and statistical approaches to measurement uncertainty, in order to deal with unknown GMOs and to provide rapid screening methods for unknown samples. There is also urgent need for harmonized decision trees common to all stakeholders and regulatory agencies.

ACKNOWLEDGMENTS

Part of this work was partially funded by the EC FP6 research project Co-Extra (contract 007158) and the EC FP6 research project PETER (contract number 031717), and is gratefully acknowledged.

REFERENCES

Akiyama, H., Watanabe, T., Wakabayashi, K., et al. (2005). Quantitative detection system for maize sample containing combined-trait genetically modified maize. *Anal. Chem.,* 77(22), 7421–7428.

Bertheau, Y., Diolez, A., Kobilinsky, A., & Magin, K. (2002). Detection methods and performance criteria for genetically modified organisms. *J. AOAC Int., 85*(3), 801–808.

COM2006 104 (2006). Communication from the Commission to the Council and the Parliament. Report on the Implementation of National Measures on the Coexistence of Genetically Modified Crops with Conventional and Organic Farming. ©European Communities, 1995–2008. Website: http://ec.europa.eu/agriculture/coexistence/com104_en.pdf

Craddock, N. (2004). Flies in the soup–European GM labeling legislation. *Nat. Biotechnol.,* 4, 383–384.

De la Vega, F. M., Lazaruk, K. D., Rhodes, M. D., & Wenz, M. H. (2005). Assessment of two flexible and compatible SNP genotyping platforms: TaqMan SNP Genotyping Assays and the SNPlex Genotyping System. *Mutat. Res., 573*(1–2), 111–135.

Eurobarometer No. 224 (2005). Europeans, Science and Technology. ©European Communities, 1995–2008. Website http://europa.eu.int/comm/public_opinion/archives/ebs/ebs_224_report_en.pdf

European Network of GMO Laboratories (ENGL). Explanatory document on the use of "Percentage of GM-DNA copy numbers in relation to taxon specific DNA copy numbers calculated in terms of haploid genomes" as a general unit to express the percentage of GMOs. ©European Communities, 1995–2008. Website: http://engl.jrc.it/docs/HGE%20release%20version%201.pdf

Fernandez-Cornejo, J., & Caswell, M. (2006). USDA Report: The first decade of genetically engineered crops in the United States. Economic Information Bulletin.

Halpin, C. (2005). Gene stacking in transgenic plants – the challenge for 21st century plant biotechnology. *Plant Biotech. J., 3*, 141–155.

Hernandez, M., Rodriguez-Lazaro, D., Zhang, D., Esteve, T., Pla, M., & Prat, S. (2005). Interlaboratory transfer of a PCR multiplex method for simultaneous detection of four genetically modified maize lines: Bt11, MON810, T25, and GA21. *J. Agric. Food Chem., 53*(9), 3333–3337.

Holst-Jensen, A., & Berdal, K. G. (2004). The modular analytical procedure and validation approach and the units of measurement for genetically modified materials in foods and feeds. *J. AOAC Int., 87*(4), 927–936.

Holst-Jensen, A., De Loose, M., & Van den Eede, G. (2006). Coherence between legal requirements and approaches for detection of genetically modified organisms (GMOs) and their derived products. *J. Agric. Food Chem., 54*(8), 2799–2809.

Joint Research Centre (JRC). KeSTE (Kernel Sampling Technique Evaluation). European Commission, (1995–2008). Website accessed 2008. http://biotech.jrc.it/home/sampling%20Keste.htm

Joint Research Centre (JRC). KelDA (Kernel Lot Distribution Assessment) Project. European Commission, (1995–2008). Website accessed 2008. http://biotech.jrc.it/home/sampling_KeLDA.htm

Kobilinsky, A., & Bertheau, Y. (2005). Minimum cost acceptance sampling plans for grain control, with application to GMO detection. *Chemom. Intel. Lab. Syst., 75*(2), 189–200.

Kuribara, H., Shindo, Y., Matsuoka, T., et al. (2002). Novel reference molecules for quantitation of genetically modified maize and soybean. *J. AOAC Int., 85*(5), 1077–1089.

Laffont, J.-L., Remund, K. M., Wright, D., Simpson, R. D., & Grégoire, S. (2005). Testing for the presence of adventitious material in GM seed or grain lots using quantitative laboratory methods: statistical procedures and their implementation. *Seed Sci. Res., 15*, 197–204.

Leimanis, S., Hernandez, M., Fernandez, S., et al. (2006). A microarray-based detection system for genetically modified (GM) food ingredients. *Plant Mol. Biol., 61*(1–2), 123–139.

Mattarucchi, E., Weighardt, F., Barbati, C., Querciand, M., & Van den Eede, G. (2005). Development and applications of real-time PCR standards for GMO quantification based on tandem-marker plasmids. *Euro. Food Res. Technol., 221*(3–4), 511–519.

Nadal, A., Coll, A., La Paz, J. L., Esteve, T., & Pla, M. (2006). A new PCR-CGE (size and color) method for simultaneous detection of genetically modified maize events. *Electrophoresis, 27*(19), 3879–3888.

Nesvold, H., Kristoffersen, A. B., Holst-Jensen, A., & Berdal, K. G. (2005). Design of a DNA chip for detection of unknown genetically modified organisms (GMOs). *Bioinformatics, 21*(9), 1917–1926.

Onishi, M., Matsuoka, T., Kodama, T., et al. (2005). Development of a multiplex polymerase chain reaction method for simultaneous detection of eight events of genetically modified maize. *J. Agric. Food Chem., 53*(25), 9713–9721.

Paoletti, C., Heissenberger, A., Mazarra, M., et al. (2006). Kernel lot distribution assessment (KeLDA): a study of the distribution of GMO in large soybean shipments. *Eur. Food Res. Technol., 224*, 129–139.

Weighardt, F. (2006). European GMO labeling thresholds impractical and unscientific. *Nat. Biotechnol., 24*(1), 23–25.

Nanotechnology in Food Applications

CHAPTER 21

Food Nanotechnology: Current Developments and Future Prospects

Carmen Moraru, Qingrong Huang, Paul Takhistov, Hulya Dogan, *and* Jozef Kokini

Contents

Abstract

Nanotechnology, one of the most promising scientific fields of research in decades, is the science and technology that has enabled the study of phenomena at the nanometer scale and the manipulation of materials at macromolecular, molecular, and atomic scales. The unusual properties displayed by materials in the nanometer range provide ground-breaking scientific, technological, and commercial opportunities. Using the tools of nanotechnology, it is now possible to study the behavior of individual molecules and the interactions in which they participate, and use these interactions to improve the functionality of synthetic or biological materials.

Nanotechnology has the potential to revolutionize the global food and agricultural system. Nanoscale control of food molecules could allow the modification of macroscale characteristics of foods such as processability, texture, sensory attributes, and shelf life. It is expected that nanotechnology will facilitate the development of lighter and more

Global Issues in Food Science and Technology
© 2009 Elsevier Inc.

ISBN 9780123741240

precise food manufacturing equipment, novel packaging materials, as well as non-polluting, cheaper, and more efficient food-processing techniques.

The food industry is currently trying to identify opportunities to use the most promising discoveries of nanotechnology to improve food processing and food products. As with any new technology, a series of challenges is foreseen. In particular, the potential dangers caused by the misuse of nanotechnology should be clearly identified and prevented. Overall though, it is generally recognized that nanotechnology has the potential to enable significant steps for the enhancement of human life, partly through the improvement of the food and agricultural system.

This chapter will provide an up-to-date, comprehensive evaluation of the existing and upcoming applications and implications of nanotechnology for the food and agricultural sector.

I. INTRODUCTION

Nanotechnology, defined by the British Standards Institution (BSI) as the 'design, characterization, production and application of structures, devices and systems by controlling shape and size at the nanoscale' (BSI, 2005), has emerged as one of the most promising scientific fields of research in decades. This new area of science and technology allows scientists to study phenomena that occur at the nanometer scale and to manipulate materials at macro-molecular, molecular, and atomic scales, where properties and functionalities differ significantly from those at macroscopic scale. 'Nanoscale' refers to the size range where one or more dimensions are of the order of 100 nm or less (BSI, 2005).

In the United States, interest in nanotechnology received a big boost after the National Nanotechnology Initiative identified it as an emerging area of national interest (Srivastava, 2000). The unusual properties displayed by materials at the nanometer-length scale and the development of tech-nologies capable of manipulating or self-assembling materials molecule by molecule, have since been shown to potentially provide world-changing scientific, technological, and commercial opportunities.

The building strategies used in nanotechnology include the 'top down' approach, in which nano-level structures are generated by breaking up bulk materials through milling, nanolithography, or precision engineering; and the 'bottom up' approach, in which nanostructures are built from individual atoms or molecules capable of self-assembling. The 'bottom up' strategy has led to the development of remarkably strong materials with precisely con-trolled structure and properties, high surface area particles, and materials that encapsulate active compounds, or smart coatings that are self-cleaning or can change color depending on environmental conditions. Nanotechnology has also opened up new ways for studying individual molecules and specifically the intra- and inter-molecular interactions in which they participate, and can improve the complexity and functionality of biological materials. Examples include elucidation of the mechanisms of catalysis and enzymatic reactions, muscular contraction, cellular transport, DNA replication and transcription, DNA unknotting and unwinding, and protein folding and unfolding.

Following the first feature article published on the topic of food nanotechnology (Moraru *et al.*, 2003), several more authors addressed the potential of nanotechnology in areas such as food safety and biosecurity (Baeumner, 2004), nanoscale properties of food materials (Weiss *et al.*, 2006; Lee *et al.*, 2007), and functional foods for health promotion (Chen *et al.*, 2006). This chapter will provide an up-to-date evaluation of the applications and implications of nanotechnology for the food and agricultural sector.

II. THE POTENTIAL OF NANOTECHNOLOGY FOR THE FOOD AND AGRICULTURAL SYSTEM

The progress made in recent years suggests that nanotechnology has the potential to revolutionize the global food and agricultural system. Nanoscale control of food molecules could allow the modification of many macroscale characteristics of foods, such as texture, sensory attributes, processability, and shelf life. The tools of nanotechnology have already allowed scientists to better understand the way in which food components are structured, and how they interact with each other; it is expected that this understanding will enable a more precise manipulation of food molecules for the design of healthier, tastier, and safer foods. Non-polluting, cheaper, and more efficient processes will likely be developed; lighter and more precise food manufacturing equipment will be built; and 'smart' packaging materials will be developed and used to safely package wholesome, nutraceutical-loaded foods. Development of novel delivery methods and the manufacture of sensors for pathogen detection are some other useful applications of nanotechnology in the food sector.

Some of the benefits of nanotechnology will be conveyed to the food system directly through agriculture and agricultural research. Newly created tools in molecular and cellular biology will enable significant advances in reproductive science and technology, disease prevention, and treatment of plants and animals, potentially boosting the production of raw food materials, while the development of biosensors for pathogen and contaminant detection in agricultural products will help ensure the safety of the food supply. Conversion of renewable agricultural materials or food waste into energy and useful byproducts is an environmentally oriented area of research that could be greatly enhanced by nanotechnology.

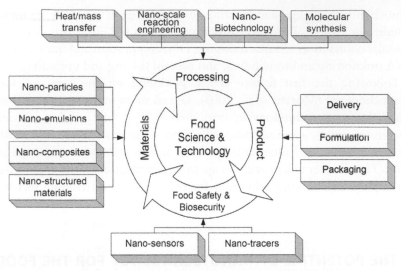

Figure 21.1 Application matrix of nanotechnology in food science and technology (Moraru *et al.*, 2003).

Four major target areas in food science and technology are expected to be significantly enhanced by nanotechnology: (1) development of materials with novel functionality, (2) micro- and nanoscale processing, (3) product development, and (4) the design of methods and instrumentation for food safety and biosecurity. A matrix of the applications of nanotechnology in the food industry, grouped by target area, is shown in Figure 21.1.

III. CHARACTERIZATION AND MANIPULATION OF FOOD BIOMOLECULES AT THE NANOSCALE

III.A. Instrumentation for studying nanoscale properties of food and biopolymer systems

Nanotechnology uses a range of tools to observe, characterize, and control phenomena at the nanoscale. Scanning probe microscopy (SPM) is a family of techniques that provides images of surface topography and, in some cases, allows quantification of surface properties down to the molecular or atomic scale. Since the invention of atomic force microscopy (AFM) and its use in structural biology of nucleic acids, this technique has been intensely used as a micromanipulation research tool (Binnig *et al.*, 1986; Viani *et al.*, 2000). Compared to other microscopic methods, AFM has the advantage of allowing imaging of non-conducting biomaterials including food materials, is one of the first and most popular applications of AFM.

AFM operates in various modes, allowing a multitude of applications. When used in the contact mode, AFM can measure properties like stiffness, hardness, friction, elasticity, or adhesion on a surface, down to the molecular level. Force–distance curves can be obtained by extending the AFM cantilever tip down to the surface of a sample, making contact between the tip and the sample surface, followed by retracting the tip from the surface (Figure 21.2). The slope of the force–deformation curve can be used as a measure of the modulus of elasticity at the sample surface. The non-contact and intermittent contact modes of AFM are used when a probability of damaging the sample surface exists due to the contact with the cantilever. In the intermittent contact mode, the tip of the cantilever oscillates at a resonant frequency and taps the sample surface lightly during each oscillation cycle. The decrease in oscillation amplitude occurring when the tip is brought closer to the sample is used as a feedback parameter to build surface morphology. The tapping mode of AFM can be used for structural investigation of delicate biopolymer structures, such as hylan (Cowman *et al.*, 2000) or xanthan gum (Camesano and Wilkinson, 2001). The non-contact mode is also very useful in imaging isolated biomacromolecules or supramolecular assemblies, since it is able to avoid sample disruption or tip contamination; several non-contact studies have been performed on food biomacromolecules such as xanthan gum, k-carageenan, gellan, and collagen (McIntire and Brant, 1997).

Some AFM measurements employ the colloidal probe technique, in which the tip of the cantilever is chemically modified by attaching silica beads functionalized with amino groups, or other molecules to it (Hilal

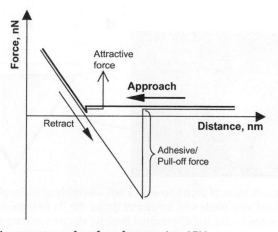

Figure 21.2   Measurement of surface forces using AFM.

et al., 2002; Sakai *et al.*, 2002). Modification of AFM probe tips by covalent linking with a variety of functional groups enables the direct probing of specific molecular interactions and imaging with chemical sensitivity. In many cases, the probe tip is ω-functionalized with alkyl thiols, which spontaneously form monolayers that terminate in well-defined functional groups (Figure 21.3). A range of ω-functionalized alkyl thiols is now commercially available, including those with methyl, amine, carboxylic acid, hydroxyl and phenyl head groups (Smith *et al.*, 2003).

By using probe tips functionalized with hydrophobic or hydrophilic molecules, Frisbie *et al.* (1994) showed that the adhesive interactions between simple hydrophobic–hydrophobic (-CH$_3$ ... -CH$_3$), hydrophobic–hydrophilic (-CH$_3$ ... -COOH), and hydrophilic–hydrophilic (-COOH ... -COOH) groups could be measured reproducibly. The difference between these adhesive interactions correlated well with the friction images of the patterned sample surface, demonstrating that the spatial distribution of hydrophilic and hydrophobic functional groups can be predictably mapped with functionalized tips. Examples of force–displacement curves recorded on -CH$_3$ and -COOH surfaces with functionalized probe tips are shown in Figure 21.4.

Schonherr (1999) determined the pull-off force distributions by using a COOH-terminated tip with self-assembled monolayers (SAMs) of octadecanethiol (COOH-terminated surface) and 11-mercaptoundecanoic acid

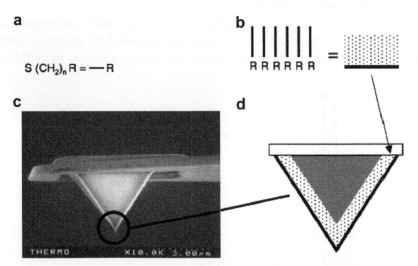

Figure 21.3 Modification of probe tip with a self-assembly of molecular monolayers. (a) ω-functionalized alkyl thiols with functional group (R); (b) formation of monolayers (R group represented by the thick horizontal line); (c) electron micrograph of a typical AFM probe; (d) schematic representation of the AFM probe (Smith *et al.*, 2003).

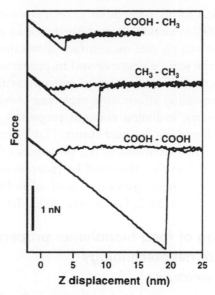

Figure 21.4 Typical force–displacement curves recorded for -COOH/-CH_3, -CH_3/-CH_3, and -COOH/-COOH tip-sample interactions (Frisbie et al., 1994).

(CH_3-terminated surface) on sputtered gold. Noy et al. (1997) provided images of a patterned SAM with 10×10 μm regions terminated with –COOH groups that repeated every 30 μm in a regular pattern. Topographical images failed to reveal this pattern since the surfaces exhibited almost flat topography across the CH_3- and COOH-terminated regions of the sample, while the chemical force maps obtained with functionalized tips readily showed chemical information about the surfaces.

The development of new SPM instrumentation, such as the near-field scanning optical microscope (NSOM) (Betzig et al., 1991), scanning thermal microscope (SThM) (Majumdar et al., 1993; Majumdar, 1999), scanning capacitance microscope (SCM) (Williams et al., 1989), magnetic force and resonance microscopes (MFM) (Hobbs et al., 1989; Rugar et al., 1992), and scanning electrochemical microscope (SECM) (Bard et al., 1991), enhanced the capability of SPM investigations. Among the applications of SPM for analysis of nanomaterials, a sensor that can measure atomic-level forces between two surfaces (as they approach each other and come into contact) has been developed (Houston and Michalske, 1992). The interfacial force microscope (IFM), which utilizes a feedback force sensor that eliminates the snap-to-contact event inherent in other scanning force microscope designs, allows more accurate measurement of the full range of adhesive interactions; it can also be used to image surfaces. Another instrument, the 'Atom Tracker,' can observe an individual atom in motion and track this motion up to 1000 times faster than a conventional STM (Swartzentruber, 1996). Continuous monitoring of an individual atom in motion as it binds to various sites on a surface allows the diffusing atom to become a probe of the surface structure and its properties.

An entire new generation of analytical instrumentation and nanoscale devices capable of providing information regarding physical, chemical, and mechanical phenomena, including material properties at nanoscale, will most likely be developed in the near future (Takhistov, 2006). As non-destructive, real-time measurements of the properties of nanoparticles and nanostructured materials evolve, there will be exceptional opportunities for fundamental and applied investigations, as well as problem-solving at the nanoscale (Brust and Kiely, 2002; Haruyama, 2003; Takhistov, 2006).

III.B. Investigation of food biopolymer properties using the tools of nanotechnology

III.B.1. Molecular structure

AFM has been increasingly used in the last decade for the structural characterization of food biopolymers such as starch (Gallant *et al.*, 1997; Baker *et al.*, 2001), hydrocolloids (Baker *et al.*, 1997; McIntire and Brant, 1997), and proteins (McMaster *et al.*, 1999 – gliadin molecules). AFM images of zein film drop, deposited onto a silicon wafer with a hydrophobic coating, helped Panchapakesan (2005) visualize the orientation of zein molecules on the surface of the film (Figure 21.5). Ellipsoidal structures with a mean height of 13.6 nm were observed on the surface of the spin-cast films.

AFM also enabled the study of rheological properties of single-polymer chains, such as proteins (Nakajima *et al.*, 2001; Rief and Grubmüller, 2002), polysaccharides (dextran, amylose, cellulose) (Rief and Grubmüller, 2002), and DNA (Marko, 1997; Clausen-Schaumann *et al.*, 2000; Strick *et al.*, 2000). Nitta *et al.* (2000) used AFM to measure the local elastic

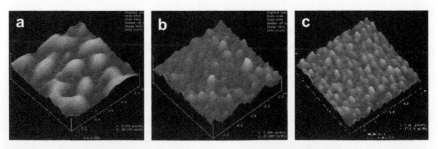

Figure 21.5 AFM images of spin-cast zein film with various nanoscale well patterns: (a) 0.1% film of 90% EtOH insoluble zein fraction, (b) 0.01% spin-cast commercial zein film, and (c) 0.1% commercial zein (Panchapakesan, 2005).

modulus of agar gels, which correlated well with the elastic modulus obtained from traditional rheological measurements.

Most recently, the conformation of bovine serum albumin (BSA) and its interactions with negatively charged surfaces in saline solutions of different pH were investigated via small-angle neutron scattering (SANS) and AFM-based chemical force microscopy (CFM) (Li et al., 2008). In this study, the contribution of elementary interactions was extracted from statistically averaged force–extension curves. A comparison of the SANS and CFM results showed that, at pH lower than the isoelectric point (pI) of BSA, both the size of the BSA chain and the stretchable extension of BSA fragments decreased with increase of pH. This suggested that when the BSA chain is attractive to the substrate, the change in chain conformation on the surface presents similar pH dependence as in bulk solution. However, at pH higher than the pI, the conformation change of the BSA chain in bulk solution is opposite to that of BSA fragments on the surface. Such an approach could be extended to the study of interactions between various food components at nanoscale.

III.B.2. Biopolymer transitions

The development of nanomaterials requires a good understanding of their reactions to stress, strain, temperature, plasticizers, and aging at the molecular and atomic levels. For nanostructured materials, it is important to be able to characterize the local properties of the matrix, which is extremely difficult, especially when interactions between the different phases occur (Oulevey et al., 1998).

The ability to manipulate individual biomolecules using AFM has made it possible to study their structural transitions and action mechanisms (Strick et al., 2000). Atomic and friction force microscopy has been used by various researchers for measuring local glass transition, and nanorheological and nanotribological properties of polymers (Terada et al., 2000; Morris et al., 2001; Nakajima et al., 2001; Boskovic et al., 2002). One drawback in such measurements is the significant friction that occurs in the system, hindering the accuracy of the data. To overcome this problem, Oulevey et al. (1998) developed a technique able to probe the properties of the submicron phases of inhomogeneous materials, called variable-temperature scanning local-acceleration microscopy (T-SLAM). This is the nanoscale equivalent of DMTA, and gives information related to the internal friction and the var-iations of the dynamic modulus of nanometer-sized volumes. By using T-SLAM, it is possible to study the defect dynamics in nanomaterials and composites, to thus determine primary and secondary relaxations as well as viscoplasticity (Oulevey et al., 1998). Micro-thermal analysis is another technique that has generated considerable interest in the polymer and pharmaceutical industries, and can provide similar information.

Non-intrusive determination of the local phase behavior is extremely valuable for understanding the miscibility of food biopolymers. The development of the above-mentioned techniques facilitates such studies, which could ultimately lead to improved control and design of the quality and stability of foods.

III.B.3. Intermolecular interactions

Most biological processes, such as DNA replication, protein synthesis, drug interaction, as well as interactions between biomolecules, are largely governed by intermolecular forces and can also be measured with AFM (Willemsen et al., 2000; Strick et al., 2001; Rief and Grubmüller, 2002). By combining SPM and AFM with other imaging, mechanical, and spectroscopic methods, it is now possible to quantitatively characterize the structures of polymers at the micron and nanometer levels, as well as the intra- and inter-molecular forces stabilizing such structures.

A research group from The Technical University of Munich measured the strength of the chemical bond between a water-soluble vitamin (biotin) and a protein (streptavidin) (Florin et al., 1994). Macromolecular adsorption and interactions at the liquid–liquid interface are of considerable importance in food processing. Detailed studies of the distribution and interactions of surfactants at interfaces in emulsions and micro-emulsions have become possible due to the development of in-situ techniques such as AFM and total internal reflection fluorescence microscopy (TIRFM) (Gajraj and Ofoli, 2000).

The gelation ability, the mechanisms of gelation, and the microstructure of the resulting gels were studied for biopolymers such as k-carageenan (Ikeda et al., 2001) and xanthan gum (Morris et al., 2001). Gajraj and Ofoli (2000) visualized protein networks by AFM and showed that their resistance to fracture determines the interfacial stability. The non-covalent interactions of the high-molecular-weight (HMW) subunits of wheat glutenin, which play an important role in the viscoelasticity of wheat flour dough, were also investigated using AFM (Humphris et al., 2000).

AFM has also been proven useful in studying the adsorption of small particles and biological molecules from an aqueous fluid onto a solid surface (Hilal et al., 2002). This could lead to a better understanding of mineral fouling of food-processing equipment. Mineral deposits are a serious problem in heat exchangers used to pasteurize liquid foods (i.e. milk or beverages), since such deposits hinder the heat transfer, leading to higher energy consumption, difficulties in cleaning, as well as reduced product quality and safety problems. A good understanding of factors controlling fouling of such surfaces could facilitate the design of heat exchangers to minimize the formation of mineral deposits and to allow easier cleaning, which would aid the processing of liquid foods significantly. The in-depth study of membrane fouling in membrane separation processes, such as microfiltration and ultrafiltration, has also been approached using AFM (James *et al.*, 2003).

IV. DEVELOPMENT OF NOVEL NANO-STRUCTURES FOR FOOD APPLICATIONS

Nanotechnology has provided unprecedented opportunities for the development of novel nanomaterials and the investigation of their properties at the nanoscale. Nanomaterials can be found in nature, such as nano-particles existing in soil (clays, zeolites, imogolite, iron, and manganese oxides), but they can also be created through nanotechnology itself (Takhistov, 2006). Nanostructured materials exhibit unique properties that open windows of opportunity for the creation of new, high-performance materials that will have a critical impact on food manufacturing, packaging, and storage. Some examples of how nanomaterials can be used in food applications will be discussed in the following section.

IV.A. Manipulation of molecules at nanoscale for improved processing

Freezing is one of the best methods known for food preservation. When properly processed, packaged, and stored, frozen foods come close to fresh foods in terms of nutrition, color, flavor, and texture; however, temperature fluctuations in the frozen product may cause the free water to experience several cycles of ice formation and melting. This can lead to a number of undesirable effects, including moisture migration, dehydration, structural breakdown, and formation of large ice crystals, imparting a gritty mouthfeel upon consumption of the product. By using purified extracellular ice nucleators (ECIN), ice nucleation temperatures can be elevated, which promotes freezing and reduced freezing time and leads to significant energy savings (Li and Lee, 1995; Zasypkin and Lee, 1999). Recently, Huang (current author) and associates at Rutgers University developed a new processing method that forms a nanoscale ECIN coating on the surface of polymer films by using the layer-by-layer (LbL) approach. LbL is widely used in the construction of multilayer films of polyelectrolytes for formation of two-dimensional structures on flat substrates. With suitable control of the charge density of the polyelectrolyte solution, it is possible to dramatically manipulate the molecular organization, surface morphology, and mechanical properties of the multilayer films (Decher, 1997). Lin *et al.* (2008) used the quartz crystal microbalance with dissipation monitoring (QCM-D) technique to monitor (in-situ) the multilayer formation of poly(diallyl-dimethylammonium chloride) (PDDA)/kappa-carrageenan. The surface morphology of the multilayer films by PDDA with kappa-carrageenan and iota-carrageenan was imaged by using AFM in the tapping mode, which indicated that the surface morphology of the final polysaccharide multilayer film depends on the conformation of the polysaccharides in solution during the multilayer buildup. Furthermore, the same group of researchers successfully prepared ECIN nano-coating on the carrageenan/PDDA multilayer polymer film surface. Preliminary results indicated that the water on the multilayer film surfaces containing ECIN was frozen, while the water on the same surfaces without ECIN remained in liquid form, suggesting that the ECIN nano-coating still maintained ice nucleation activity.

Nanotechnology tools could also be used to functionalize microfiltration or ultrafiltration polymeric membranes by filling their pores with polymeric or oligomeric liquids with an affinity for compounds of interest (Jelinski, 1999). An Australian–American team of researchers has discovered a nanoparticle-enhanced membrane that combines organic polymers with inorganic silica nanoparticles, enabling large molecules to pass through more readily than smaller molecules (Kingsley, 2002). The addition of silica resulted in over 200% improvement in the flux and increased the permeability of the membrane. The increased permeability was due to the nanoparticles, which, according to the authors, pushed the polymer chains apart, creating larger openings in the membrane's structure. The reported use of such membranes is to purify ethanol and methanol inexpensively, but food-related applications could also become possible.

These are just a couple of examples of how manipulation at the nanoscale could be used directly to enhance food-processing operations, and the future will probably bring about a multitude of such examples.

IV.B. Nanotubes and high surface area nanomaterials

Nanotubes, a relatively new generation of nanomaterials, were discovered in 1991 by the Japanese electron microscopist Sumio Iijima of NEC Corporation. Nanotubes are made by 'winding' single sheets of graphite with honeycomb structures into very long, thin tubes that have a stable, strong, and flexible structure. Nanotubes are the strongest fibers known today; one nanotube is estimated to be 10–100 times stronger than steel per unit weight. Researchers are trying to exploit their extraordinary strength and flexibility, and use, in making nanotube-reinforced composites with high fracture and thermal resistance. Such new materials could replace conventional ceramics, alumina, or even metals in the building of aircraft, gears, bearings, car parts, medical devices, sports equipment, and industrial food-processing equipment (Spice, 1999; Gorman, 2003; Zhan et al., 2003).

Recent studies suggest the use of carbon nanotubes for biological purposes, such as crystallization of proteins, building of bioreactors and biosensors (Huang et al., 2002). However, for biological applications, the insolubility of carbon nanotubes in aqueous media needs to be overcome. A research group at University of California solubilized single-wall carbon nanotubes (SWNT) in aqueous iodine–starch solutions (Dagani, 2002), while a research group in Israel obtained a similar result using aqueous solutions of gum arabic (Bandyopadhyaya et al., 2002). Other solutions for solubilization of SWNT consist of functionalizing the tubes with glucos-amine (Pompeo and Resasco, 2002) or bovine serum albumin (Huang et al., 2002). In another example, nanotubes are formed by the self-assembly of phospholipid bilayers capable of entrapping active compounds. Due to their biocompatibility, such tubules are ideal for delivery in biological systems (Jelinski, 1999). Carbon nanotubes, particularly multi-wall nanotubes (MWNT) with well-defined nanostructures, can also be used to build sensors.

Manufacturing of nanotube membranes is an area with significant potential for use in food systems. High-selectivity nanotube membranes can be used for analytical purposes, as part of the sensors for molecular recognition of enzymes, antibodies, various proteins and DNA, or for the membrane separation of biomolecules, such as proteins (Lee and Martin, 2002; Rouhi, 2002). The selectivity and yield of membranes currently used in the food industry are not fully satisfactory, mainly due to the limited control of their structure and chemical affinity. By functionalizing nanotubes in a desired manner, membranes could be tailored to efficiently separate molecules both on the basis of their molecular size and shape, and on their chemical affinity. Lee and Martin (2002) developed membranes that contain monodisperse gold nanotubes with inside diameters smaller than 1 nm; such membranes can be used either for the separation of molecules or for the transport of ions. The same authors were able to make the interior of the nanotubes hydrophobic, which allowed the nanotube membrane to preferentially extract and transport neutral hydrophobic molecules. While such technologies are still too expensive for commercial-scale food applications, as the technology is further developed such membranes might be used in the future for the isolation of food biomolecules with unique functional properties.

Another area of carbon-nanotube-based applications is the development of electrically conductive membranes. The high length per diameter ratio of carbon nanotubes can be used to turn ordinary synthetic polymers, which are typically electrical insulators, into conducting polymers. In addition to

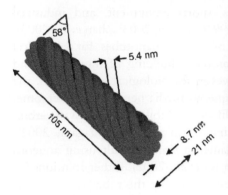 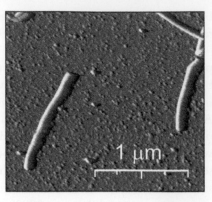

Figure 21.6 α-Lactalbumin nanotubes. Left: model of α-lactalbumin nanotube with outer diameter 21 nm, inner diameter 8.7 nm, pitch 105 nm, pitch angle 58°, and inter-strand spacing of 5.4 nm; Right: microscopic image of nanotubes self-assembled after partial proteolysis of α-lactalbumin at 50°C for 1.5 h (Graveland-Bikker et al., 2006).

their use in the electronic and automobile industries, such polymers could also be utilized to develop novel membranes that enhance the separation and energy efficiency of valuable, delicate molecules. The basic idea of this development would be to incorporate minuscule carbon nanotubes uniformly into polymer substrates that could then be developed into membranes. Membrane pervaporation of food flavors, dehydration of alcohols by pervaporation or membrane distillation, and temperature-swing absorption of volatile liquid foods are just some examples of the possible applications of such conducting membranes in food and associated industries (Moraru et al., 2003). Conducting polymers could also be used to develop food-packaging films for new generations of quick and easy ready-to-eat meals.

Recent developments indicate that nanotubes can also be obtained by controlled self-assembly of biomolecules (i.e. proteins). Graveland-Bikker (2006) discovered that partially hydrolyzed α-lactalbumin organizes into a ten-start helix, forming tubes with diameters of 21 nm (Figure 21.6). They probed the mechanical strength of these nanotubes by AFM, and observed that these artificial helical protein self-assemblies yield very stable, strong structures. The authors suggested that such nanotubes can function either as a model system for artificial self-assembly, or as a nanostructure with potential for practical applications (Graveland-Bikker, 2006).

IV.C. Nanodelivery systems

Many functional food ingredients, such as vitamins, antimicrobials, anti-oxidants, flavorings, colorants, or preservatives, are often incorporated in foods by means of a carrier or delivery system (Weiss *et al.*, 2006). In order for delivery systems to be effective, the encapsulated active compounds must be delivered to the appropriate sites, their concentration maintained at suitable levels for long periods of time, and their premature degradation prevented (Jelinski, 1999). By facilitating a precise control of properties and functionality at the molecular level, nanotechnology enables the development of a number of novel systems that can be used to deliver functional ingredients. Examples include nanometer-sized association colloids such as surfactant micelles, vesicles, bilayers, reverse micelles, and liquid crystals. The major advantages of association colloids are that they form spontaneously, are thermodynamically favorable, and are typically a transparent solution; their major disadvantage is that they require a large quantity of surfactant, which can lead to problems related to flavor, cost, or legality (Weiss *et al.*, 2006).

IV.C.1. Nanoemulsions

Emulsions with nanometer-size droplet diameters, produced using high-pressure valve homogenizers or microfluidizers, can be used as effective nanodelivery systems (Weiss *et al.*, 2006). Encapsulating functional components in the droplets of such nanoemulsions can enable the slowdown of chemical degradation by engineering the properties of the interfacial layer surrounding them (McClements and Decker, 2000).

Compounds with low water solubility have poor bioavailability. The use of nanoemulsions could allow the effective delivery of target compounds through biological barriers. Recently, Huang and collaborators at Rutgers University successfully prepared curcumin emulsions with controllable sizes ranging from 200 nm to 5000 nm, with the purpose of creating an efficient delivery system capable to improve the bioavailability of curcumin (Wang *et al.*, 2008). Takhistov and Paul (2006) successfully investigated the formation and stability of nanoemulsions for food functionalization and drug delivery using the effect of electrochemical instability and surface tension drop at the polarized liquid–liquid interface. When a potential difference of higher than the critical voltage of emulsification was applied to an oil–water interface, the interfacial tension was reduced to almost zero and spontaneous emulsification occurred (Takhistov and Paul, 2006). The authors were able to control the structure and size of these emulsions, as shown in Figure 21.7.

Sometimes it is desirable to combine multiple bioactive compounds within a single formulation due to potential synergistic benefits. A possible challenge would be the fact that different functional compounds are often incompatible with each other within a formulation. The use of multiple emulsions, for example oil-in-water-in-oil (O/W/O) and water-in-oil-in-water (W/O/W) emulsions could represent a solution to this problem. Multiple emulsions can

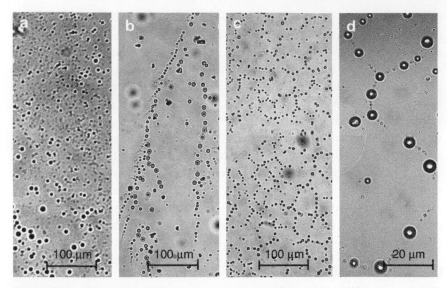

Figure 21.7 Controlled O/W emulsion structures formed by Takhistov and Paul (2006): (a) irregular patterns, (b) chains of polarized droplets, (c) oil droplets network, and (d) internal organization of the droplets' network.

be used to encapsulate functional food ingredients within their individual phases, thus allowing the development of single delivery systems that contain multiple functional components. Multilayer emulsions, consisting of oil droplets (the core) surrounded by nanometer-thick layers (the shell) comprised of different polyelectrolytes, represent another example of a novel delivery system (Weiss *et al.*, 2006). These layers are formed using a layer-by-layer electrostatic deposition method that involves sequential adsorption of poly-electrolytes onto the surfaces of oppositely charged colloidal particles. A functional component trapped within the core of a multilayer emulsion delivery system could be released in response to a specific environmental trigger by designing the response of the shell to the environment (Weiss *et al.*, 2006). One of the big advantages of this technology is that it could use food ingredients (proteins, polysaccharides, phospholipids) and established food-processing methods (homogenization, mixing), which would increase its chances of being easily adapted by the food industry.

IV.C.2. Biopolymeric nanoparticles

Nanoparticles and nanospheres allow better encapsulation and release efficiency as compared to traditional encapsulation systems; they are particularly attractive since they are small enough to even be injected directly into the circulatory system (Bodmeier *et al.*, 1989; Roy *et al.*, 1999). Roy *et al.* (1999) showed that complex coacervates of DNA and chitosan, a natural polysaccharide from crustacean shells, could be used as delivery vehicles in gene therapy and vaccine design. Their work resulted in immunization of mice against the peanut allergen gene, which indicates that oral immunization using DNA-functionalized nanoparticles could become an effective treatment for food allergies, a serious problem affecting many consumers around the world.

Many nanostructured particles are made from biodegradable polymers, such as polylactic acid (PLA). PLA is quickly removed from the bloodstream and sequestered in the liver and the kidneys, making it ideal for the treatment of intracellular pathogens. One of the significant limitations of PLA is that it breaks down in intestinal fluid, which limits its use as a carrier for oral delivery (Weiss *et al.*, 2006). These problems can be overcome by associating a hydrophilic compound, such as polyethylene glycol (PEG). Nanoparticles made from PLA-PEG diblock copolymers form a micellar-like assembly that can entrap a compound to be delivered (Riley *et al.*, 1999).

The efficiency of delivery systems can be enhanced using dendrimer-coated particles. They act as high-capacity loadable nanoreservoirs, sequestering and subsequently releasing charged molecules from solution due to the presence of an oppositely charged dendrimer (Khopade and Caruso, 2002). Dendrimers are monodisperse macromolecules with a regular, highly branched three-dimensional structure, and have a large number of functionalities due to the high local density of active groups. This characteristic makes them usable in a wide range of applications, such as for sensors, catalysts, or agents for controlled release and site-specific delivery (Khopade and Caruso, 2002).

A very stable and precise delivery system is represented by cochleates, which are stable phospholipid-divalent cation precipitates composed of naturally occurring materials. Cochleates have a multilayered structure consisting of a large, continuous, solid, lipid bilayer sheet rolled up in a spiral (Figure 21.8), and deliver their contents to target cells through the fusion between the outer layer of the cochleate and the cell membrane (Santangelo *et al.*, 2000; Gould-Fogerite *et al.*, 2003). Cochleates resist environmental attack; their solid-layered structure provides protection from degradation of the 'encochleated' molecules, even when exposed to harsh environmental conditions or enzymes, including protection from digestion in the stomach. Cochleates can be used for the encapsulation and delivery of many bioactive materials, including compounds with poor water solubility, protein and peptide drugs, and large hydrophilic molecules (Gould-Fogerite *et al.*, 2003). It is possible that such systems will be used in the future for the encapsulation and targeted delivery and controlled release of functional food molecules.

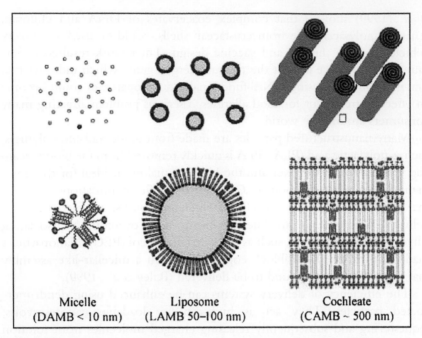

Micelle   Liposome   Cochleate
(DAMB < 10 nm)   (LAMB 50–100 nm) (CAMB ~ 500 nm)

Figure 21.8 Schematic representation of physical states of AMB delivery suspensions, including dispersed detergent micelles, ordered liposomal vesicles, and rolled crystalline cochleates (Santangelo *et al.*, 2000).

Nanoparticles can also be produced using food biopolymers such as proteins or polysaccharides, by promoting self-association or aggregation of single biopolymers, or by inducing phase separation in mixed biopolymer systems. Such nanoparticles can be used to encapsulate functional ingredients, which are then released in response to specific environmental triggers. A research group from Wageningen (The Netherlands) has reported the development of starch-based micro- and nano-particles that behave like colloids in aqueous solution (Dziechciarek *et al.*, 1998). These biodegradable starch nanoparticles can be used in food applications such as mixing, emulsification, for imparting specific rheology to foods, or in non-food applications such as manufacturing of paints, inks, and coatings (Dziechciarek *et al.*, 1998).

Salvona Technologies developed and patented a multicomponent delivery system called MultiSal™, which delivers multiple active immiscible ingredients such as nutrients and flavors and releases them consecutively (Shefer and Shefer, 2005). The system consists of solid hydrophobic nanospheres composed of a blend of food-grade hydrophobic materials encapsulated in moisture-sensitive or pH-sensitive bioadhesive microspheres. Nanospheres with a diameter of about 0.01–0.5 μm, generated using a proprietary suspension technology, are encapsulated in microspheres of about 20–50 μm in diameter. When the microsphere encounters water, such as in saliva, it dissolves, releasing the nanospheres and other ingredients.

IV.C.3. Nanocomposites

Nanostructuring adds value to traditional materials by enhancing their mechanical strength, conductivity, and ability to incorporate and efficiently deliver active substances into biological systems, at low cost, and limited environmental impact. A promising class of new materials is represented by nanocomposites, which are made from nanoscale structures with unique morphology, and have increased modulus and strength as well as good barrier properties.

The development of nanocomposites was inspired by biomineralization, the process in which an organic substance (protein, peptide, or lipid) interacts with an inorganic substance (i.e. calcium carbonate) and forms materials with increased toughness. One example is a packaging material made out of potato starch and calcium carbonate; the material has good thermal insulation properties and is lightweight and biodegradable; it also has been developed to replace the polystyrene 'clam-shell' used for fast food (Stucky, 1997).

The report, 'Nanocomposites for Packaging: New Frontiers and Future Opportunities,' prepared by BRG Townsend and Packaging Strategies (U.S.), predicted that by 2011 the total amount of nanocomposites will reach almost 100 million pounds, with about a half being used for carbonated soft drinks packaging, followed by beer, and equally by meats, as well as other foods and condiments (http://nanotechweb.org/). Nano-composites are regarded as the potentially ideal solution for plastic beer bottles, since previous attempts to use plastic for this application have resulted in spoilage and flavor problems. A Japanese company (Nano Material Inc.) developed a microgravure process for coating plastic films such as PET with a nanocomposite barrier material, which proved to be a better-performing, transparent, alternative to silica- and alumina-coated food-packaging films, and has great potential for the manufacture of beer bottles (Moore, 1999).

Nanostructures can be also built from natural materials. Natural smectite clays, particularly montmorillonite, a volcanic material that consists of nanometer-thick platelets, are a popular source for producing nanoclays (Quarmley and Rossi, 2001). Some U.S. companies (Nanocor Inc., Illinois; Southern Clay Products, Texas) are using montmorillonite as an additive in nanocomposite production. Only a 3–5% montmorillonite addition makes the plastic lighter, stronger, and more heat-resistant, along with improved oxygen, carbon dioxide, moisture, and volatile barrier properties. Mathew and Dufresne (2002) prepared and examined nanocomposites made from a matrix of waxy maize starch plasticized with sorbitol and a stable aqueous suspension of tunicin whiskers as the reinforcing phase. Park *et al.* (2003) developed thermoplastic starch/clay nanocomposites using a melt intercalation method. With a 5% by weight inclusion of the clays, higher tensile properties and a lower water vapor transmission rate than the thermoplastic starch alone were obtained.

Chitosan, a nontoxic natural polysaccharide found in shellfish that has unique antifungal or antimicrobial properties, has a wide range of applications, from food preservation, cosmetics, and wastewater treatment to wound healing and production of artificial skin (Risbud *et al.*, 2000; Juang and Shao, 2002). Unfortunately, chitosan's hydrophilic character and poor mechanical properties in the presence of water and humidity limit its application, although the addition of exfoliated hydroxyapatite layers was observed to significantly enhance the mechanical and barrier properties of chitosan films in humid environments, while maintaining their antimicrobial activity (Weiss *et al.*, 2006).

The barrier properties of nanocomposites, particularly those obtained from natural materials, could be extremely useful for food-packaging applications. The use of such materials could enhance considerably the shelf life of foods such as processed meats, cheese, confectionery, cereals, or boil-in-the-bag foods. Intense scientific and technologic efforts are being made to develop nanocomposites, but due to the expected commercial impact, many of these efforts are surrounded by confidentiality.

IV.C.4. Nanolaminates

Nanolaminates, consisting of two or more layers of material with nanometer dimensions, physically or chemically bonded to each other, could be used in the future to make edible coatings or films for food packaging or carriers for functional agents such as colors, flavors, antioxidants, nutrients, and antimicrobials (Weiss *et al.*, 2006). Nanolaminates can be made using the layer-by-layer (LbL) deposition technique, in which charged surfaces are coated with interfacial films consisting of multiple nanolayers of different materials. The deposition could be facilitated by electrostatic attraction, which causes charged substances to be deposited onto oppositely charged surfaces. The LbL technology allows precise control of the thickness and properties of nanolaminates, and layers of 1–100 nm thickness can be obtained (Weiss *et al.*, 2006). According to Kotov (2003), nanolaminates are more likely to be used as coatings instead of self-standing films because they are very fragile. Food items could be encased in nanolaminates by dipping them into a series of solutions containing substances that would adsorb to the surface of the food, or by spraying these solutions onto the food surface

(Weiss *et al.*, 2006). Adsorbing substances that could be used to create the different layers include natural polyelectrolytes (proteins, polysaccharides), charged lipids (phospholipids, surfactants), and colloidal particles (micelles, vesicles, droplets). This procedure could also be used to encapsulate various active functional components such as antimicrobials, antibrowning agents, antioxidants, enzymes, flavors, or colors into the films, thus increasing the shelf life and quality of the coated foods (Weiss *et al.*, 2006).

IV.C.5. Nanofibers

Nanofibers are polymeric strands of sub-micrometer diameters and are produced by interfacial polymerization and electrospinning. Electrospinning is used to make thin polymer strands from solution by applying a strong electric field to a spinneret with a small capillary orifice. Polymer filaments are formed from solution between two oppositely charged electrodes, with one electrode submerged in the polymer solution and the other one connected to a collector. When the electrical force at the interface of a polymer liquid overcomes the surface tension, a charged jet is ejected (Fong and Reneker, 2001). The jet initially extends in a straight line, then undergoes a vigorous whipping motion caused by the electrohydrodynamic instability (Kim *et al.*, 2006). This reduces the effective fiber diameter and aligns the polymer molecules, thereby improving the mechanical properties of the fibers. As the solvent evaporates, the polymer filament is collected onto a grounded mesh in the form of a non-woven mat (Figure 21.9) with high surface area to mass ratio ($10-1000 \text{ m}^2/\text{g}$) (Kim *et al.*, 2006).

Electrospun polymer fibers have unique mechanical, electrical, and thermal properties, and have applications in filtration, manufacture of protective clothing, and biomedical applications. The limited application of

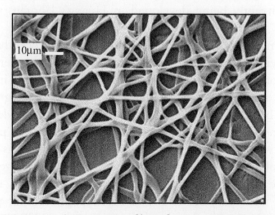

Figure 21.9 SEM image of electrospun fibers from 9 wt% DP210 cellulose/NMMO/water solution with a rotating collector at 1.2 rpm and a flow rate of 0.03 mL/min (Kim *et al.*, 2006).

electrospun fibers in food and agricultural systems is in part due to the fact that they are made primarily from synthetic polymers. Therefore, production of nanofibers from food biopolymers is likely to increase their use in the food industry. Potential food applications of nanofibers include the reinforcement of food-packaging materials and the fabrication of nanostructured and microstructured scaffolding for bacterial cultures (Weiss *et al.*, 2006).

V. NANOTOOLS FOR FOOD AND BIO-SAFETY

The biosecurity of the food and water supply has become a serious concern in recent years, and novel solutions are required for the development of fast, reliable, and highly sensitive biosensors for the detection of biological agents in food or water. A research group from Northwestern University has developed a method that produces nanoparticles with a triangular prismatic shape that can be used in detecting biological threats such as anthrax, smallpox, and tuberculosis, and a wide range of genetic and pathogenic diseases (Fellman, 2001). Chip-based sensing for rapid detection of biological pathogens is another new area with tremendous potential for application in food handling and processing, and in early warning systems for exposure to air- and water-borne bacteria, viruses, and other antigens. DNA microarrays, which consist of short strands of DNA sequences patterned on solid supports, are the primary choice for such systems. The 'target' DNA extracted from pathogens can be analyzed to obtain detailed sequence information, by sequence-specific binding (hybridization) to the microarrays. Fluorescence microscopy is usually used to detect the hybridization of fluorescent-labeled 'target' DNA in the microarrays. A novel method of making simple, linker-free, high-capacity DNA microarrays, based on highly porous organosilicate supports, is currently being developed in Huang's laboratory at Rutgers University.

A team from Clemson University in South Carolina recently investigated the ability of synthetic adhesin-specific nanoparticles to irreversibly bind to targeted types of bacteria, inhibiting them from binding to and infecting their host (Latour *et al.*, 2003). This research was aimed at reducing the infection capability of human foodborne enteropathogens (*Campylobacter jejuni*, *Campylobacter coli*, *Salmonella enteritidis*, and *Escherichia coli*) in poultry products and used two types of nanoparticle systems. One type was based on the self-assembly of organic polymers with controlled structures (i.e. polystyrene), while the other system was based on inorganic nanoparticles functionalized with polysaccharides and polypeptides that promote the adhesion of the targeted bacterial cells (Latour *et al.*, 2003).

Even a low infectious dose of *E. coli* O157:H7 (less than 100 cells) requires the development of rapid and sensitive detection methods to prevent widespread outbreaks. Traditional detection methods for trace amounts of bacteria require amplification of the target organism. Recent progress in nanotechnology has led the way for development of novel detection methods with higher sensitivity to eliminate the laborious and time-consuming amplification process. Many efforts have been made in developing new probe materials with high stability and quantum yield, to improve the stability, sensitivity, and reproducibility of various bioassays for pathogenic cell detection. For example, semiconductor quantum dots (QDs) are promising materials that can be used to improve the bacterial detection limit. Huang's group at Rutgers University has successfully developed a simple bioconjugation procedure that allows the attachment of water-soluble cadmium tellurium QD to anti-*E. coli* antibodies (Kuo *et al.*, 2008). Known for their unique optical profiles, QDs have size-dependent tunable photoluminescence with broad excitation spectra and narrow emission bandwidths. These characteristics allow simultaneous excitation of particles of different sizes at a single wavelength. In addition, their high photobleaching threshold renders possible the continuous or long-term monitoring of slow biological processes. Today, QDs have become promising probe materials in the development of antibody-based immunosensors, demonstrating high stability, sensitivity, and reproducibility. The results of Huang and associates at Rutgers University suggest that the use of QDs make the detection of a single pathogenic cell possible.

Many of the microbial safety problems encountered in the food industry are related to the contamination of food-processing equipment and surfaces with bacteria, and bacterial and fungal spores. Microorganisms and their spores adhere to surfaces, especially to those that are porous or have cracks and crevices. Spores are of particular concern, since they can survive extreme temperatures and adverse conditions. One of the most important spore properties related to the contamination of surfaces is adhesion, which has been identified as one of the principal virulence factors in microorganisms like *Bacillus*. Spore adhesion is a significant problem, particularly for pipelines in the dairy industry, where spores can germinate and resulting bacteria can multiply and resporulate (Andersson *et al.*, 1995; Bowen *et al.*, 2000). The quantification of spore adhesiveness has been facilitated by the development of tools capable of analyzing single molecular mechanics. Bowen *et al.* (2000) used AFM to study the adhesive properties of *Aspergillus niger*, one of the most common molds to contaminate food and produce mycotoxins in cereals and cereal products. Camesano and Logan (2000) developed an AFM method to probe the effects of pH, ionic strength, and bacterial surface polymers on the electrosteric repulsion between negatively charged bacteria and AFM silicon nitride tips, finding that bacterial surface polymers were the dominating factor. Using a similar approach, Bowen *et al.*

(2000) obtained adhesion properties and structural information about bacterial polysaccharides, which form biofouling layers on food-processing equipment and other surfaces. The adhesion of lactic acid bacteria to the intestinal epithelium, one of the most important properties determining probiotic activity, has been studied using AFM by Zammaretti and Ubbink (2003). Understanding such interaction between contaminated surfaces and microorganisms allowed the design of materials that are resistant to bacterial adhesion (Husmark and Ronner, 1990; Razatos et al., 1998). Today, researchers are making efforts to develop a new generation of 'self-cleaning' materials that are loaded with antimicrobial compounds, which can then be released under certain environmental conditions to kill contaminant microflora on food surfaces and equipment.

Although such developments are much anticipated by the food industry, they are a rather long-term solution. Therefore, intervention strategies such as disinfection remain important 'weapons' in fighting harmful microorganisms in the food industry. The corrosive nature of disinfectants and biocides commonly used in food-processing environments makes their use unsuitable for the decontamination of more sensitive equipment. In research by Baker and associates (University of Michigan), nanoemulsions were developed that are able to fuse with and subsequently disrupt the membrane of a variety of different pathogens, such as bacteria, spores, enveloped viruses, and fungal spores (Chepurnov et al., 2003). Since nanoemulsions can be produced cheaply and rapidly with high efficiency and stability, they are likely to be used extensively in the food industry.

VI. PROSPECTS AND CHALLENGES FOR THE FUTURE

In 2000, the National Science Foundation (NSF) estimated that in less than 10 years, products made from nanotechnology will have a $1 trillion impact on the global economy, and that the nanotechnology industry alone will employ two million workers (Rocco et al., 2000). Currently, many of the world's leading food companies such as Kraft, Unilever, Heinz, Nestlé, and Hershey are investing heavily in nanotechnology applications (Wolfe, 2005).

In 2006, participants at the First International Food Nanotechnology Conference (organized by the Institute of Food Technologists) agreed that nanotechnology is still in its infancy, with food applications more in a pre-infancy state; however, there is a great deal of enthusiasm and anticipation surrounding this technology (Bugusu et al., 2006).

At present, the food industry is trying to identify opportunities and the most promising discoveries in nanotechnology that can significantly enhance food processing and food products. A limited number of applications that deal with matter at the nanoscale and are relevant to the food sector are currently available, and can be categorized in two groups: (1) functionalized membranes, which could be used to isolate and purify highly sensitive bioactive compounds and (2) nano-structuring technologies, such as emulsification, dispersion and nano-aeration, or the encapsulation of active compounds in food polymer matrices. The fabrication of food biopolymer-based nano-structured materials with unique mechanical and functional properties; the synthesis of liposomes, which can be used for the encapsulation and targeted delivery of bioactive compounds through foods; or the development of nanosensors for pathogen detection, are all viewed as a starting point for the integration of nanotechnology into food science and food technology.

Despite the slow onset of nanotechnology-based applications in the food industry, the participants at the above-mentioned meeting agreed that the future holds great promise, particularly in the following areas: (1) health promotion through foods, by using food matrices as delivery tools for bioactive compounds, (2) development of ingredients with enhanced functionality, which could be used to manufacture foods with novel and unique flavors and textures, (3) development of novel, nano-structured packaging materials with enhanced barrier or self-sealing characteristics, (4) materials with microbial-repellant characteristics, which could be used to manufacture food contact surfaces with self-sanitizing properties, (5) novel processing technologies, and (6) advanced tools (i.e. nanosensors) for food safety shelf life or quality monitoring (Bugusu et al., 2006).

As with any new technology, a series of challenges can be expected. One of the big questions associated with the prospect of commercializing 'nano' applications in the food sector is whether or not this technology will be cost-effective in the near future, and if the benefits will outweigh the costs in the low profit margin scenario of the food industry. It must also be considered that the implementation of some nanotechnology developments could change the way in which the food industry operates at the moment, prompting the need for significant changes in current food regulations and legislation. Presently, in the United States, there are no special regulations for the utilization of nanotechnology products in foods; in the European Union (EU), although recommendations for special regulations have been made, laws have yet to be changed (Weiss et al., 2006).

Media reports regarding the risks posed by some nanoparticles clearly demonstrate that significant research is still required to evaluate the toxicity of nanotechnology products, particularly nanoparticles, and to assess the environmental and personal safety aspects of their use. The risks could be practically zero, or significant, depending on the properties of a particular product and exposure levels (Kuzma and VerHage, 2006). Few risk assessments have been done that allow one to predict what happens when such extremely small particles, some designed to be biologically active, enter the human body or are dispersed in the environment.

The Food and Drug Administration (FDA), the government agency responsible for regulating food, dietary supplements, and drugs in the U.S., states that it regulates products, not technologies, and that 'nanotechnology products will be regulated as "Combination Products" for which the regulatory pathway has been established by statute' (http://www.fda.gov/nanotechnology/regulation.html). The European Commission has taken a more proactive approach and in February 2008 released a Code of Conduct for Responsible Nanosciences and Nanotechnologies Research that contains general principles and guidelines for all member states. This Code of Conduct 'invites all stakeholders to act responsibly and cooperate with each other (…) in order to ensure that N&N research is undertaken in the Community in a safe, ethical and effective framework, supporting sustainable economic, social and environmental development' (http://cordis.europa.eu/nanotechnology/).

It is very difficult to predict the long-term impact of any technology, and nanotechnology is no exception. While one should not underestimate any potential dangers caused by its misuse, it is almost generally recognized that nanotechnology has the potential to make significant contributions for the improvement and even extension of life, partly through the improvement of the food and agricultural system. Both agricultural producers and food manufacturers could gain a competitive position through the application of nanotechnology, while consumers may benefit from those advances in nanotechnology that contribute to the improved safety and nutritional value of food.

REFERENCES

Andersson, A., Ronner, U., & Granum, P. E. (1995). What problems does the food industry have with the spore-forming pathogens *Bacillus cereus* and *Clostridium perfringens*? *Int J. Food Microbiol., 28*, 145–155.

Baeumner, A. (2004). Nanosensors identify pathogens in food. *Food Technol., 58*(8), 51–55.

Baker, A., Helbert, W., Sugiyama, J., & Miles, M. J. (1997). High resolution atomic force microscopy of native Valonia cellulose I microcrystals. *J. Struct. Biol., 119*, 129–138.

Baker, A., Miles, M. J., & Helbert, W. (2001). Internal structure of the starch granule revealed by AFM. *Carbohydrate Res., 330*, 249–256.

Bandyopadhyaya, R., Nativ-Roth, E., Regev, O., & Yerushalmi-Rozen, R. (2002). Stabilization of individual carbon nanotubes in aqueous solutions. *Nano Letters, 2*(1), 25–28.

Bard, A., Fan, F., Pierce, D., Unwin, P., Wipf, D., & Zhou, F. (1991). Chemical imaging of surfaces with the scanning electrochemical microscope. *Science, 254*, 68–74.

Betzig, E., Trautman, J., Harris, T., Weiner, J., & Kostelak, R. (1991). Beating the diffraction barrier: Optical microscopy on a nanometer scale. *Science, 251*, 1468–1470.

Binnig, G., Quate, C. F., & Gerber, Ch. (1986). Atomic force microscope. *Phys. Rev. Lett., 56*, 930–933.

Bodmeier, R., Chen, H. G., & Paeratakul, O. (1989). A novel approach to the oral delivery of micro- or nanoparticles. *Pharma. Res., 6*, 413–417.

Boskovic, S., Chon, J. W. M., Mulvaney, P., & Sader, J. E. (2002). Rheological measurements using cantilevers. *J. Rheology, 46*(4), 891–899.

Bowen, R., Lovitt, R., & Wright, C. (2000). Direct quantification of *Aspergillus niger* spore adhesion in liquid using an atomic force microscope. *J. Colloid Interface Sci., 228*, 428–433.

British Standards Institution (BSI). (2005). PAS 71:2005. Vocabulary – Nanoparticles. Available online at http://www.bsi-global.com (accessed April 2008).

Brust, M., & Kiely, C. J. (2002). Some recent advances in nanostructure preparation from gold and silver particles: a short topical review. *Colloids and Surfaces A: Physicochem. Engineer. Aspects, 202*, 175–186.

Bugusu, B., Bryant, C., Cartwright, T.T., et al. (2006). Report on the First IFT International Food Nanotechnology Conference. June 28–29, 2006, Orlando, FL. Available online at: http://members.ift.org/IFT/Research/ConferencePapers/firstfoodnano.htm (accessed March 2008).

Camesano, T. A., & Logan, B. E. (2000). Probing bacterial electrosteric interactions using atomic force microscopy. *Environ. Sci. Technol., 34*(16), 3354–3362.

Camesano, T. A., & Wilkinson, K. (2001). Single molecule study of xanthan conformation using atomic force microscopy. *Biomacromolecules, 2*, 1184–1191.

Chen, H., Weiss, J., & Shahidi, F. (2006). Nanotechnology in nutraceuticals and functional foods. *Food Technol., 60*(3), 30–36.

Chepurnov, A. A., Bakulina, L. F., Dadaeva, A. A., Ustinova, E. N., Chepurnova, T. S., & Baker, J. R., Jr. (2003). Inactivation of Ebola virus with a surfactant nanoemulsion. *Acta. Trop., 87*(3), 315–320.

Clausen-Schaumann, H., Rief, M., Tolksdorf, C., & Gaub, H. E. (2000). Mechanical stability of single DNA molecules. *Biophys. J., 78*(4), 1997–2007.

Cowman, M. K., Li, M., Dyal, A., & Balasz, E. A. (2000). Tapping mode atomic force microscopy of the hyaluronan derivative, hylan A. *Carbohydrate Polymers, 41*, 229–235.

Dagani, R. (2002). Sugary ways to make nanotubes dissolve. *Chem. Engineer. News, 80*(28), 38–39.

Decher, G. (1997). Fuzzy nanoassemblies: Toward layered polymeric multicomposites. *Science, 277*, 1232–1237.

Dziechciarek, Y., van Schijndel, R. J. G., Gotlieb, K. F., Feil, H., & van Soest, J. J. G. (1998). Development of starch-based nanoparticles: Structure, colloidal and rheological properties. Presented at the Meeting of the Dutch Society of Rheology, October 22, 1998. Abstract available online at: http://www.mate.tue.nl/nrv/ede/dziechciarek.html (accessed June 10, 2003).

The European Commission (EC). (2008). A code of conduct for responsible nanosciences and nanotechnologies research. A Commission Recommendation of 07/02/2008, Brussels. Available online at: http://cordis.europa.eu/nanotechnology/ (accessed May, 2008).

Fellman, M. (2001). Nanoparticle Prism Could Serve as Bioterror Detector. Available online at: http://unisci.com/stories/20014/1204011.htm (accessed May 28, 2002).

Florin, E. L., Moy, V. T., & Gaub, H. E. (1994). Adhesion forces between individual ligand–receptor pairs. *Science, 264*, 415–417.

Fong, H., & Reneker, D. H. (2001) In D. R. Salem (Ed.), *Structure Formation in Polymeric Fibers*, Ch. 6 (pp. 225–246). Munich: Hanser Gardner Publications, Inc.

Frisbie, C. D., Rozsnyai, L. F., Noy, A., Wrighton, M. S., & Lieber, C. M. (1994). Functional group imaging by chemical force microscopy. *Science, 265*, 2071–2074.

Gajraj, A., & Ofoli, R. (2000). Quantitative technique for investigating macromolecular adsorption and interactions at the liquid–liquid interface. *Langmuir, 16*, 4279–4285.

Gallant, D. J., Bouchet, B., & Baldwin, P. M. (1997). Microscopy of starch: evidence of a new level of granule organization. *Carbohydrate Polymers 32(3/4):* 177–191.

Gorman, J. (2003). Fracture protection: Nanotubes toughen up ceramics. *Science News, 163*, 1.

Gould-Fogerite, S., Mannino, R. J., & Margolis, D. (2003). Cochleate delivery vehicles: Applications to gene Therapy. *Drug Delivery Technol., 3*(2), 40–47.

Graveland-Bikker, J. F., Schaap, I. A. T., Schmidt, C. F., & de Kruif, C. G. (2006). Structural and mechanical study of self-assembling protein nanotubes. *Nano Letters, 6*, 616–621.

Haruyama, T. (2003). Micro- and nanobiotechnology for biosensing cellular responses. *Adv. Drug Delivery Rev., 55*, 393–401.

Hilal, N., & Bowen, R. (2002). Atomic force microscope study of the rejection of colloids by membrane pores. *Desalination, 150*, 289–295.

Hobbs, P., Abraham, D., & Wickramasinghe, H. (1989). Magnetic force microscopy with 25 nm resolution. *Appl. Phys. Lett., 55*, 2357–2359.

Houston, J., & Michalske, T. (1992). Interfacial-force microscope. *Nature, 356*, 266.

Huang, W., Taylor, S., http://pubs.acs.org/cgi-bin/article.cgi/nalefd/2002/2/i04/html/ - nl010095iAF2 Fu, K., Lin, Y., Zhang, D., Hanks, T.W., Rao, A.M., http://pubs.acs.org/cgi-bin/article.cgi/nalefd/2002/2/i04/html/ - nl010095iAF4 and Sun, Y.P. (2002). Attaching proteins to carbon nanotubes via diimide-activated amidation. *Nano Letters* 2(4): 311–314.

Humphris, A., McMaster, T., Miles, M., Gilbert, S., Shewry, P., & Tatham, A. (2000). Atomic force microscopy study of interactions of HMW subunits of wheat glutenin. *Cereal Chem., 77*(2), 107–110.

Husmark, U., & Ronner, U. (1990). Forces involved in adhesion of *Bacillus cereus* spores to solid surfaces under different environmental conditions. *J. Appl. Bacteriol., 69*(4), 557–562.

Ikeda, S., Morris, V., & Nishinari, K. (2001). Microstructure of aggregated and non aggregated k-carrageenan helices visualized by atomic force microscopy. *Biomacromolecules, 2*, 1331–1337.

James, B. J., Jing, Y., & Chen, X. D. (2003). Membrane fouling during filtration of milk: a microstructural study. *J. Food Eng., 60*, 431–437.

Jelinski, L. (1999). Biologically related aspects of nanoparticles, nanostructured materials and nanodevices. In: Siegel, R.W., Hu, E., and Roco, M.C. (eds) Nanostructure Science and Technology. A Worldwide Study. Prepared under the guidance of the National Science and Technology Council and The Interagency Working Group on NanoScience, Engineering and Technology. Available online at http://www.wtec.org (accessed May 2002).

Juang, R. S., & Shao, H. J. (2002). A simplified equilibriu1m model for sorption of heavy metal ions from aqueous solutions on chitosan. *Water Res., 36*(12), 2999–3008.

Khopade, A. J., & Caruso, F. (2002). Electrostatically assembled polyelectrolyte/dendrimer multilayer films as ultrathin nanoreservoirs. *Nano Letters, 2*(4), 415–418.

Kim, C. W., Kim, D. S., Kang, S. Y., Marquez, M., & Joo, Y. L. (2006). Structural studies of electrospun cellulose nanofibers. *Polymer, 47*, 5097–5107.

Kingsley, D. (2002). Membranes show pure promise. ABC Science Online, May 1, 2002. Available online at: http://www.abc.net.au (accessed August 22, 2003).

Kotov, N. A. (2003). Layer-by-layer assembly of nanoparticles and nanocolloids: intermolecular interactions, structure and materials perspective. In G. Decher, & J. B. and Schlenoff (Eds), *Multilayer thin films: sequential assembly of nanocomposite materials* (pp. 207–243). Weinheim, Germany: Wiley-VCH.

Kuo, Y. C., Wang, Q., Ruengruglikit, C., & Huang, Q. R. (2008). Antibody-conjugated CdTe quantum dots for *E. coli* detection. *J. Phys. Chem. C, 112*, 4818–4824.

Kuzma, J., & VerHage, P. (2006). New Report on Nanotechnology in Agriculture and Food Looks at Potential Applications, Benefits and Risks. Available online at http://www.nanotechproject.org/news/archive/new_report_on_nanotechnology_in/ (accessed April 2008).

Latour, R. A., Stutzenberger, F. J., Sun, Y. P., Rodgers, J., & Tzeng, T. R. (2003). Adhesion-specific nanoparticles for removal of *Campylobacter jejuni* from poultry. CSREES Grant (2000–2003), Clemson University (SC). Available online at http://www. clemson.edu (accessed June 10, 2003).

Lee, J. Y., Wang, X. Y., Ruengruglikit, C., & Huang, Q. R. (2007). Nanotechnology in food materials research. In J. M. Aguilera, & P. J. and Lillford (Eds), *Topics in Food Materials Science* (pp. 123–144). New York: Springer.

Lee, S. B., & Martin, C. R. (2002). Electromodulated molecular transport in gold-nanotube membranes. *J. Am. Chem. Soc., 124*(40), 11850–11851.

Li, J., & Lee, T.-C. (1995). Bacterial ice nucleation and its potential application in the food industry. *Trends Food Sci. Technol., 6*, 259–265.

Li, Y. Q., Lee, J. Y., Wang, X. Y., Lal, J., An, L. J., & Huang, Q. R. (2008). Effects of pH on the interactions and conformation of bovine serum albumin: Comparison between chemical force microscopy and small-angle neutron scattering. *J. Phys. Chem. B, 112*, 3797–3806.

Lin, Y., Su, Z.H., & Huang, Q. R. Assembly of poly(diallyldimethylammonium chloride)/carrageenan multilayer films using carrageenans of different charge densities. Unpublished, (2008).

Majumdar, A. (1999). Scanning thermal microscopy. *Ann. Rev. Materials Sci., 29*, 505–585.

Majumdar, A., Carrejo, J., & Lai, J. (1993). Thermal imaging using the atomic force microscope. *Appl. Phys. Lett., 62*, 2501–2503.

Marko, J. F. (1997). Twist and shout (and pull): Molecular chiropractors undo DNA. *Proc. Natl. Acad. Sci. USA, 94*, 11770–11772.

Mathew, A. P., & Dufresne, A. (2002). Morphological investigation of nanocomposites from sorbitol plasticized starch and tunicin whiskers. *Biomacromolecules, 3*(3), 609–617.

McClements, D. J., & Decker, E. A. (2000). Lipid oxidation in oil-in-water emulsions: impact of molecular environment on chemical reactions in heterogeneous food systems. *J. Food Sci., 65*(8), 1270–1282.

McIntire, T. M., & Brant, D. A. (1997). Imaging of individual biopolymers and supramolecular assemblies using noncontact atomic force microscopy. *Biopoly, 42*, 133–146.

McMaster, T., Miles, M., Kasarda, D., Shewry, P., & Tatham, A. (1999). Atomic force microscopy of A-gliadin fibrils and in-situ degradation. *J. Cereal Sci., 31*, 281–286.

Moore, S. (1999). Nanocomposite achieves exceptional barrier in films. *Modern Plastics, 76*(2), 31–32.

Moraru, C. I., Panchapakesan, C. P., Huang, Q., Takhistov, P., Liu, S., & Kokini, J. L. (2003). Nanotechnology: A new frontier in food science. *Food Technol., 57*(12), 24–29.

Morris, V., Mackie, A., Wilde, P., Kirby, A., Mills, C., & Gunning, P. (2001). Atomic force microscopy as a tool for interpreting the rheology of food biopolymers at the molecular level. *Lebensm-Wiss.u-Technol., 34*, 3–10.

Nakajima, K., Mitsui, K., Ikai, A., & Hara, M. (2001). Nanorheology of single protein molecules. *Riken Review, 37*, 58–62.

Nanotechweb.org. (2002). Technology Update: Nanocomposites set to wrap up the packaging market. Available online at: http://nanotechweb.org/articles/news/1/8/18/1 (accessed April 2008).

Nitta, T., Haga, H., Kawabata, K., Abe, K., & Sambongi, T. (2000). Comparing microscopic with macroscopic elastic properties of polymer gel. *Ultramicroscopy, 82*, 223–226.

Noy, A., Vezenov, D. V., & Lieber, C. M. (1997). Chemical force microscopy. *Annu. Rev. Mater. Sci., 27*, 381–421.

Oulevey, F., Gremaud, G., & Semoroz, A. (1998). Local mechanical spectroscopy with nanometer-scale lateral resolution. *Rev. Scientific Instruments, 69*(5), 2085–2094.

Panchapakesan, C. P. (2005). Analysis of the topography, molecular organization and phase properties of films formed from zein and its fractions. M.S. thesis, New Brunswick, NJ: Rutgers University.

Park, H. M., Lee, W. K., Park, C. Y., Cho, W. J., & Ha, C. S. (2003). Environmentally friendly polymer hybrids. Part I: mechanical, thermal, and barrier properties of the thermoplastic starch/clay nanocomposites. *J. Mater. Sci., 38*, 909–915.

Pompeo, F., & Resasco, D. E. (2002). Water solubilization of single-walled carbon nanotubes by functionalization with glucosamine. *Nano Letters, 2*(4), 369–373.

Quarmley, J., & Rossi, A. (2001). Nanoclays. Opportunities in polymer compounds. *Industrial Minerals, 400*, 47–49, 52–53,

Razatos, A., Ong, Y., Sharma, M. M., & Georgiou, G. (1998). Molecular determinants of bacterial adhesion analyzed by atomic force microscopy. *Proc. Natl Acad. Sci. USA, 95*, 11059–11064.

Rief, M., & Grubmüller, H. (2002). Force spectroscopy of single biomolecules. *Chemphyschem, 3*, 255–261.

Riley, T., Govender, T., Stolnik, S., Xiong, C. D., Garnett, M. C., Illum, L., & Davis, S. S. (1999). Colloidal stability and drug incorporation aspects of micellar-like PLA-PEG nanoparticles. *Colloids Surf. B, 16*, 147–159.

Risbud, M., Hardikar, A., & Bhonde, R. (2000). Growth modulation of fibroblasts by chitosan-polyvinyl pyrrolidone hydrogel: implications for wound management? *J Biosci., 25*(1), 25–31.

Roco, M. C., Williams, R. S., & Alivisatos, P. (2000). Nanotechnology Research Directions: IWGN Workshop Report, pp. (iii–iv). Berlin, Germany: Springer.

Rouhi, M. (2002). Novel chiral separation tool. *Chem. Engineer. News, 80*(25), 13.

Roy, K., Mao, H. Q., Huang, S. K., & Leong, K. W. (1999). Oral gene delivery with chitosan–DNA nanoparticles generates immunologic protection in a murine model of peanut allergy. *Nature Med., 5*(4), 387–391.

Rugar, D., Yannoni, C., & Sidles, J. (1992). Mechanical detection of magnetic resonance. *Nature, 360*, 563–566.

Sakai, K., Sadayama, S., Yoshimura, T., & Esumi, K. (2002). Direct force measurements between adlayers consisting of poly (amidoamine) dendrimers with primary amino groups or quaternary ammonium groups. *J. Colloid Interface Sci., 254*(2), 406–409.

Santangelo, R., Paderu, P., Delmas, G., Chen, Z. W., Mannino, R., Zarif, L., & Perlin, D. S. (2000). Efficacy of oral cochleate-amphotericin B in a mouse model of systemic candidiasis. *Antimicrob. Agents Chemother., 44*(9), 2356–2360.

Shefer, A., & Shefer, S. (2005). The Application of Nanotechnology in the Food Industry. Available online at: http://www.foodtech-international.com/papers/application-nano. htm (accessed March 2008).

Shonherr, H. (1999). From functional group assembles to single molecules: Scanning force microscopy of supramolecules and polymer systems. Ph.D. Thesis, The Netherlands: Twente University.

Smith, D. A., Connel, S. D., Robinson, C., & Kirkham, J. (2003). Chemical force microscopy: Applications in surface characterization of natural hydroxyapatite. *Analytica Chimica Acta, 479*, 39–57.

Spice, B. (1999). Tiny molecules called nanotubes have scientists dreaming big. Post-Gazette. October 11, 1999. Available online at: http://www.post-gazette.com/ (accessed June 5, 2003).

Srivastava, D. (2000). Editorial. Nanotechnology 11 doi: 10.1088/0957–4484/11/2/001. Available online at: http://www.iop.org/EJ/abstract/0957-4484/11/2/001/ (accessed March 2008).

Strick, T., Allemand, J., Croquette, V., & Bensimon, D. (2000). Stress-induced structural transitions in DNA and proteins. Annu. Rev. Biophys. Biomol. Struct., 29, 523–543.

Strick, T., Allemand, J., Croquette, V., & Bensimon, D. (2001). The manipulation of single biomolecules. *Physics Today, 54*, 46–52.

Stucky, G. D. (1997). Oral Presentation. WTEC Workshop on R&D Status and Trends in Nanoparticles, Nanostructured Materials, and Nanodevices in the United States, May 8–9, Rosslyn, VA.

Swartzentruber, B. (1996). Direct measurement of surface diffusion using atom-tracking scanning tunneling microscopy. *Phys. Rev. Lett.*, 76, 459.

Takhistov, P., & Paul, S. (2006). Formation of oil/water emulsions due to electrochemical instability at the liquid/liquid interface. *Food Biophysics*, 1(2), 57–73.

Takhistov, P. (2006). Nanotechnology and its applications for the food industry. In: Handbook of Food Science, Technology, and Engineering, Vol. 3, pp. 127.1–127.18.

Terada, Y., Harada, M., & Ikehara, T. (2000). Nanotribology of polymer blends. *J. Appl. Phys.*, 87(6), 2803–2807.

Viani, M. B., Pietrasanta, L. I., Thompson, J. B., Chand, A., Gebeshuber, I. C., Kindt, J. H., Richter, M., Hansma, H. G., & Hansma, P. K. (2000). Probing protein–protein interactions in real time. *Nature Struct. Biol.*, 7, 644–647.

Wang, X. Y., Jiang, Y., Wang, Y.-W., Huang, M. T., Ho, C. T., & Huang, Q. R. (2008). Enhancing anti-inflammation activity of curcumin through O/W nanoemulsions. *Food Chem.*, 108, 419–424.

Weiss, J., Takhistov, P., & McClements, D. J. (2006). Functional materials in food nanotechnology. Scientific Summary. *J. Food Sci.*, 71(9), R107–R116.

Willemsen, O. H., Snel, M. M. E., Cambi, A., Greve, J., De Grooth, B. G., & Figdor, C. G. (2000). Biomolecular interactions measured by atomic force microscopy. *Biophys. J.*, 79(6), 3267–3281.

Williams, C., Slinkman, J., Hough, W., & Wickramasinghe, H. (1989). Lateral dopant profiling with 200 nm resolution by scanning capacitance microscopy. *Appl. Phys. Lett.*, 55, 1663–1664.

Wolfe, J. (2005). Safer and Guilt-Free Nano Foods.Forbes.com, August 10, 2005. Available online at: http://www.forbes.com (accessed April, 2008).

Zammaretti, P., & Ubbink, J. (2003). Imaging of lactic acid bacteria with AFM – elasticity and adhesion maps and their relationship to biological and structural data. *Ultramicroscopy*, 97(1), 199–208.

Zasypkin, D. V., & Lee, T. C. (1999). Extracellular ice nucleators from *Pantoea ananas*: Effects on freezing of model foods. *J. Food Sci.*, 64, 473–478.

Zhan, G. D., Kuntz, J., Wan, J., & Mukherjee, A. K. (2003). Single-wall carbon nanotubes as attractive toughening agents in alumina-based nanocomposites. *Nature Materials*, 2, 38–42.

CHAPTER 22

Nanotechnology and Applications in Food Safety

Bita Farhang

Contents

Abstract

Nanotechnology involves research and technology development at the atomic, molecular, and macromolecular levels, aimed at creating and using structures, devices, and systems with novel properties and functions based on their small size. It is strategically a very important research field with considerable industrial potential.

Over the past few years nanotechnology has rapidly become a significant component in the food industry, applications ranging from smart packaging to interactive foods. Major food producers are presently using nanotechnology to improve food quality but the future belongs to new products and processes and customization of such products, as the nanofood market worldwide is expected to increase monetarily to over $20.4 billion in 2010.

Indeed, nanotechnology could provide innovative answers to current problems concerning food safety. Recent developments show many applications of nanotechnology that improve the quality and safety of food products; for example, nanosensors to detect pathogens and contaminants and nanodevices to identify preservation and tracking. In food packaging, nanomaterials are being developed with enhanced mechanical and thermal properties to improve protection of foods against exterior mechanical, thermal, chemical, or microbiological effects. Some potential uses in food packaging include modifying permeation behavior of foils, increasing barrier properties, improving mechanical and heat-resistance properties, developing active antimicrobial and antifungal

Global Issues in Food Science and Technology
© 2009 Elsevier Inc.

surfaces, and sensing and signaling microbiological and biochemical changes.

Given the importance of this novel technology in making food safe to eat, this chapter will provide insights into the potential benefits of nanotechnology in food safety.

I. INTRODUCTION

Recent trends in global food production, processing, distribution, and preparation are creating an increasing demand for food safety research to ensure a safer global food supply. Food safety is therefore an increasingly important public health issue and governments all over the world are intensifying their efforts to improve food safety procedures. These efforts are also in response to an increasing number of food safety problems and rising consumer concerns (WHO, 2006).

One of the new technologies to improve food safety is nanotechnology, which is the manipulation or self-assembly of individual atoms, molecules, or molecular clusters into structures, the purpose of which is to create materials and devices with new or vastly different properties (Joseph and Morrison, 2006).

This new technology deals with controlling the properties of matter with dimensions between 1 and 100 nanometers. One nanometer is equal to one billionth of a meter, and is about the size of a small molecule (ElAmin, 2005). Therefore it can be said that nanotechnology focuses on the characterization, fabrication, and manipulation of biological and non-biological structures smaller than 100 nanometers. Structures on this scale have been shown to have unique and novel functional properties (Weiss et al., 2006).

Nanotechnology will enable the manufacture of high-quality products at a very low cost and fast pace. It is commonly referred to as a generic technology that offers better-built, safer, longer-lasting, cheaper, and smarter non-food products with wide applications in consumer households, the communications industry, and medicine field, however there is even more potential in the agriculture and food industry (Warad and Dutta, 2006).

According to a recent report by the consulting company, Cientifica, the value of nanotechnologies in the food industry reached $410 million in 2006, with applications restricted primarily to food packaging for improved gas barrier protection and improved nutraceutical delivery systems. Today more than 200 companies around the world are active in research and development. The United States is the leader followed by Japan and China. The Asian market, comprising more than 50% of the world's population, will be the biggest market for nanofood by 2010 with China leading (Moraru *et al.*, 2003).

Nanotechnology can be used in food processing and storage as nano-composites for plastic film coatings in food packaging, as antimicrobial nanoemulsions for decontamination of food equipment, in packaging, and as food nanotechnology-based antigens to detect biosensors for identification of pathogen contamination (Salamanca-Buentello *et al.*, 2005).

II. POTENTIAL FOOD APPLICATIONS

The application of nanotechnology and nanoparticles in food processing is emerging rapidly (ElAmin, 2006). According to reports, five out of ten of the world's largest food and beverage companies are currently investing in nanotechnology research and development (R&D) (Nanotechwire report, 2005).

Nanotechnology has the potential to impact many aspects of the world's food and agricultural systems. Some examples of the important links that nanotechnology has to the science and engineering of these agricultural and food systems are (Weiss *et al.*, 2006):

- Food security
- Disease treatment delivery methods
- New tools for molecular and cellular biology
- New materials for pathogen detection
- Protection of the environment.

According to a study by Mallika *et al.* (2005), uses for nanotechnology in the food industry, in addition to food security, include design of new food products, nutrient and flavor encapsulation, microbiological food safety, food biotechnology, beverage filtration, nano biosensors, restoring cellular damage in foods, and food packaging. Thus, nanotechnology has the potential to improve food quality and safety significantly. There are many new applications of this novel technology for improving food safety in particular (to be discussed in this chapter) such as:

- Food packaging
- Detection of foodborne pathogens and their toxins
- Detection of chemicals and contaminants
- Identity preservation and tracking.

II.A. Food packaging

Packaging is a critical key to ensuring the safety of food products. Nano-technology can improve packaging material and its functionality and consequently ensure food safety and consumer protection. Today, food packaging and monitoring of products are a major focus of food-industry-related nanotechnology R&D. Packaging that incorporates nanomaterials can be 'smart,' which means it can respond to environmental conditions or repair itself, or even alert a consumer about contamination and/or the presence of pathogens. According to an industry analyst the current U.S. market for active, controlled, smart packaging of foods and beverages is an estimated $38 billion, and will surpass $54 billion by 2008 (ETC Group Report, 2004). Nanotechnology makes food packaging intelligent, smart and long-lasting, providing better safety against bacteria and microorganisms than traditional market packaging methods. It is predicted that nanotechnology will change 25% of the worldwide food-packaging business in the next decade, which means over $30 billion annually in the market (Nanotechwire report, 2005).

Developing smart packaging to optimize product shelf life has been the goal of many companies. Such packaging systems would be able to repair small holes/tears, respond to environmental conditions (e.g. temperature and moisture changes), and alert the customer if the food is contaminated. For example, nanomaterials, which are materials with a microstructure of characteristic length on the order of a few, typically 1–100 nanometers, could be developed to modify the permeation behavior of foils, increase barrier properties (mechanical, thermal, chemical, and microbial), improve mechanical and heat-resistance properties, develop active antimicrobial and antifungal surfaces, and sense and signal microbiological and biochemical changes (De Jong, 2005; Joseph and Morrison, 2006).

Plastics are increasingly being used in food packaging, but there are concerns about their ability to allow the exchange of oxygen, carbon dioxide, water, and aroma components, which could compromise the quality and safety of packaged foods. However nanocomposites, which are high-barrier materials, are a potential solution to this problem (Lagaron, 2006). High-barrier materials are good barriers against oxygen, water vapor, and aromas and also offer good barrier properties under different packing, handling, shipping, and storage conditions.

For instance the polymer–clay nanocomposite has emerged as a novel food-packaging material due to several benefits, such as its enhanced mechanical, thermal, and barrier properties (Ray et al., 2006). Its use has improved plastic packaging of food products. For example, the Bayer Company produces a transparent plastic film containing nanoparticles of clay. The nanoparticles are dispersed throughout the plastic and are able to block oxygen, carbon dioxide, and moisture from reaching fresh meats or other foods. The nanoclay also makes the plastic lighter, stronger, and more heat-resistant (Brody, 2006). According to Avella et al. (2005), biodegradable starch/clay nanocomposite films can also be used for food packaging while in a similar study Lagaron *et al.* (2006) showed that nanocomposites can improve the quality and safety of packaged food.

As mentioned, researchers are even experimenting with materials that will change their properties to address outside environmental factors such as temperature or humidity. For instance, imagine an ice cream carton that would tighten its existing molecular structure to prevent heat from affecting the contents if left in the back of an automobile on a hot summer day (ETC Group Report, 2004).

Zinc oxide and magnesium oxide nanoparticles, two in a range of nano ingredients produced by advanced nanotechnology, could also have anti-microbial properties. Nanotechnology researchers in the UK reported that both zinc oxide and magnesium oxide are effective in killing microorganisms (Patton, 2006). According to another research group in the UK (University of Leeds), nanoparticles of magnesium oxide and zinc oxide are indeed highly effective at destroying microorganisms and therefore could have applications in food packaging (Food Production Daily Report, 2005).

With different nanostructures, plastics can achieve various levels of gas/water vapor permeability to fit the requirements of preserving fruits, vegetables, beverages, and other foods. By adding nanoparticles, food processors can also produce bottles and packages that are more resistant to light, have stronger mechanical and thermal performance, and are less gas absorptive. These properties can significantly increase the shelf life, efficiently preserve flavor and color, and facilitate transportation and usage. Further, nanostructured film can effectively protect the food from invasion of bacteria and microorganisms and ensure food safety. With embedded nanosensors in the packaging, consumers will be able to 'read' the condition of the food inside (Asadi and Mousavi, 2006).

Meanwhile, scientists at the University of Strathclyde (Glasgow) have developed 'intelligent ink' by using light-sensitive nanoparticles that only detect oxygen. The patented ink could be used for labels on any food, and because it is inexpensive, it is suitable for use in large numbers. The ink could also be used to indicate if the original modified atmosphere inside a package has changed (Dunn, 2004).

At Kraft foods Inc, as well as at Rutgers University in the U.S., scientists are developing an electronic tongue for inclusion in packaging. This consists of an array of nanosensors that are extremely sensitive to gases released by food as it spoils, causing the sensor strip to change color as a result, giving a clear visible signal of whether the food is fresh or not (Joseph and Morrison, 2006).

II.B. Detection of foodborne pathogens and their toxins

Foodborne diseases are a widespread and growing public health problem, both in developed and developing countries. In industrialized countries, the percentage of people suffering from foodborne diseases each year has been reported to be up to 30%. In the U.S., for example, there are around 76 million cases of foodborne diseases estimated to occur each year, resulting in 325,000 hospitalizations and 5,000 deaths (WHO report, 2006).

The presence of microorganisms in food is a natural and unavoidable occurrence. Researchers are continuously searching for sensitive tools that are fast, accurate, and ultrasensitive. In recent years, there has been much research activity in the area of sensor development for detecting pathogenic microorganisms (Bhunia and Lathrop, 2003). The more exciting of nanotechnology applications is the development of nanosensors that can be placed in food production and food distribution facilities, and in the packaging itself to detect the presence of everything from *E. coli* and *Listeria* to *Campylobacter* and *Salmonella* (Russell, 2005; Sage, 2007).

The ever-present need for rapid and sensitive assay methods to detect foodborne pathogens, particularly *Salmonella*, has led to increased incorporation of biosensor technology into microarray and other platforms. The use of mimetics and aptamers has been added to these procedures. Incorporating nanoparticles, particularly fluorophores and quantum dots in various procedures, has decreased the size of instrumentation while increasing automation, sensitivity, and rapidity of results (Goldschmidt, 2006).

This exciting possibility of combining biology and nanoscale technology into sensors not only holds the potential of increased sensitivity, it significantly reduces the response-time to sense potential problems. A bioanalytical nanosensor would be able to detect a single virus particle long before the virus multiplied and long before symptoms are evident in plants or animals. Some examples of potential applications for bioanalytical nanosensors are detection of pathogens, contaminants, environmental characteristics (light/dark, hot/cold, wet/dry), heavy metals, and particulates or allergens (Scott and Chen, 2003).

Nanocantilevers, which look like tiny diving boards made of silicon, are one of the types of nanoscale materials being studied as part of general nanotechnology research. Cantilever structures are the simplest of microelectro-mechanical systems (MEMS) and can be easily micro-machined and mass produced. The ability to detect extremely small displacements makes the cantilever beams an ideal device for detection of extremely small forces and stresses (Datskos *et al.*, 2004). Nanocantilevers could be used as future detectors because they vibrate at different frequencies when contaminants stick to them, revealing the presence of dangerous substances. Nanocantilevers coated with antibodies to detect certain viruses attract different densities – or a quantity of antibodies per area – depending on the size of the cantilever. The nanocantilever devices are immersed into liquid containing antibodies to allow the proteins to stick to the nanocantilever surface (ElAmin, 2006). Also, a nanocantilever device developed using DNA biochips, to detect pathogens in different food products, has been applied successfully (Sage, 2007).

II.C. Detection of chemicals and contaminants

The contamination of food due to chemical hazards is a worldwide public health concern and a leading cause of trade problems internationally. Contamination may occur through environmental pollution of the air, water, and soil, as in the case of toxic metals, dioxins, or the intentional use of various chemicals, such as pesticides, animal drugs, and other agrochemicals. Also, there are some naturally occurring toxins such as mycotoxins, marine biotoxins, and cyanogenic glycosides that can contaminate food (WHO, 2006). Therefore the detection of these chemicals in foods is critical, whereupon nanotechnology can provide a rapid means to detecting such materials through biosensors (Hanna, 2006).

II.D. Nanodevices for identity preservation and tracking

Identity preservation (IP) is a system that creates increased value by providing customers with information about the practices and activities used to produce a particular crop or other agricultural product. Quality assurance of agricultural products' safety and security could be significantly improved through IP at the nanoscale. Nanoscale IP holds the possibility of continuous tracking and recording of history that a particular agricultural product experiences and, further, can improve identity preservation of food and agricultural products (Scott and Chen, 2003).

In another study by Lachance (2004), use of traceability systems for ensuring safety and quality of nutraceutical foods and pharmaceutical products is discussed. A composite safety system, termed intelligent product delivery system (IPDS), is proposed for preventing introduction of chemical and/or microbiological hazards into these products by bioterrorism. IPDS is based on the integration of three existing technologies:

1. Global positioning systems for location of crop field or place of manufacture
2. Bar codes
3. HACCP (Hazard Analysis and Critical Control Point).

It is further suggested that the IPDS can be coupled by developing rapid nanotechnology marker assays for ensuring traceability throughout the processing and packaging process.

III. REGULATION

The safety of nanoproducts has become the focus of increasing attention. Despite the rapid commercialization of nanotechnology, no nano-specific regulations exist anywhere in the world. Most regulatory agencies remain in an information-gathering mode, lacking the legal and scientific tools, information, and resources they need to adequately oversee exponential nanotechnology market growth (Azonano report, 2008). The U.S. Food and Drug Administration (FDA) claims that it regulates 'products, not technologies.' Nevertheless, the FDA expects that many products of nanotechnology will come under the jurisdiction of many of its centers; thus, the Office of Combination Products will likely absorb any relevant responsibilities (Tarver, 2007). The FDA regulates a wide range of products, including foods, cosmetics, drugs, devices, and veterinary products, some of which are produced utilizing nanotechnology or contain nanomaterials. Acting Commissioner of the FDA initiated the Nanotechnology Task Force (Task Force) in 2006 to help address the questions regarding adequacy and application of regulatory authorities (FDA Nanotechnology Task Force, 2007).

On the other hand, the European Commission aims at reinforcing nanotechnology and, at the same time, boosting support for collaborative R&D on the potential impact of nanotechnology on human health and the environment via toxicological and ecotoxicological studies. The Commission is performing a regulatory inventory, covering EU regulatory frameworks that are applicable to nanomaterials (chemicals, worker protection, environmental legislation, product specific legislation, etc.). The purpose of this inventory is to examine and, where appropriate, propose adaptations of EU regulations in relevant sectors (Cordis, 2008).

IV. CONCLUSION

Nanotechnology has emerged as one of the most innovative technologies to occur in decades and has the potential to improve food quality and safety. Currently a lot of research has been done on nanotechnology application in the areas of food packaging, and the detection of pathogens and contaminants. Nanotechnology will change the packaging industry by modifying the structure of packaging materials at the molecular level. As mentioned, the most beneficial application of nanotechnology is in food packaging. A self-cleaning coating at the nanoscale and an antimicrobial packaging coating has already been applied successfully as well as a polymer nanocomposite for barrier protection. For food manufacturers, using nanotechnology can mean gaining a more competitive position. While in the long term, consumers may benefit from advances in nanotechnology as new methods for improving the safety and quality of food products are developed and applied, the products still need to be regulated to ensure safety and consumer protection.

REFERENCES

Asadi, G., & Mousavi, M. (2006). Application of nanotechnology in food packaging. Proceedings of 13th IUFoST World Congress 2006. Nantes, France. Website: http://iufost.edpsciences.org

Avella, M., De Vlieger, J. B., Emanuela Errico, M. A., Fischer, S. B., Vacca, C. P., & Grazia Volpe, M. C. (2005). Biodegradable clay nanocomposite films for food packaging applications. *Food Chem., 93*(3), 467–474.

Azonano report (2008). Federal Agency Officials Meet to Discuss the Regulation of Nanotechnology, Foods, Drugs, Cosmetics, Devices and Biologics. Website: www.azonano.com/news.asp?newsID=5699

Bhunia, A. K., & Lathrop, A. (2003). Foodborne pathogen detection. In *McGraw-Hill Yearbook of Science & Technology*, pp. 320–323. New York: The McGraw-Hill Companies, Inc.

Brody, A. L. (2006). Nano and food packaging technologies converge. *Food Technology, 60*(3), 92–94.

Cappello, J. (2000). Overview of Nanotechnology: Risks, Initiatives and Standardization. Website: http://www.asse.org/practicespecialties/articles/nantechArticle.php

Cientifica Report. (2006). Nanotechnologies in the Food Industry. Website: www.cientifica.com/www/details.php?id=47

Cordis report on safety aspects of nanotechnology. (2008). Available from http://cordis.europa.eu/nanotechnology/src/safety.htm

Datskos, P. G., Thundat, T., & Lavrik, V. N. (2004). Micro and Nanocantilever Sensors. Encyclopedia of Nanoscience and Nanotechnology, Vol. X: 1–10. Website: www.mnl.ornl.gov/Publications/Datskos-2004a.pdf

De Jong, L. (2005). Nanotechnology in action. *Intern. Food Ingredients, 5*, 107–108.

Dunn, J. (2004). A mini revolution. *Food Manufacture, 79*(9), 49–50.

ElAmin, A. (2005). UK nanotechnology research directed at food industry. Website: http://www.nutraingredients.com/news/ng.asp?id=72148

ElAmin, A. (2006). Nanocantilevers studied for quick pathogen detection. Website: http://www.foodproductiondaily.com/news-by-product/news.asp

ETC Group Report. (2004). Down on the farm: The Impact of nano-scale Technologies on Food and Agriculture. Website: http://www.etcgroup.org/en/

Food Production Daily Report. (2005). Nanotech discovery promises safer food packaging. Website: http://www.foodproductiondaily.com/news/ng.asp?n=59980-nanotech-discovery-promises

Goldschmidt, M. C. (2006). The use of biosensor and microarray techniques in the rapid detection and identification of salmonellae. *J. AOAC Intern., 89*(2), 530–537.

Hanna, L. (2006). Sensors and sensing – opportunities for the food industry. *Food Manufacturing Efficiency (Special theme: Modeling and simulation), 1*(1), 55–56.

Helmut Kaiser Consultancy. (2005). Study: nanotechnology in food and food processing industry worldwide. Websites: http://www.hkc22.com/nanofood.html; http://www.who.int/foodsafety/en/; http://www.who.int/foodsafety/chem/en/; http://www.who.int/mediacentre/factsheets/fs237/en/

Joseph, T., & Morrison, M. (2006). Nanotechnology in agriculture and food, Institute of nanotechnology. Website: http://www.nanoforum.org/dateien/temp/nanotechnology%20in%20agriculture%20and%20food.pdf?20032007152346

Lachance, P. A. (2004). Nutraceutical/drug/anti-terrorism safety assurance through traceability. Toxicology Letters *(Nutraceutical and functional food industries: aspects on safety and regulatory requirements), 150*(1), 25–27.

Lagaron, J. M. (2006). Higher barriers and better performance. *Food Engineer. Ingredients,* *31*(2), 50–51.

Mallika, C. (2005). Nano-technology: applications in food industry. *Indian Food Industry,* *24*(4), 19–21, 31,

Moraru, C. I., Panchapakesan, C., Huang Q., Takhistov, P., Liu S., & Kokini J. (2003). Nanotechnology: A new frontier in food science. *Food Technol.* 57: 25–27. Website: http://members.ift.org/NR/rdonlyres/5F641E00-290A-4EB0-93D7-C02171FF5D17/0/1203moraru.pdf

Patton, D. (2006). Australian nanotech firm promises better food packaging film. Website: http://www.foodproductiondaily.com/news

Ray, S., Quek, S. Y., Easteal, A., & Chen, X. D. (2006). The potential use of polymer-clay nanocomposites in food packaging. *Intern. J. Food Engineer.* 2(4): 22–25. Website: http://www.bepress.com/ijfe/vol2/iss4/art5/

Russell, E. (2005). The nuts & bolts of nanoscience. *Intern. Food Ingredients,* *5*, 103–105.

Sage, L. (2007). Detecting pathogens on produce. *Analytical Chemistry,* *79*(1), 7–8.

Salamanca-Buentello, F., Persad, D. L., Court, E. B., Martin, D. K., Daar, A. S., & Singer, P. A. (2005). Nanotechnology and the Developing World. PLoS Med 2(5): e97 doi:10.1371/journal.pmed.0020097.

Scott, N., & Chen, H. (2003). Nanoscale science and engineering of agriculture and food systems. Report submitted to Cooperative State research, Education and Extension Service (CSREES), U.S. Dept. of Agriculture. Website: http://www.nseafs.cornell.edu

Taver, T. (2007). IFT scientific summery on Food nanotechnology. Website: http://members.ift.org/NR/rdonlyres/E725D811-3620-4CC1-8AAD-40E2BF66CE7E/0/1106Nano.pdf

U.S. Food and Drug Administration, Nanotechnology Task Force report. (2007). Website: http://www.fda.gov/nanotechnology/taskforce/report2007.pdf

Warad, H. C., & Dutta, J. (2006). Nanotechnology for agriculture and food systems–A view. Asian Institute of nanotechnology. Website: http://www.foresight.org/publications/weekly0026.html

Weiss, Jochen, Takhistov, Paul, & McClements, D. Julian (2006). Functional Materials in Food Nanotechnology. *J. Food Sci.,* *71*(9), R107–RR116.

CHAPTER 23

Nanotechnology for Foods: Delivery Systems

Eyal Shimoni

Contents

Abstract

Nanotechnology and nanosciences, and their related fields, are bound to make a tremendous impact on most aspects of our lives in the coming years. The broad spectrum starts from the tools provided to us for study of the materials we work with at the molecular and atomic scale. These analytical capabilities enable food scientists and technologists to attain precise control over the assembly of food biomaterials, and also food-related materials, as ones used for packaging. Moreover, new tools for precise analysis and detection will ensure a safer food chain. The focus of this chapter is on the combination of 'know-how' accumulated from understanding the nanoscale phenomenon in biomaterials, and the design and application of nano-derived assemblies as tools for improved delivery and bioavailability of bioactive nutrients.

I. INTRODUCTION

Nanotechnology and nanosciences stem from our ability to understand physical and chemical phenomena and processes at molecular and atomic resolution. The very basic concept leading this field is the notion that by governing structure at the atomic level, we can achieve a new dimension of tools and machines, which once controlled efficiently, and assembled in an orchestrated manner, can open a tremendous array of opportunities for new and innovative properties of materials at the macro–scale.

While going from macro- to micron scale can be done by diminishing the size of the tools used to assemble the parts of the system we build, it is

Global Issues in Food Science and Technology
© 2009 Elsevier Inc.

important to note that such downscaling cannot be done when we reduce the size of the assembled machine to three orders of magnitude at the nanoscale level. Here, processes are governed by different forces: molecular interactions mediated by weak chemical bonds such as van der Waals forces, ionic and hydrophobic interactions. Therefore, the construction of nanoscale structures relies in most cases on the self-assembly of the materials. This is true for lipids, proteins, and polysaccharides. Thus, the physicochemical properties of food components are the major players in the use of nanoscale architectures in food science and technology.

The study of food nanostructure and the application of nanotechnology to the various aspects of food science and technology picked up speed at the beginning of the third millennium, and it is now being quickly adopted as a legitimate field of research in the food science community. In this chapter we will focus on the use and application of nanotechnology concepts for the design of oral delivery systems, for the protection, release, and improved bioavailability of nutrients and bioactive food compounds.

Challenges facing introduction of bioactives into foods are not limited solely to their inclusion in free-flowing powder or solution. Such compounds are often ultrasensitive to thermal and oxidative stress and are degraded during processing or storage. Moreover, as many bioactives are practically insoluble in water, they require a vehicle to increase their solubility or dispersability in the commonly hydrophilic food matrix/liquid. Upon their consumption, enzymatic degradation, acid conditions in the stomach, as well as changes in the osmolality of the intestinal fluids, might affect the properties of the nutrients and their functionality. If we consider the fact that for every compound there is preferred loci in the gastro intestine (GI) for its absorption into the blood stream, it is evident that a carrier of functional biochemicals in food should also bare some controlled release properties.

As noted above, the ideal delivery system for functional bioactives in foods encompasses a wide array of controlled release properties, which are probably too wide to be all answered by the same system. Considering the fact that in food, unlike in pharmaceuticals, one can use only food-grade materials, meeting these requirements is exceptionally difficult. Additionally, the fact that nanosize systems call for new processes of formation, as well as new ways of design, one can ask: Why go nano? The reasons can be found in new properties stemming from the use of matter being organized at the molecular level. Such properties include the precise control of material behavior under temperature, pH, water activity, enzymatic environment, etc., as described in the following sections, including several examples of delivery systems designed at the molecular, nanoscale level.

II. LIPID-BASED NANOENCAPSULATION SYSTEMS

Lipid-based nanoencapsulation systems are among the most rapidly developing fields of nanotechnology application in food systems. Lipid-based nanoencapsulation systems have several advantages, including the ability to entrap material with different solubilities, the use of natural ingredients on an industrial scale, and the possibility to target them (Brummer, 2004; Mozafari, 2004; Yurdugul and Mozafari, 2004; Mozafari and Mortazavi, 2005; Taylor *et al.*, 2005). By protecting the bioactive ingredient from free radicals, metal ions, pH, and enzymes these systems practically prevent the degradation of the food ingredient. Lyposomes induce stability in water–soluble material, particularly in high–water–activity applications (Gouin, 2004). Lipid-based nanocarriers can be also used for targeted delivery of their content to specific areas within the GI or food matrix (Mozafari and Mortazavi, 2005). In this section we will focus on two types of lipid-based nanocarriers: liposomes and nano-sized self-assembled liquids.

Liposomes are spherical lipid vesicles, which may incorporate a wide range of bioactive compounds in their hydrophilic interior. Due to their bilayer structure they often serve as a model for biological membrane, however, they may consist of more than one bilayer membrane. Their resemblance to cellular membranes makes them an interesting candidate for delivering bioactives into cells through interaction with the membrane. While liposomes are used extensively in the pharmaceutical industry, their introduction to food systems is growing rapidly, but it is still in its infancy.

A liposome composed of a number of concentric bilayers is called a multilamellar vesicle (MLV). Others composed of many small non-concentric vesicles entrapped in a single bilayer are called multivesicular vesicles (MVV). A third type of liposome is called a unilamellar vesicle (ULV), which contains a single lipidic bilayer (Figure 23.1). Liposomes can be used for delivery and release of both water- and lipid-soluble materials.

When referring to nanoscale lipid vesicles, the term nanoliposome has recently been introduced (Mozafari and Mortazavi, 2005) to describe lipid

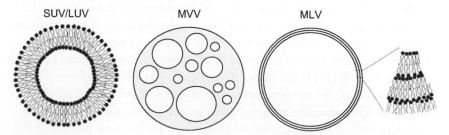

Figure 23.1 The three liposome types (left to right): small and large unilamellar vesicles (SUV/LUV), multivesicular vesicle (MVV), and multilamellar vesicle (MLV).

vesicles, which have diameters that range from tens of nanometers. These so-called nanoliposomes have similar structural, physical, and thermodynamic properties to the liposomes described previously. An important point is that, while in nanoscience and technology we often refer to self-assembly phenomenon as the preferred tool for structuring matter, the manufacture of nanoliposomes (e.g. liposomes) requires high energy for the dispersion of lipid/phospholipid molecules in the aqueous medium (Mozafari and Mortazavi, 2005; Mozafari, 2005). The underlying mechanism for the formation of liposomes and nanoliposomes is basically the hydrophilic–hydrophobic interaction between phospholipids and water molecules. As do all dynamic entities, vesicles prepared in nanometric size tend to aggregate or fuse and may end up growing into micron-size particles during storage.

For food applications, probably the most important issue is the food-grade status of the materials making the liposome. Therefore, the study of food-grade liposomes made down to the nanoscale level is an essential component. The effect of edible lipid composition on size, stability, and entrapment efficiency of polypeptide antimicrobials in liposomal nanocapsules was investigated by Were et al. (2003). A mixture of phosphatidylcholine (PC), phosphatidylglycerol (PG), and cholesterol was used to form liposomes with antimicrobial peptides. With calcein and nisin, the entrapment reached 54–70%, and the size of the liposomes 85–233 nm. The highest concentration of antimicrobials was encapsulated in 100% PC liposomes. Their results show that stable nanoparticulate aqueous dispersions of polypeptide antimicrobials for food products depend on the selection of suitable lipid–antimicrobial combinations.

In order to evaluate the feasibility of using lecithins for nanocapsules, including functional food materials, Takahashi et al. (2007) prepared liposomes from different lecithins and examined their physicochemical properties. There was little difference in the trapping efficiency among the three types of liposomes. In all cases, the trapping efficiency clearly increased with an increase of the lecithin concentration up to 10% wt, and the maximal efficiency reached 15%. Confocal laser scanning microscopy (CLSM) showed that the particle size of liposomes prepared from SLP-WHITE was significantly smaller than that of other lecithins. This liposomal solution remained well dispersed for at least 30 days. After using a homogenizer and microfluidizer to improve the efficiency and stability, the particle size of the SLP-WHITE liposomes decreased and reached between 73 and 123 nm based on the measurement with dynamic light scattering (DLS). Using these liposomes, the authors demonstrated the trapping of curcumins up to over 85%. Thus, their results show that the method may have the potential for manufacturing nanocapsules, which serve as novel carriers of functional food materials.

An interesting advantage of liposomes and nanoliposomes is their ability to incorporate and release two materials with different solubilities simultaneously. These 'bifunctional liposomes' (Suntres and Shek, 1996) were used to incorporate two antioxidant agents: (1) tocopherol (a lipid-soluble molecule) and glutathione (a water-soluble molecule) (Suntres and Shek, 1996; Mozafari et al., 2004); and (2) ascorbic acid and tocopherol (Kirby, 1993). By delivering both vitamin E and ascorbic acid to the oxidation site a synergistic effect can be achieved. Some studies have actually reported that when tocopherol is entrapped in liposome, it is more effective than in its free form (Arnaud, 1995).

One of the most promising lipid-based nanodelivery systems for food applications is the development of nano-sized self-assembled liquids (NSSL), as studied by Garti et al. (2005). These carriers are intensively being studied by Garti's group at the Hebrew University (Jerusalem, Israel). They developed NSSL vehicles to tackle shortcomings in the microemulsion system. They found that a unique mixture of food-grade oils, in which two or more food-grade nonionic hydrophilic emulsifiers, cosolvents (polyol), and coemulsifiers self-assemble to form mixed reverse micelles (the concentrate), can be inverted into oil-in-water nanodroplets. This system is transformed into bicontinuous structures by dilution with an aqueous phase, progressively and continuously, without phase separation. These reversed micelles can solubilize compounds that are poorly soluble in water or in the oil phase. Garti's group demonstrated that the system provides 10–20 times more solubility capacity to nutraceuticals, antioxidants, and others than in any food-grade oils or water phase.

The detailed studies of Garti et al. (2005) demonstrated that NSSLs can be used to solubilize phytosterols up to 12 times more than the dissolution capacity of the oil (R-(+)-limonene) for the same compounds. Also, the solubilization of lycopene in the concentrate was found to be about ten times more than in the corresponding oil. These systems were shown to be superior in protecting the solubilizates from environmental reactivity (oxidation). Lycopene did not oxidize even after 75 days in an open vessel if solubilized in the microemulsion medium, while if left unformulated, it was totally oxidized. This method is most likely to be implemented in a wide range of food and pharma products.

III. THE USE OF PROTEINS IN NANOSCALE DELIVERY SYSTEMS

The use of the protein–polysaccharide interaction to form encapsulation systems based on coacervation was downsized to nanoscale by Huang and

Jiang (2004). Their study on health-promoting flavonoids focused on the most active form contained in green tea, epigallocatechin gallate (EGCG). Green tea has long been used as a beverage, but low bioavailability is always a problem. Therefore, this group suggested encapsulation of the tea catechins to enhance their stability and bioavailability. They used coacervates formed by protein–polysaccharide complexes as an inexpensive encapsulation method for green tea catechins at the micro- and nanoscale level.

Yu *et al.* (2005) applied peptide nanotubes as supports for enzyme immobilization. They encapsulated a lipase inside peptide nanotubes and found that the catalytic activity of the nanotube-bound enzymes was actually higher (>33%) than the free lipase at room temperature. Interestingly, at 65°C, the activity of lipase in the nanotubes was 70% higher as compared to the free lipases. They showed that the enzyme inside the nanotube can be recycled by fabricating magnetic nanotubes using FePt superparamagnetic nanocrystals immobilized on the outside of the peptide nanotubes. They suggested that the activity enhancement of lipases in the peptide nanotubes is likely induced by the conformation change of lipases to the open form (enzymically active structure), as lipases are adsorbed on the inner surfaces of peptide nanotubes.

Nohiro *et al.* (2006) used casein to form nano-sized protein micelles to hold hydrophobic substances. The group used transglutaminase to form ANS-encapsulated casein micelles with particle size 36 nm, which retained $\geq 50\%$ ANS when treated with trypsin. This method is useful for manufacturing transparent supersaturated solutions by solubilization of hydrophobic substances in functional foods and pharmaceuticals.

One of the more recent concepts introduced by Semo *et al.* (2006; led by Y. Livney) is the use of self-assembled casein micelles as nanocapsular vehicles. The authors realized that casein micelles (CM) are in effect nanocapsules created by nature to deliver nutrients, such as calcium, phosphate, and protein to the neonate. Thus, they suggested using CM as a self-assembled system for nanoencapsulation and stabilization of hydrophobic nutraceutical substances for enrichment of food products. Vitamin D2 was used as a model for hydrophobic nutraceutical compounds. D2-rCM and rCM had similar morphology, which was also typical in naturally occurring CM, as demonstrated in Cryo-TEM micrographs (Figure 23.2).

The reassembled micelles had average diameters of 146 and 152 nm with and without vitamin D2, respectively, similar to normal CM, which are typically 150 nm on average. It was demonstrated that a nutraceutical compound can be loaded into CM, by utilizing the natural self-assembly tendency of bovine caseins. The vitamin concentration in the micelle was about 5.5 times more than in the serum. Even in the serum, vitamin D2 was

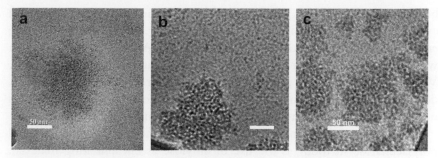

Figure 23.2 Cryo-TEM images: (a) reassembled CM (rCM), (b) Vitamin D2 containing rCM (D2-rCM), (c) naturally occurring CM in skim milk (unpublished sets; courtesy of Dr. Livney and Dr. Danino, Faculty of Biotechnology and Food Engineering).

only present as bound to residual soluble caseins. A very interesting observation was that the morphology and average diameter of the reassembled micelles were similar to those of naturally occurring CM. Partial protection against UV light-induced degradation of vitamin D2 was afforded by the micelles. This study therefore demonstrated that CM may be useful as nanovehicles for entrapment of hydrophobic nutraceuticals within food products. Such nanocapsules may be incorporated in dairy products without modifying their sensory properties.

Recently, Livney and Ron (2007) demonstrated the use of β-lactoglobulin–polysaccharide complexes to carry hydrophobic nutraceuticals. The authors took advantage of the ability of beta-lactoglobulin to bind hydrophobic molecules, and added a secondary protective layer by its complexation with charged polysaccharide. They showed that by carefully controlling the zeta potential of the particle, a stable nanosize vehicle can be produced. It should be noted, however, that in such instances the load of the bioactive in the vesicle was low and the stabilization effect has yet to be demonstrated. Yet, this clever control of the physical properties of the system is an excellent example for assembling a nanosize encapsulation vesicle based on controlling the physical properties.

IV. POLYSACCHARIDE-BASED NANOCAPSULES

A new beverage product, with particle size around 100 nm, was produced by ultra-high-pressure homogenization and stabilized by microencapsulation with a food-grade starch (Chen and Wagner, 2004). The vitamin E nanoparticle product was stable in the beverage and did not alter its appearance. Particles were produced by dissolving starch sodium octenyl succinate in distilled water with vitamin E acetate slowly added and homogenized with a high shear mixer until the emulsion droplet size was below 1.5 μm. The crude emulsion was then further homogenized and cycled through the homogenization process for several passes until the emulsion droplets reached the target particle size. Afterwards, the emulsion was spray-dried to yield a powder containing about 15% vitamin E acetate. While providing desired properties, this method represents the traditional downsizing of conventional encapsulation methodologies. In principle, it is not different in concept from conventional encapsulation methods.

One interesting use of starch, probably the most abundant functional polysaccharide in foods, is the use of amylose as encapsulating material. Here, molecular entrapment by amylose is based on the interaction between amylose and lipids, characterized by amylose chains forming V-crystalline structure. In this form, the amylose chain forms a helix with a large cavity in which low-molecular-weight chemicals can be situated. The size of the ligand determines the number of glucosyl residues per turn (6, 7, or 8) (Snape et al., 1998). This crystalline state of amylose–fatty acid complex involves the V-amylose six-fold single chain left-handed helix with the fatty acid as a 'stem' (planar zigzag) inside the helix (Godet, 1993). The V-helical complex segments are interrupted by short sections of uncomplexed amylose that permit random orientation of the helical segments (Karklas and Raphaelides, 1986; Biliaderis and Galloway, 1989; Karkalas et al., 1995). The various studies on complexes produced from mono- and di- glycerides (Eliasson and Krog, 1985; Biliaderis and Galloway, 1989; Seneviratne and Biliaderis, 1991; Tufvesson and Eliasson, 2000; Tufvesson et al., 2001), saturated fatty acids (Godet et al., 1993, 1995; Karkalas et al., 1995; Lebail et al., 2000; Tufvesson et al., 2003), as well as unsaturated fatty acids (Szejtli and Banky-Elod, 1975; Eliasson and Krog, 1985; Karkalas et al., 1995; Snape et al., 1998; Tufvesson et al., 2003), have shown that they have high melting temperatures; that the complex fatty acid is efficiently protected from oxidation; and that the digestibility of starch is influenced by complex formation, which decreases the digestibility of starch. In light of these studies, it was suggested that amylase–lipid complexes could be used as a delivery system for hydrophobic bioactive nutrients. It was hypothesized that these complexes can protect polyunsaturated fatty acids (PUFA) during processing and storage, and that they release them in the intestine following amylolytic hydrolysis. The concept was examined using conjugated linoleic acid (CLA) as a model (Lalush et al., 2005).

Amylose–CLA complexes were produced to provide optimal stability to oxidation and thermal treatments, to dissolution in the stomach, and efficient release by mammalian amylases (Lalush et al., 2005). Inclusion complexes were produced using the method of Eliasson and Krog (1985) in a water/DMSO mixture, as well as the method of Karkalas et al. (1995). The DSC thermal analysis of the complexes showed a transition temperature ranging from 88–95°C, suggesting stability of the complexes during food processing. AFM scanning showed that the complexes have a globular structure of heterogeneous nature with an average diameter of 152 nm ± 39 nm. Complexes created by KOH/HCl solution exhibited mainly rod-like structures of heterogeneous nature with a width of about 40 nm and length ranging from 0.35 to 3.2 μm. As shown in Figure 23.3, the amylase–CLA complexes produced by the water/DMSO complexation system yield nano-particulated molecular capsules. Stability tests showed that, regardless of the complexation method and temperature, the complexes protect and inhibit the oxidation of the ligand. Enzymatic digestion of amylase–ligand complexes by mammalian pancreatic amylase and various microbial enzymes showed that amylolytic enzymes can digest the complex. Following the digestion experiments, the ligands were released only when complexes

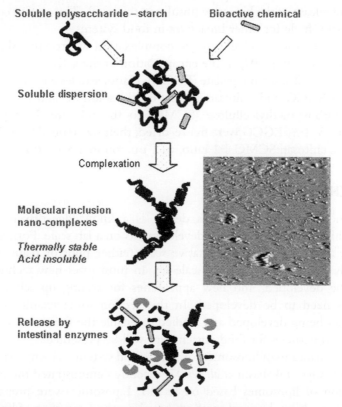

Figure 23.3 Schematic description of the assembly and release of molecular inclusion complexes made of amylose and hydrophobic bioactive molecules. The AFM image on the right-hand side shows nanosize spherical clusters of inclusion complexes, 80–150 nm diameter.

were digested by amylases. These results demonstrate the potential of the system for molecular encapsulation of hydrophobic functional materials. Currently, this method is being further developed to encapsulate a wide range of bioactives using continuous processes. A continuous method was developed recently and used to produce complexes with isoflavones, PUFA, and other substances (Shimoni *et al.*, 2007).

One very interesting complex system has been developed for oral insulin delivery by Pinto Reis *et al.* (2007), however the system is composed of food-grade materials and thus could be used to carry other components. The authors used nanoemulsion triggering in situ alginate gelation with dextrin to form the core. This was followed by polyelectrolyte complex coating, using chitosan as the first layer, followed by albumin. By using this technique the albumin provided protection against proteolysis, and the chitosan acted as mucoadhesive to bind the nanocapsules to the intestine wall. In using these nanocapsules, the authors demonstrated efficient delivery and release of the bioactive (insulin in this case). This system could be a potential vehicle for other bioactives in food systems.

Indeed, the use of polyelectrolyte complexation was recently shown by Seangsitthisak *et al.* (2007) in the encapsulation of the active component in green tea, epigallocatechin gallate (EGCG). Nanoparticles were prepared by dissolving EGCG and chitosan in acetic acid solution, and then adding sodium carboxymethylcellulose (SCMC) to the mixture. Nanoparticles containing 5–10% EGCG were freeze-dried; their size ranged from 200 nm when in a chitosan:SCMC 1:1 ratio, to 1 micron in a 5:1 ratio.

V. TECHNOLOGIES

The nanoscale delivery system's design is a knowledge-driven process achieved in other technological developments on a lab scale. Formation of nanosize assemblies on a quantitative scale (either pilot or plant scale) is obviously not a straightforward scale up. In most cases new technologies should be developed, and new approaches for scaling up self-assembly processes need to be developed. In this section some examples of the techniques being developed and studied to enable the formation of nano-carriers and particles for food enrichment will be presented.

The preparation of liposomes for use in food systems is being performed in various ways. Takahashi *et al.* (2007) recently demonstrated the efficient preparation of liposomes based on lecitin. Liposomes were prepared by using the mechanochemical method, based on the formation of liposomal suspension by processing with a microfluidizer. To manufacture phospholipid liposomes containing antimicrobial peptides Were *et al.* (2003) used a modified method of Pinnaduwage and Bruce (1996). PC lipid stocks were dissolved in chloroform and dried by a stream of nitrogen followed by overnight drying under vacuum. The dried lipid combinations were then rehydrated and ultrasonicated to initiate encapsulation and liposome formation. Vesicles collected by centrifugation were then separated from unencapsulated antimicrobials using size exclusion chromatography. The authors produced liposomes from different formulations and showed that indeed, for each antimicrobial peptide, there is an optimal composition based on compatibility of the lipid and the antimicrobial.

As for nanoparticles, the number of methods that could be utilized for their formation is as high as the various formulations proposed. In one case, phase separation was used by Parris *et al.* (2005) to form zein nanoparticles. The authors used the method to form nanospheres to carry essential oils of oregano, thyme, and cassia. Here, the particles were formed by dissolving the oil and the coating/matrix material in 85% ethanol. Particles were formed by rapid dispersion of the solution during high-speed mixing into water. The dry form of the particles was obtained by freeze-drying the opaque solution. The authors reported a yield of 65–75%. In principle, this approach can be used to form starch-inclusion nanocomplexes (Lalush *et al.*, 2005; Shimoni *et al.*, 2006, U.S. patent). Starch and ligand could be co-dissolved in DMSO or alkali solution, and then either diluted in water or acidified, respectively. The result in both cases is the formation of conditions that promote the sedimentation (e.g. phase separation) of the starch while forming inclusion complexes. The continuous dual feed jet homogenization developed for the process provides sub-micron size particles. A similar approach was used by Chen and Wagner (2004) to form vitamin E nano-particles. They used a microfluidics homogenizer to form the nanoparticles, which were further dried with a spray dryer.

Fibers are also potent tools for protection, delivery, and controlled release systems. Electrospinning was demonstrated by Jiang *et al.* (2006) in the formation of biodegradable core-shell fibers. By applying high voltage to a spinneret, an electric jet of viscous polymer solution can be formed. By stretching the jet prior to its reaching the target, it is possible to evaporate the solvent quickly. The resulting fibers have a tremendously high surface of porosive matrix. Spinnerets, composed of two coaxial capillaries, can simultaneously electrospin two different solutions to form core and shell nanofibers. Jiang *et al.* (2006) demonstrated the use of the technique to incorporate and control the release of proteins from a biodegradable nanofiber. An interesting point is their use of polyethylene glycol (PEG) as the shell, and of dextran as a core polymer. These two are compatible with food systems, and thus one can clearly see the potential of the method for future food applications.

VI. CONCLUDING REMARKS

The use of nano-sized vehicles for the protection and controlled release of nutrients and bioactive food ingredients is a growing area of interest in the food science and technology community. In this chapter, we presented only very few examples to demonstrate the potential of some of the techniques and concepts that have evolved in recent years. These methods as well as others not mentioned here are likely to change the way we deliver functional bioactive food ingredients in the years to come. They have the potential to provide new opportunities for use in new food systems. Furthermore, the use of these platforms may very well fit into the new world of personalized nutrition.

REFERENCES

Arnaud, J. P. (1995). Pro-liposomes for the food industry. *Food Technol. Eur.,* 2, 30–34.

Biliaderis, C. G., & Galloway, G. (1989). *Carbohydrate Research, 189,* 31–48.

Bummer, P. M. (2004). Physical chemical considerations of lipid-based oral drug delivery – solid lipid nanoparticles. *Crit. Rev. Ther. Drug Carrier Syst., 21,* 1–20.

Chen, C.-C., & Wagner, G. (2004). Vitamin E nanoparticle for beverage applications. *Chem. Engineer. Res. Design, 82*(A11), 1432–1437.

Eliasson, A. C., & Krog, N. J. (1985). Physical properties of amylose-monoglyceride complexes. *Cereal Sci., 3,* 239–248.

Garti, N., Spernath, A., Aserin, A., & Lutz, R. (2005). Nano-sized self-assemblies of nonionic surfactants as solubilization reservoirs and microreactors for food systems. *Soft Matter, 1,* 206–218.

Godet, M. C., Buleon, A., Tran, V., & Colonna, P. (1993). Structural features of fatty acid–amylose complexes. *Carbohydrate Polymers, 21,* 91–95.

Godet, M. C., Tran, V., Colonna, P., Buleon, A., & Pezolet, M. (1995). Inclusion/exclusion of fatty acids in amylose complexes as a function of the fatty acid chain length. *Int. J. Biol. Macromol., 17*(6), 405–408.

Godet, M. C., Bouchet, B., Colonna, P., Gallant, D. J., & Buleon, A. (1996). Crystalline amylose–fatty acid complexes: Morphology and crystal thickness. *J. Food Sci., 61*(6), 1196–1201.

Gouin, S. (2004). Micro-encapsulation: industrial appraisal of existing technologies and trends. *Trends Food Sci. Technol., 15,* 330–347.

Huang, Qingrong and Jiang, Yan. (2004). Enhancing the stability of phenolic antioxidants by nanoencapsulation. Abstracts of Papers. 228th ACS National Meeting, Philadelphia, PA, United States, August 22–26, 2004.

Jiang, Hongliang, Hu, Yingqian, Zhao, Pengcheng, Li, Yan, & Zhu, Kangjie (2006). Modulation of protein release from biodegradable core-shell structured fibers prepared by coaxial electrospinning. *J. Biomed. Mater. Res. B Appl. Biomater., 79B*(1), 50–57.

Karkalas, J., & Raphaelides, S. (1986). Quantitative aspects of amylose-lipid interactions. *Carbohydrate Res., 157,* 215–234.

Karkalas, J., Ma, S., Morrison, W., & Pethrick, R. A. (1995). Some factors determining the thermal properties of amylose inclusion complexes with fatty acids. *Carbohydrate Res., 268,* 233–247.

Kirby, C. J. (1993). Controlled delivery of functional food ingredients: opportunities for liposomes in the food industry. In G. Gregoriadis (Ed.), *Liposome Technology* (pp. 215–232). London: CRC Press.

Lalush, I., Bar, H., Zakaria, I., Eichler, S., & Shimoni, E. (2005). Utilization of amylose-lipid complexes as molecular nanocapsules for conjugated linoleic acid. *Biomacromolecules, 6*, 121–130.

Lebail, P., Buleon, A., Shiftan, D., & Marchessault, R. H. (2000). Mobility of lipid in complexes of amylose–fatty acids by deuterium and 13C solid state NMR. *Carbohydrate Polymers, 43*(4), 317–326.

Livney, Y. D., & Ron, Nadav. (2007). Beta-lactoglobulin (ß-Lg)–polysaccharide complexes as nanovehicles for hydrophobic nutraceuticals. Proceedings of XV International workshop on Bioencapsulation, Vienna, Austria, September 6–8, 2007.

Mozafari, M. R. (2004). Micro and nano carrier technologies: high quality production within pharmaceutical standards. *Cell Mol. Biol. Lett., 9*(Suppl. 2), 44–45.

Mozafari, M. R. (2005). Liposomes: an overview of manufacturing techniques. *Cell Mol. Biol. Lett., 10*, 711–719.

Mozafari, M. R., & Mortazavi, S. M. (2005). Nanoliposomes: from Fundamentals to Recent Developments. Oxford: Trafford.

Mozafari, M. R., Reed, C. J., & Rostron, C. (2004). Ultrastructural architecture of liposome-entrapped glutathione: a cryo-SEM study. *Cell Mol. Biol. Lett., 9*(Suppl. 2), 101–103.

Mozafari, M. R., Flanagan, J., Matia-Merino, L., Awati, A., Omri, A., Suntres, Z. E., & Singh, H. (2006). Recent trends in the lipid-based nanoencapsulation of antioxidants and their role in foods. *J. Sci. Food Agriculture, 86*(13), 2038–2045.

Noriho, K., Goto, M., Shioashi, Y., & Nio, T. (2006). Manufacture of nano-sized protein micelles holding and absorbing hydrophobic substances inside of the micelles. *Jpn.* Kokai Tokkyo Koho, 2006. Patent Application: JP 2004-306546 20041021.

Parris, N., Cooke, P. H., & Hicks, K. B. (2005). Encapsulation of essential oils in zein nanospherical particles. *J. Agric. Food Chem., 53*, 4788–4792.

Pinnaduwage, P., & Bruce, B. D. (1996). In vitro interaction between a chloroplast transit peptide and chloroplast outer envelope lipids is sequence-specific and lipid class-dependent. *J. Biol. Chem., 51*, 32907–32915.

Pinto Reis, C., Sarmento, B., Erdinc, B., Bowey, K., Ribeiro, A., Veiga, F., Ferreira, D., Damge, C., & Neufeld, R. J. (2007). Effective oral delivery of insulin by a nanoplex composite delivery system containing alginate as core material. Proceedings of XV International workshop on Bioencapsulation, Vienna, Austria, September 6–8, 2007.

Seangsitthisak, B., Duangrat, C., and Okonogi, S. (2007). Nano-Size Encapsulation of Bioactive Compound of Green Tea. Proceedings of XV International workshop on Bioencapsulation, Vienna, Austria, September 6–8, 2007.

Semo, E., Kesselman, E., Danino, D., & Livney, Y. D. (2006). Casein micelle as a natural nano-capsular vehicle for nutraceuticals. *Food Hydrocolloids, 21*(5–6), 936–942.

Seneviratne, H. D., & Biliaderis, C. G. (1991). Action of α-amylases on amylose-lipid complex superstructures. *J. Cereal Sci., 13*(2), 129–143.

Shimoni, E., Lesmes, U., & Ungar, Y. (2006). *Non-covalent nanocomplexes of bioactive agents with starch for oral delivery,* U.S. Provisional Patent Application, 2006.

Shimoni, E., Lesmes, U., Cohen, R., & Ades, H. (2007). Using starch molecular complexes as carriers for therapeutics and nutrients. Proceedings of XV International workshop on Bioencapsulation, Vienna, Austria, September 6–8, 2007.

Snape, C. E., Morrison, W. R., Maroto-Valer, M. M., Karkalas, J., & Pethrick, R. A. (1998). *Carbohydrates Polymers, 36*, 225–237.

Suntres, Z. E., & Shek, P. N. (1996). Alleviation of paraquat-induced lung injury by pretreatment with bifunctional liposomes containing alpha-tocopherol and glutathione. *Biochem. Pharmacol., 52*, 1515–1520.

Szejtli, J., & Banky-Elod, E. (1975). Inclusion complexes of unsaturated fatty acids with amylose and cyclodextrin. *Starch, 27*, 368–376.

Takahashi, M., Inafuku, K., Miyagi, T., Oku, H., Wada, K., Imura, T., & Kitamoto, D. (2007). Efficient preparation of liposomes encapsulating food materials using lecithins by a mechanochemical method. *J. Oleo Sci., 56*(1), 35–42.

Taylor, T. M., Davidson, P. M., Bruce, B. D., & Weiss, J. (2005). Liposomal Nanocapsules in Food Science and Agriculture. *Crit. Rev. Food Sci. Nutrition, 45,* 587–605.

Tufvesson, F., & Eliasson, A. C. (2000). Formation and crystallization of amylose–monoglyceride complex in a starch matrix. *Carbohydrate Polymers, 43*(4), 359–365.

Tufvesson, F., Skrabanja, V., Bjorck, I., Elmstahl, H. L., & Eliasson, A. C. (2001). Digestibility of starch systems containing amylose–glycerol monopalmitin complexes. *Lebensmittel-Wissenschaft und -Technologie, 34*(3), 131–139.

Tufvesson, F., Wahlgren, M., & Eliasson, A. C. (2003). Formation of amylase–lipid complexes and effects of temperature treatment Part 2, Fatty acids. *Starch, 55,* 138–149.

Were, L. M., Bruce, B. D., Davidson, P. M., & Weiss, J. (2003). Size, stability, and entrapment efficiency of phospholipid nanocapsules containing polypeptide antimicrobials. *J. Agricultural Food Chem., 51*(27), 8073–8079.

Yu, L., Banerjee, I., Gao, X., & Matsui, H. (2005). Fabrications and applications of enzyme-incorporated peptide nanotubes and magnetic nanocrystal-coated peptide nanotubes. *Abstracts of Papers* San Diego, CA, USA: 229th ACS National Meeting, March 13-17, 2005.

Yu, L., Banerjee, I. A., Gao, X., Nuraje, N., & Matsui, H. (2005). Fabrication and application of enzyme-incorporated peptide nanotubes. *Bioconjugate Chem., 16*(6), 1484–1487.

Yurdugul, S., & Mozafari, M. R. (2004). Recent advances in micro- and nano-encapsulation of food ingredients. *Cell Mol. Biol. Lett., 9*(Suppl. 2), 64–65.

CHAPTER 24

Nanostructured Encapsulation Systems: Food Antimicrobials

Jochen Weiss, Sylvia Gaysinsky, Michael Davidson, *and* Julian McClements

Contents

Abstract

This chapter provides an overview of the current development of the design and production of nanostructured encapsulation systems, including nanoemulsions, microemulsions, solid-lipid nanoparticles, and liposomes. Nanoscalar particulate systems have been shown to differ substantially in terms of their physicochemical properties from larger microscopic systems due to their submicron particle diameters. Properties affected by the small particle diameters include particle–particle interactions, interaction with electromagnetic waves (e.g. light), interaction with biological tissue, crystallization processes, catalytic activities, and others. For food

Global Issues in Food Science and Technology
© 2009 Elsevier Inc.

ISBN 9780123741240
All rights reserved.

manufacturers, introduction of these nanoscalar systems into foods can result in favorable changes in food quality attributes such as appearance, texture, flavor, and aroma. In the case of biologically or biochemically active compounds, such as nutraceuticals, antimicrobials, and antioxidants, substantial alterations in the functionalities of the encapsulated compounds may be seen. In this chapter, we focus on the use of these nano-encapsulation systems to deliver food antimicrobials with the specific goal to improve the quality and safety of foods. To this purpose, each of the four above-mentioned encapsulation systems is first defined, their functionalities are discussed, and manufacturing and fabrication techniques are introduced. Lastly, studies utilizing each of the encapsulation systems to inhibit food pathogens or spoilage organisms are reviewed. In addition to the four above-mentioned established nanoencapsulations systems, new nanostructures that are currently under active development, the so-called second-generation nanocapsules, are discussed. Finally, some selection criteria to rationally select nanoencapsulation systems for food antimicrobials are highlighted.

I. INTRODUCTION

Nanofabrication technologies are rapidly being adapted and developed for use in the food industry for a number of applications: to produce novel packaging materials; to fabricate improved sensors to detect pathogens, allergens, contaminants, and quality degradation indicators in foods; to create novel nanostructured processing aids such as filters, membranes, and reactors; and to produce new food ingredients such as particles and fibers and assembled aggregates thereof (Weiss *et al.*, 2006). Progress in this new field of 'food nanotechnology' has been particularly rapid in the U.S., where governmental funding was first disseminated through the National Nanotechnology Initiative (NNI) and the United States Department of Agriculture (USDA) to facilitate a transfer of nanoscalar sciences to food applications (NNI, 2008). In the context of food applications, the 2008 USDA National Research Initiative listed as its specific long-term goals for research in food nanotechnology the following statement: '... to effectively aid the design of nano-based devices and systems, which are highly sensitive and specific for monitoring, detection, and intervention of *food quality, safety, and biosecurity*; to develop a mature knowledge base of nanoscale processing, product formulation, and *shelf-stability of food* with enhanced nutrition value as a part of individualized health management (IHM) practices ...' (Cooperative State Research, 2007). These goals, formulated by the primary food nanotechnology research funding agency in the U.S., recognize the tremendous potential that nanofabrication technologies have in improving the safety and quality of future food systems. Research in the area of food safety has become particularly urgent due to a lack of new preventative measures to control foodborne illnesses (Doores, 1999). Two high-profile recalls occurred in 2006; a recall of 100% carrot juice prompted by three cases of botulism following consumption of temperature-abused juice (CDC, 2006b), and the removal of all fresh bagged spinach from store shelves after 199 people across 26 states became infected with *E. coli* O157:H7 (Anonymous, 2006a, b; CDC, 2006a). Laboratory-confirmed cases of foodborne illness in the U.S. from all food products was 16,614 cases, comprising 15% of the population (37.4 cases per 100,000) in 2005, with *Salmonella*, *Campylobacter*, and *Shigella* being the most commonly reported pathogens (Allos *et al.*, 2004; Anonymous, 2006b). These incidences are further evidence that new and improved tools need to be developed that can be used to better control foodborne pathogens.

Currently, there are four basic approaches to food preservation: (a) aseptic handling to prevent entry of microorganisms into food, (b) mechanical removal of microorganisms through washing or filtration, (c) destroying microorganisms with heat, pressure, irradiation, or chemical sanitizers, and (d) inhibiting growth of food pathogens and spoilage organisms through environmental control (Davidson *et al.*, 2005). Studies over the past 7 years have shown that the latter approach in particular may benefit from recent discoveries in food nanotechnology. Growth inhibition through environmental control can be achieved by adding chemicals (food antimicrobials or preservatives) that exhibit inhibitory (bacteriostatic or fungistatic), bactericidal, or fungicidal activity. These compounds may be synthesized or extracted from natural sources. Naturally occurring antimicrobial agents in particular have attracted the attention of food processors and researchers because of the changes in consumer preference that have led to an increasing rejection of traditional antimicrobials. Unfortunately, the list of available antimicrobial compounds that are approved by FDA for use in foods is very limited (Table 24.1), and regulatory hurdles make it unlikely that a large number of new food antimicrobials will be approved for use in foods in the near future. It is thus imperative to make better use of available approved antimicrobial compounds. Food scientists have thus focused on the development of nanoencapsulation technologies to help improve the functionality of approved food antimicrobials.

As with most other bioactive compounds (e.g. flavors, antioxidants, nutraceuticals), antimicrobial agents are chemically diverse. This raises considerable problems when attempting to introduce these compounds into a complex food system. For example, addition of bioactive compounds may negatively affect the physical stability of the food system and/or alter the chemical integrity and biological activity of the bioactive compounds. For food antimicrobials, the consequence of low activity is that high concentrations of antimicrobials must be used to effectively control the growth of

Table 24.1 List of FDA approved antimicrobials for use in foods (adapted from Institute of Food Technologists, 2006)

Compound	Microbial target	Food application
Acetic acid, acetates, diacetates, dehydroacetic acid	yeasts, bacteria	baked goods condiments, confections dairy products, fats and oils meats, sauces
Benzoic acid, benzoates	yeasts, molds	beverages, fruit products margarine
Dimethyl dicarbonate	yeasts	beverages
Lactic acid, lactates	bacteria	meats, fermented foods
Lactoferrin	bacteria	meats
Lysozyme	*Clostridium botulinum*, other bacteria	cheese, casings for frankfurters, cooked meat and poultry products
Natamycin	molds	cheese
Nisin	*Clostridium botulinum*, other bacteria	cheese, casings for frankfurters, cooked meat and poultry products
Nitrite, nitrate	*Clostridium botulinum*	cured meats
Parabens	yeasts, molds	beverages, baked goods, syrups, dry sausage
Propionic acid, propionates	molds	bakery products, dairy products
Sorbic acid, sorbates	yeasts, molds and bacteria	most foods, beverages, wines
Sulfites	yeasts, molds	fruits, fruit products, potato products, wines

microorganisms. Required concentrations may in fact be so high as to exceed the regulatory levels or adversely affect the flavor of the food products. Moreover, complex mass transport phenomena, coupled with potential chemical reactions of antimicrobials with food constituents, may reduce the concentration and effectiveness of the antimicrobials in the vicinity of food pathogens or spoilage organisms. In order to inhibit growth of microorganisms, antimicrobial compounds must directly interact with the target organisms and partition into either the microbial membranes or the microbial intracellular space. Physical and chemical processes can alter the structure and functionality of antimicrobials, thereby preventing interaction of antimicrobials with target pathogens or spoilage organisms. Ingredient interactions may thus have a profoundly negative impact on the ability of antimicrobials to successfully disrupt membrane integrity, as they can interrupt functionality of protein-bound membrane complexes, change intracellular pH, and interfere with the metabolic or genetic systems of microorganisms.

Encapsulation of functional ingredients has been shown to reduce ingredient interactions and allow for better control over mass transport phenomena and chemical reactions. Encapsulation of biologically active components such as nutraceuticals may also improve bioactivity, e.g. by improving adsorption and uptake of components during digestion. In the case of food antimicrobials, encapsulation may increase the effectiveness of concentration of bioactive compounds in areas of the food system where target microorganisms are preferentially located (e.g. in water–rich phases or at solid–liquid interfaces). Novel nanoencapsulation systems have been shown to be particularly suited to the task of protecting ingredients and controlling the delivery to sites of action. This is because, compared to larger microcapsules, they often have superior physical and chemical stability, better compatibility with food matrices, and allow for targeting of bacterial surfaces through tailoring of interfacial properties (interfacial engineering), and enable high concentrations of very lipophilic functional components that are evenly dispersed in aqueous phases.

In this chapter, we will provide an overview of the fabrication, as well as the structural and functional characteristics, of a number of recently developed nanocapsules that have been successfully used to deliver a variety of antimicrobials in different foods. Initially, a brief definition and overview of food antimicrobials (the target of nanoencapsulation) is given, and their current limitations in food applications are discussed. Four emerging nanoencapsulation systems for food antimicrobials, namely nanoemulsions, surfactant micelles, phospholipid liposomes, and solid lipid nanoparticles, are introduced and some results in terms of their functionality and efficacies in model microbiological and food systems are presented. Finally, some future developments in the area of nano-encapsulation are presented.

II. FOOD ANTIMICROBIALS AS TARGETS OF NANOENCAPSULATION

II.A. Definition of antimicrobials

Antimicrobial agents have been defined by the U.S. Food and Drug Administration as 'chemical compounds that when present in or added to foods, food packaging, food contact surfaces, or food processing environments inhibit the growth of, or inactivate pathogenic or spoilage microorganisms' (Davidson *et al.*, 2005). These agents may be classified as either traditional or naturally occurring antimicrobials (Brul and Coote, 1999; Branen and Davidson, 2000; López-Malo *et al.*, 2000). In recent years consumer demand and preference has shifted towards use of naturally occurring antimicrobials derived from microbial, plant, or animal sources, while products containing traditional antimicrobials have increasingly been rejected by consumers (Menon and Garg, 2001). This has led food manufacturers to either begin replacing traditional with naturally occurring compounds or finding alternative methods that allow for reduced concentrations of traditional antimicrobials, e.g. through the use of additional hurdles such as the combination of antimicrobials with other preservation techniques. The need to develop new single antimicrobial formulations, or combinations of multiple antimicrobials, has spurred renewed research efforts in this area and led to the early adoption of a variety of nanoencapsulation approaches.

II.B. Traditional versus naturally occurring antimicrobials
II.B.1. Traditional antimicrobials

Traditional or regulatory approved antimicrobials in the U.S. consist of a diverse list of compounds that include well-known preservatives such as benzoic acid and benzoates, acetic acids and acetates, nitrite and nitrate, sorbic acids and sorbates, sulfites, propionic acid and propionates, parabens, dimethyl dicarbonate, and more complex compounds such as lactoferrin, lysozyme, natamycin, and nisin (Davidson *et al.*, 2005). In Europe, the list of approved compounds is quite similar to that in the U.S., but specific salts or organic acids are listed rather than the general class of organic acid salts, e.g. calcium, sodium, or potassium acetate rather than just acetates. The great advantage of existing traditional compounds is that their functionality in a large number of foods is well known and documented and, most importantly, that their use is economical. Additionally, as an in-depth report has recently illustrated, some bacterial pathogens and spoilage organisms may develop a temporary tolerance to traditional antimicrobials, which is another reason the search for alternative antimicrobials is gaining momentum (Institute of Food Technologists, 2006).

II.B.2. Naturally occurring antimicrobials

Naturally occurring antimicrobials can be obtained from animal, plant, microbial, and mineral sources. Highly active naturally occurring antimicrobials can be found in spices, herbs, or their essential oils, and include phenolic compounds and their subclasses such as terpenes, coumarins, flavonoids (Gaysinsky and Weiss, 2007). These compounds are thought to constitute an essential part of various biological systems' defense mechanisms against predators (Hill et al., 1997), parasites (Acamovic and Brooker, 2005), and microorganisms (Hill et al., 1997; Acamovic and Brooker, 2005; Kong et al., 2008). Of particular interest have been plant-derived antimicrobials, e.g. in herbs and spices. An astonishingly high number (> 1340) of plants are known to be sources of antimicrobial compounds, including lemongrass, oregano, clove, cinnamon, palmarose, and others (Chao and Young, 2000; Velluti et al., 2003). For example, studies in the 1980s confirmed the growth inhibition of Gram-negative and Gram-positive bacteria, yeasts, and molds by garlic, onion, cinnamon, cloves, thyme, savory, sage, and others (Shelef, 1983; Conner and Beuchat, 1984; Deans and Svoboda, 1989). The antimicrobial compounds in plants appear to be mostly present in their essential oils. These lipophilic fractions of oils can be extracted from leaves (rosemary and sage), flowers and flower buds (clove), bulbs (garlic, onion), rhizomes (asafetida), fruits (pepper, cardamom) or other parts of the plant by expression, fermentation, extraction, or steam distillation (López-Malo et al., 2000; Burt, 2004). Essential oils are composed of a heterogenic mixture of organic compounds such as phenylpropanes, terpenes, and aliphatic compounds of low molecular weight such as alkanes, alcohols, aldehydes, ketones, esters, and acids (Oka, 1964; Brul and Coote, 1999). Generally, essential oils possessing the strongest antibacterial properties are those that contain phenolic compounds such as carvacrol, eugenol, and thymol (Hirasa and Takemasa, 1998; Rota et al., 2004).

Essential oils from oregano and thyme, for example, exhibit antimicrobial activity against several foodborne bacteria (Paster et al., 1990). Friedman et al. (2002) reported that the concentration level needed in some spices such as clove, oregano, cinnamon and others, to impart bactericidal activity against E. coli, ranged from 0.046 to 0.14%. The concentration needed to exhibit bactericidal activity with oregano, thyme, clove bud, and others ranged from 0.057 to 0.092%. Clove oil restricted the growth of Listeria monocytogenes in meat and cheese (Menon and Garg, 2001). Thyme essential oils inhibited mycelial growth of fungi at a concentration of 4 ppm while only 2 ppm of oregano essential oil was needed (Paster *et al.*, 1990).

II.C. Mechanism of action of antimicrobials

Fundamentally, antimicrobials may act in two principal ways (Figure 24.1). They can either act as membrane perturbers or act as inhibitors of proton transfer dynamics. Examples of the latter are weak organic acids such as benzoic and sorbic acid, which have been widely used as preservatives to inhibit the growth of microorganisms in beverages (Brul and Coote, 1999; Hazan *et al.*, 2004; Lambert and Stratford, 2004). The efficacy of these weak organic acids is a function of the environmental pH. For example, both benzoic and sorbic acid become more active with decreasing pH (Davidson *et al.*, 2005). This is because some of the weak acid molecules dissociate in solution to form the negatively charged acid anions, while other molecules remain undissociated. Only the undissociated organic acid molecules are able to penetrate the bacterial cell envelope, whereas the dissociated anions are repelled due to electrostatic interactions. Since the concentration of the undissociated acid increases as the pH decreases, higher concentrations of active compounds can migrate into the intracellular space at lower pH to inhibit essential metabolic reactions (Brul *et al.*, 2002). Alternatively, antimicrobials may act as membrane perturbers by inserting themselves into the

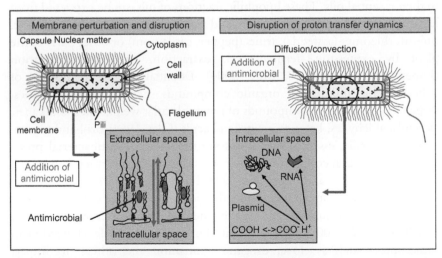

Figure 24.1 Illustration of the two basic mechanisms of action of antimicrobials. (Left) addition of antimicrobial leads to insertion into the membrane of bacteria, resulting in perturbation and disruption of the integrity of the membrane followed by leakage of intracellular content and permeation of extracellular content into the cell interior. (Right) Diffusion of an antimicrobial into the cell interior followed by dissociation and disruption of the proton transfer dynamics of the cell.

microbial membrane, thereby disrupting the structure and functionality of the phospholipid bilayer. In turn, the membrane loses its capability to act as a barrier, preventing exchange of material between the cell interior and exterior and its ability to host and regulate the functionality of protein complexes that are suspended in the bilayer.

While the basis of antimicrobial activity of many traditional antimicrobials has been well studied, the mechanism of action of newly discovered naturally occurring antimicrobials is still under investigation. For example, it has been proposed that the antibacterial activity of many essential oil compounds is predominantly due to membrane disruption (Moleyar and Narasimham, 1992). Cyclic hydrocarbons such as aromatics and terpenes from plant sources have been shown to strongly interact with biological membranes (Sikkema et al., 1994) with compounds typically more active against Gram-positive than Gram-negative bacteria (Smith et al., 1998; Friedman et al., 2000; Bagamboula et al., 2003). Several authors have demonstrated that the presence of a hydroxyl group in cyclic hydrocarbon antimicrobials enhances the antimicrobial activity (Aziz et al., 1998; Brul and Coote, 1999; Cowan, 1999; López-Malo et al., 2000; Ultee and Smith, 2002). Phenolic compounds are also known to disrupt enzyme functionality essential to the survival of microorganisms, thereby inducing a loss of homeostasis.

II.D. Limitations of food antimicrobials and considerations for selecting a suitable nanoencapsulation system

When applying antimicrobials in foods, one frequently encounters the situation wherein antimicrobials are substantially less active in complex food matrices compared to their activity in model microbiological systems. Minimum inhibitory concentration (MIC), i.e. the minimum concentration required to inhibit growth for a specific time period (typically 24 hours), generally increases when the antimicrobial is tested in a select food system. The increases in MIC depend on the nature of the antimicrobial and the composition and structure of the food system. Several factors governing this limitation should be carefully considered when designing and selecting a suitable nanoencapsulation system for antimicrobials:

- **Polarity of antimicrobials.** Many antimicrobials are partially or completely lipophilic and are only sparingly soluble in the aqueous phase. It appears that in many cases this lipophilicity is essential to cause compounds to migrate into or through the microbial cell membranes. However, for food manufacturers, this creates a significant application challenge since compounds may migrate not only into the microbial cell membrane but also into any other phase that is thermodynamically more favorable than the aqueous phase. For example, in food products where significant amounts of other lipid phases or interfaces are present, such as high-fat milk or cream, the lipid droplets act as a 'sink' resulting in a loss of concentration and efficacy when compared to a lipid-free food system (e.g. skim milk) (Rico-Muñoz and Davidson, 1983). In terms of selecting a suitable nanoencapsulation system, polarity is thus a key criterion since the capsules should provide a thermodynamically favorable environment for the antimicrobials to prevent migration to other lipid phases while not inhibiting interaction with microbial cells. Nanoencapsulation systems, as will be discussed later, are typically only able to incorporate components that have specific polarities. For example, for highly hydrophobic compounds, nanoemulsions and solid lipid nanoparticles may be the only systems of choice, while for amphiphilic and low-molecular-weight hydrophobic compounds, microemulsions, nanoemulsions, and solid lipid nanoparticles may be suitable. On the other hand, for hydrophilic antimicrobials such as nisin or lysozyme, liposomes may be a highly effective carrier system.

- **Temperature and pH dependence of antimicrobial activity.** Wide varieties of antimicrobials such as organic acids require a specific pH environment in order to be effective. Their activity may also depend on temperature since the dissociation equilibrium can be altered if temperature increases or decreases. In high-pH environments, some antimicrobial compounds may completely lose their efficacies. Other antimicrobials such as lysozyme may lose activity after exposure to a critical temperature or pH due to irreversible structural changes. It is thus important for many antimicrobial encapsulation systems to maintain, to the extent possible, the pH environment required for optimal antimicrobial activity and to protect the antimicrobial compounds from degradation upon temperature-induced changes.

- **Sensory effects of antimicrobials.** A significant limitation of many of the emerging naturally occurring antimicrobials is their sensory threshold level. Moreover, essential oil extracts have been predominantly used as flavors, and therefore introduction of essential oil compounds in food products will undoubtedly alter the flavor profile of the food. While some flavor alterations can be masked, compatibility with the food matrix in terms of sensory characteristics is nevertheless essential. Moreover, the high levels of essential oil compounds required to inhibit or inactivate microorganisms may exceed regulatory levels. At these levels, the health of the consumer upon repeated exposure/consumption may be compromised. Therefore, two essential goals of any antimicrobial nanoencapsulation system should be (1) to reduce the potential negative sensory effects of antimicrobials (e.g. by suppressing volatility) and (2) to not increase (or possibly even decrease) the minimum inhibitory concentrations of antimicrobials to allow for a reduction of required concentrations.

- **Economics of antimicrobials.** Use of food antimicrobials must be economically feasible. If the antimicrobial at the concentration required is too expensive or the achievable shelf life extension is too small, the antimicrobial will not be used by the food industry. In the context of the development of suitable nanoencapsulation systems, the delivery system must be capable of being economically manufactured, one that uses inexpensive ingredients so that the benefits gained from encapsulating the antimicrobial components (e.g. increased and prolonged activity or reduced sensory impact) outweigh the additional cost associated with the encapsulation process.

III. NANOENCAPSULATION SYSTEMS

III.A. Nanoemulsions

III.A.1. Definition

Emulsions are liquid–liquid dispersions of two completely or partially immiscible liquids (e.g. oil and water) with one phase being dispersed in the second phase in the form of droplets. Emulsions are a widely used class of food products; examples include milk, cream, salad dressing, mayonnaise, beverages, and sauces. In addition to the food industry, emulsions are heavily used in other industries in products such as pharmaceuticals, petrochemicals, and cosmetics (Schramm, 1992; McClements, 2005a). Emulsions are not only used to change the texture, flavor, and appearance of foods, but they are increasingly used to delivery biologically functional components that are lipophilic such as antioxidants, flavors, and nutraceuticals. Traditionally, emulsions could only be manufactured on a commercial scale with droplets in the micron-size range (i.e. d = 0.1 to 100 µm) (McClements, 1999; Friberg *et al.*, 2004; McClements, 2005a; Leal-Calderon *et al.*, 2007). However, in recent years due to the introduction of new ultra-high-pressure homogenizers (e.g. microfluidizers), food manufacturers have been able to produce droplets in the nano-size range (i.e. d < 100 nm). These so-called nanoemulsions differ appreciably from conventional emulsions in their functional performances due to the decreased size (Friberg *et al.*, 2004; McClements, 2005a). For example, these emulsions may not scatter light strongly in the visible region and can thus be transparent. Moreover, encapsulation of lipid bioactive compounds in nanoemulsions has been shown to increase bioactivity. In a study by Wang and associates, curcumin-loaded nanoemulsions were topically applied on mouse ear edema to reduce growth of tumors. The authors of the study reported that application of nanoemulsion (d ~ 70 nm) led to an 85% reduction in tumor tissue compared to a 43% reduction seen in the topical application of a macroemulsion (d ~ 500 nm) (Wang et al., 2008).

III.A.2. Functional properties of nanoemulsions

As mentioned, nanoemulsions may have functional properties such as appearance, physical and chemical stability, texture, and activity of encapsulated bioactive compounds that differ from that of conventional emulsions. This has raised interest in using nanoemulsions as carriers of functional lipids (e.g. antioxidants, flavors, colors, and antimicrobials). It should be noted though that no one single critical size exists at which all these functional properties change simultaneously. Instead, each one has a specific size dependence that may differ from that of other functional properties. In some cases, the functionality may change dramatically over a narrow size range, as is the case for optical properties of the emulsion (i.e. emulsion changes from white-milky opaque to completely transparent when mean droplet diameters decrease from 120 to ca. 80 nm). For other properties, such as stability of the emulsion to gravitational separation, this change is more gradual and substantial improvements may be seen at sizes below 300 nm (Figure 24.2).

- **Appearance.** The scattering intensity of droplets decreases as their size decreases relative to the wavelength of light (McClements, 2002a, b). Consequently, the appearance of emulsions is strongly dependent on droplet size, and emulsions become transparent when the size of the droplet falls below a critical diameter (d < 90–100 nm). Nanoemulsions are thus easily distinguishable from conventional emulsions and may be quite attractive to beverage manufacturers trying to avoid introduction of turbidity with addition of antimicrobial-carrying emulsions.
- **Gravitational stability.** The stability of emulsions to gravitational separation (creaming or sedimentation) increases as droplet size decreases, with creaming velocity proportional to d^2. When droplet size falls below a critical value (d ~ 100 nm), emulsions become completely stable to creaming or sedimentation because the effects of Brownian motion dominate gravitational effects (McClements, 2005a). Nanoemulsions can thus be kinetically stable for many years, a property that makes them again very attractive to food manufacturers.
- **Aggregation stability and rheology.** Emulsion stability to flocculation and coalescence depends strongly on droplet size because of the size dependence of the droplet collision (e.g. frequency and efficiency). The

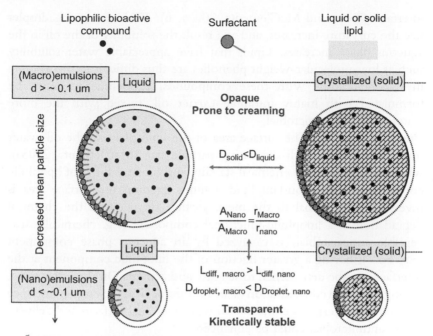

Figure 24.2 Comparison of structures and key properties of (left) liquid and (right) solid (top) macro and (bottom) nanoemulsions. D denotes the diffusion coefficients, A the particle surfaces, r the particle radii, and L the diffusion pathlengths.

magnitude of repulsive interactions (e.g. steric, electrostatic) and attractive interactions (e.g. van der Waals, depletion) tend to decrease with decreasing droplet size. The droplet collision frequency tends to increase with decreasing droplet size, which may promote droplet aggregation (McClements, 2005a). The rheology of emulsions may also be strongly dependent on droplet size, particularly when the size becomes small. For example, a significant increase in viscosity may be observed in emulsions stabilized by ionic surfactants when particle size is reduced since this alters the ratio between the thickness of the Debye layer and the droplet diameter, causing a 'virtual' increase in droplet concentration. Nano-emulsions with an oil droplet concentration of 25 wt%, stabilized by sodium dodecyl sulfate (SDS), had rheological behavior similar to that of conventional emulsions containing more than 70 wt% oil (Weiss and McClements, 2000a).

- **Ostwald ripening.** Emulsion instability to Ostwald ripening, i.e. the growth of larger droplets at the expense of smaller ones, greatly increases as the droplet size decreases. This is because Ostwald ripening is driven by a molecular mass transport of the oil through the intervening aqueous phase, which depends on the solubility of the oil above the droplet interface (Weiss and McClements, 2000a, b). With decreasing droplet size the curvature increases, and as a result the solubility of the oil in the aqueous phase increases. Lipids that have appreciable water solubility such as low-molecular-weight phenolics are thus difficult to incorporate in nanoemulsions. With these compounds, nanoemulsions must be formulated with highly insoluble carrier oil, mixed with the more water-soluble antimicrobial.

- **Bioactive exposure.** The surface area of the lipid phase at the oil–water interface increases with decreasing particle size (McClements, 2005a), which may impact the chemical stability and bioavailability of lipophilic components. At a constant lipid volume fraction, the surface area is inversely proportional to the mean particle diameter. If the chemical degradation of a lipophilic functional component (e.g. chemically susceptible antimicrobial) is catalyzed by an aqueous phase component (e.g. Fe^{2+}), having a greater fraction of the lipophilic component at the interface may be detrimental to stability and activity. On the other hand, having increased concentration of an antimicrobial at the droplet interface could be beneficial since interaction with microbial surfaces and delivery of the antimicrobial could be enhanced.

- **Bioactive solubility.** The solubility of encapsulated lipid antimicrobials in the surrounding aqueous phase increases with decreasing particle size due to curvature effects (McClements, 2005a), which again may negatively impact the chemical stability of the antimicrobial but possibly improve bioactivity (Lombardi *et al.*, 2005; Dias *et al.*, 2007; Tagne *et al.*, 2008; Wang *et al.*, 2008).

III.A.3. Fabrication of nanoemulsions

Conventionally, oil-in-water emulsions are prepared by homogenizing the oil, water, and emulsifier together using a mechanical device known as a homogenizer, e.g. high-shear mixer, high-pressure homogenizer, colloid mill, sonicator or membrane homogenizer (Walstra, 1993; Walstra, 2003; McClements, 2005a). To manufacture nanoemulsions, mean droplet diameter should be below 100 nm. The manufacturing device of choice has proven to be the so-called microfluidizer (Figure 24.3). In contrast to high-pressure homogenization, microfluidization diverts the flow of the emulsion in the homogenization chamber to create fluid jets that impinge in a mixing chamber, thereby creating additional forces through cavitation and droplet disruption in addition to pressure changes. Since the microfluidizer has no moving parts, maintenance is typically lower. On the negative side, there is an increased risk of blockage at the exit of the homogenization chamber, which may be quite difficult to resolve.

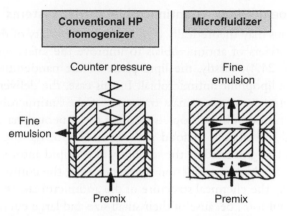

Figure 24.3 Schematic drawing of the generalized design of (left) a conventional high-pressure homogenization valve and (right) a microfluidizer.

The factors determining the final droplet size of the nanoemulsion include emulsifier type and concentration, volumetric energy input during the homogenization process, composition of component phases, and temperature. In the homogenizer two processes occur simultaneously, resulting in the generation of small droplets from a premix that contains larger droplets. Firstly, the volumetric energy input induces droplet deformation and disruption, generating smaller droplets with new interfaces. Generally, the larger the volumetric energy input the smaller the droplets. The droplet disruption also depends on the viscosity ratio between the two phases, which may depend on the temperature of the system. Secondly, surfactants must rapidly absorb at the new interfaces, creating an interfacial layer that reduces the interfacial tension (thereby facilitating droplet disruption) and induces repulsive interactions (thereby retarding re-coalescence). The droplet size thus depends on the molecular properties of the emulsifiers governing the rate of adsorption, reduction in surface tension, and degree of induced repulsive and attractive colloid interactions, e.g. van der Waals, steric, and electrostatic (McClements, 2005a).

In nanoemulsions, the choice of surfactants is very critical since emulsifiers have to rapidly cover the many new surfaces that are formed. Generally, in food emulsions two classes of surface-active species are used: (1) small-molecule surfactants such as monoglycerides, sucrose esters, and others and (2) macro-molecular emulsifiers such as protein or modified starches (Dickinson, 2003). Since small-molecule surfactants adsorb much more rapidly to newly formed interfaces than macro-molecular surfactants, they are typically better suited for manufacture of nanoemulsions.

III.A.4. Nanoemulsions as antimicrobial carrier systems

Nanoemulsions may theoretically be designed in a variety of different ways to serve as carriers of antimicrobials to improve the safety and quality of foods (Figure 24.4). Firstly, the lipid phase of the nanoemulsion may be loaded with a lipophilic antimicrobial. In this case, the delivery of the antimicrobial may occur via a mass transport of the antimicrobial from the inside of the nanoemulsion droplets through the aqueous phase of the food system to the membrane of food pathogens or spoilage organisms. This process would be governed by the solubility of the lipid antimicrobial in the aqueous phase, which is a function of temperature, the composition of the aqueous phase, the chemical structure of the antimicrobial, and the droplet size. Nanoemulsions, because of their small size and large curvature, can be expected to have better activity since the driving force for the mass transport process, i.e. the concentration difference of the antimicrobial in the vicinity of the oil droplet and in the bulk phase, is much higher due to the Laplace effect (McClements, 2005a). Secondly, antimicrobial nanoemulsions can be constructed from a surface-active antimicrobial that stabilizes the nanoemulsion and an inert lipid. Examples of surface-active antimicrobials include lysozyme, nisin, and lauric arginate. Additional emulsifiers may be required to enhance emulsion formation and to achieve the small droplet diameter needed to produce nanoemulsions, as well as to improve stability of nanoemulsions for breakdown of mechanisms such as coalescence or flocculation. Additional emulsifiers also allow for the interaction of droplets

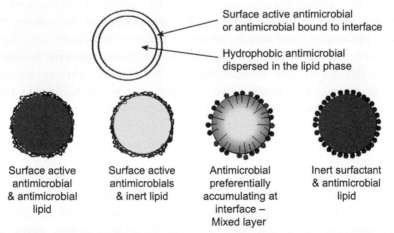

Surface active antimicrobial or antimicrobial bound to interface

Hydrophobic antimicrobial dispersed in the lipid phase

| Surface active antimicrobial & antimicrobial lipid | Surface active antimicrobials & inert lipid | Antimicrobial preferentially accumulating at interface – Mixed layer | Inert surfactant & antimicrobial lipid |

Figure 24.4 Structures of antimicrobial nanoemulsions. From left to right: nanoemulsions with antimicrobial lipid and surfactant; nanoemulsions with antimicrobial surfactant and inactive lipid; nanoemulsions with preferentially accumulated lipid antimicrobial and inactive surfactant; and nanoemulsions with inert surfactant and antimicrobial lipid homogeneously dispersed throughout the droplet phase.

with the microorganisms, which can be improved by creating an electrostatic attraction between droplets and negatively charged microbial surfaces. Nanoemulsions may have an advantage over conventional emulsions because the frequency of collision with microorganisms is higher due to the smaller size. Thirdly, antimicrobial nanoemulsions could be constructed by combining the two approaches outlined above, i.e. using one antimicrobial that is part of the dispersed droplet phase and a second antimicrobial that is surface active and part of the emulsifier layer. It should be noted though, that some antimicrobials (e.g. simple and more complex phytophenols) may not strictly fall into either category. While antimicrobials are typically added to the lipid phase of nanoemulsions, they have a tendency to accumulate in higher concentrations near the vicinity of the interface. In the case of phenolic antimicrobials, this is because the hydroxyl group facilitates an interaction with the headgroup of the surfactant.

Interestingly, while the above-outlined approaches are very promising, little systematic research has been done recently to evaluate the antimicrobial activity of nanoemulsions or emulsions in general. Marinating food products with food emulsions is a technique used in the meat industry to enrich flavor, increase moisture and tenderness, and preserve color, while simultaneously increasing shelf life of products (Carlos and Harrison, 1999; Bjorkroth, 2005). Marinades are emulsions containing salt, sugars, spices, stabilizers (gums), and antimicrobial agents. The pH is often acidic and less than 5 (Bjorkroth, 2005). In many cases, spices and their extracts are added to the marinade, not just to impart flavor, but also as a preservation technique, contributing to the long-term microbial stability of different meat products such as jerky (Buege et al., 2006; Busquets et al., 2006). Marinades are usually added to meat products at concentrations ranging from 20 to 30 wt% (Bjorkroth, 2005). In a completely different product – beer – addition of hop oil emulsions has been shown to rapidly inactivate spoilage organisms to below detectable levels within 24 hours. However, no information about the exact composition and physicochemical properties of the emulsion (size and charge) was given (Biatwright, 1976).

Conversely, in another study, the activity of an emulsion composed of castor oil and medium-chain triglycerides stabilized by Poloxamer 188 (mean particle diameter of 1 μm) in the presence of the antimicrobial compound chitosan was investigated. Researchers stated that the emulsion served as a carrier for chitosan but reported that the emulsion was in fact slightly less active than the free chitosan; however, it conformed to preservation efficacy requirements for topical formulations. No explanation was given as to the underlying physicochemical effects. Most likely the positively charged chitosan adsorbed on the surface of the emulsion droplets, thereby decreasing the concentration of chitosan in the aqueous phase and therefore decreasing its overall activity (Jumaa *et al.*, 2002). Similarly, formation of submicron oil-in-water emulsions that contained a large variety of preservatives (including benzalkonium chloride, methyl and propyl parabens, and chlorocresol) was problematic. Introduction of these compounds induced, in many cases, instability of the emulsion and led to insufficient activities against *Staphylococcus aureus*, *Escherichia coli* O157:H7, and *Pseudomonas aeruginosa*. Again, relatively little information was given on the physicochemical properties of the emulsions and the location of the antimicrobials in the emulsion system (Sznitowska *et al.*, 2002).

We conducted studies in which the stability and antimicrobial efficacy of oil-in-water nanoemulsions formulated with phytophenol eugenol and other lipids were evaluated (Gaysinsky *et al.*, 2006, 2007b). Eugenol and lipids (hexadecane, dodecane, tetradecane, or corn oil) were mixed in eugenol:lipid ratios and varied from 0:1 to 1:0. Oil-in-water emulsions were prepared by homogenizing 5 wt% of the lipid mixture with 95 wt% of a 0.5 wt% aqueous solution of Tween 20. The particle size distribution was measured after 0, 1, 3, 6, 12, 24, 48, 96, 168, 264, 336, and 504 h. Stable nanoemulsions were tested for antimicrobial activity against four strains of *E. coli* O157:H7 and *L. monocytogenes*, using a spot inoculation test and growth curves by plate counting. Formulation of oil-in-water nanoemulsions was found to be challenging as emulsions composed of phytophenols above a critical loading ratio broke down in less than 1 h. The appreciable water solubility of the antimicrobials appeared to promote rapid breakdown by Ostwald ripening. Below this critical concentration, rate of emulsion particle size increase varied depending on type of carrier lipid and concentration of eugenol. Corn-oil-in-water emulsions loaded with eugenol were inhibitory against *E. coli* O157:H7 strains depending on loading ratio, but failed to inhibit growth of *L. monocytogenes* strains. Direct addition of eugenol to the aqueous phase at concentrations similar to the lowest loaded emulsions was inhibitory against all strains of both pathogens. However, addition of eugenol, at a concentration attained after partitioning equilibrium between the droplets and the aqueous phase had been established, was not inhibitory, suggesting that the emulsion in fact served as a reservoir for the antimicrobial. In another study, we tested the antimicrobial activity of a fine-disperse oil-in-water emulsion that had been formulated with lauric arginate as the emulsifier and corn oil as the droplet phase. Lauric arginate emulsions were not stable to coalescence unless Tween 20 was added to the emulsion. However, submicron emulsions manufactured with mixtures of Tween 20 and lauric arginate were stable to aggregate and coalescence and emulsions displayed similar activities against *E. coli* O157:H7 and *L. monocytogenes* on an absolute antimicrobial concentration basis as the free (water-soluble) antimicrobial.

The above-outlined studies emphasize that significantly more research is needed to answer the important question of how the composition and structure of a nanoemulsion composed of one or more antimicrobials should be designed, in order to maximize the antimicrobial efficacy of the system. These reviewed studies suggest that it is critical that the underlying physicochemical principles driving assembly of the structure, release of the associated antimicrobial and its interaction with microorganisms are thoroughly understood.

III.B. Solid lipid nanoparticles (SLN)
III.B.1. Definition
Conventional emulsion-based delivery systems contain lipid droplets that are completely liquid, e.g. the droplets in flavor emulsions or in ω-3 fatty acid delivery systems. Recent work in the pharmaceutical area has shown that *controlled crystallization* of lipid droplets in emulsions can be used to greatly increase the performance of this type of delivery system. When nanoemulsions are manufactured with melted lipids prone to crystallization at or below room temperature, such as triacylglycerides (TAGs), and are then cooled under controlled conditions, small solid particles are formed. These so-called SLN have been shown to be ideal carriers for functional lipids. The physical stability, chemical stability, and bioavailability of hydrophobic drugs, for example, can be greatly improved by encapsulating them within these small solid particles, rather than liquid droplets (Müller *et al.*, 1996a, 1997; zur Mühlen *et al.*, 1998; Mehnert and Mader, 2001; Wissing *et al.*, 2004; Helgason *et al.*, 2008). To date these systems have been predominantly used in the pharmaceutical industry as carrier systems for highly hydrophobic, chemically unstable drugs. They have been reported to combine the advantages of the parent liquid nanoemulsions of high disso- lution velocities associated with high permeability of the active compound through cell walls, while simultaneously solving existing problems associ- ated with physical stability of the dispersion and chemical stability of the encapsulated compound. This is because mobility of bioactive compounds and particle aggregation can be controlled by controlling the physical state of the lipid matrix (Videira *et al.*, 2002; Sivaramakrishnan *et al.*, 2004; Zhang *et al.*, 2004; Wang and Wu, 2006; Yang *et al.*, 2006). The food industry may benefit from the utilization of this approach for the encapsulation and delivery of lipophilic antimicrobials (e.g. parabens).

III.B.2. Crystallization in emulsions
Solid lipid nanoparticles have radically different physical properties compared to liquid emulsions since the mobility of droplet components (surfactant, lipid,

lipophilic bioactive compounds) and external reagents (e.g. oxygen) into and out of the droplet may be altered upon crystallization (see Figure 24.2). In addition, morphological changes can occur that give rise to altered optical and rheological properties. It is therefore important to understand the mechanism of lipid droplet physical state transformations (crystallization) on the stability and performance of solid lipid nanoparticles.

In almost all bulk liquids some impurities in the melt will act as the starting point for heterogeneous nucleation, before a degree of supercooling can be reached, where new nuclei would be formed by homogeneous nucleation. However, when the liquid is finely dispersed as emulsion droplets the number of droplets may greatly exceed the number of impurities. Then, a majority of lipid droplets may be impurity free and hence crystallize by an apparently homogeneous mechanism. Thus, not only composition and purity of the lipid, but also size of the emulsion droplets, will play a critical role in the crystallization process since homogeneous nucleation will only become the dominant process if the particle size is reduced to a point where the possibility of finding a nucleus within a droplet is < 1. Interfacial composition and packing (which may depend on curvature of droplets) may also influence emulsion crystallization. For example, if the hydrophobic portion of surfactant adsorbed at the lipid–water phase has a similar molecular structure to that of the to-be-crystallized lipid, nucleation rates may be increased (Hartel, 2001). In this case the nucleation rate is proportional to the interfacial area and not the volume; in other words, the crystallization proceeds by a surface-mediated heterogeneous nucleation process. Finally, the presence of solid droplets in a liquid emulsion can induce nucleation in the liquid droplets, presumably through a collision mechanism (i.e. interdroplet nucleation), a mechanism that again is influenced by the droplet size of the emulsion since collision frequency and number of droplets increase as size of droplets decreases.

III.B.3. Functional properties of solid lipid nanoparticles

Fundamentally, solid lipid nanoparticles are composed of solidified nano-emulsions. Their unique properties are a direct consequence of the re-duction in particle size that leads, firstly, to increased surface curvatures and, secondly, to an increase in the number of molecules per particle that actively interact with molecules in the interface (e.g. surfactants). With increasing surface curvature, lipid molecules are restricted in their ability to assume the shape of a perfect crystal. Consequently, smaller particles have increasing concentrations of α- and β'-crystals, while bulk fat or larger emulsion droplets tend to crystallize predominantly in the form of β-crystals (Westesen et al., 1993; Jenning et al., 2000). While this crystal form is thermodynamically unstable, it appears to be much better suited to carry hydrophobic compounds. For example, studies have shown that when nanoemulsions crystallize in the α-form, the hydrophobic compound is not expelled from the matrix, and instead becomes part of a mixed crystal structure. However, while some in the pharmaceutical industry have reported particles to be kinetically stable for up to 12 months upon selection of appropriate cooling conditions, surfactants, and surfactant concentrations (Siekmann and Westesen, 1994; Jenning et al., 2000; Palanuwech and Coupland, 2003), polymorphic transformation from α-form to the more stable β-form may pose a significant problem to the overall stability of the system (Awad et al., 2008; Helgason et al., 2008). Compared to liquid nanoemulsions, solid lipid nanoparticles differ in their functional properties in a number of ways.

- **Appearance.** The scattering intensity of droplets increases as their refractive index increases (McClements, 2002a, b). Solid fats have a higher refractive index than liquid oils. Consequently, the appearance of emulsions is dependent on the physical state of the lipid droplets. In practical terms, this means that solidified nanoemulsions may be turbid despite the fact that their size is the same as that of a liquid nanoemulsion.
- **Gravitational stability.** The stability of emulsions to gravitational separation (creaming or sedimentation) depends on the density contrast between the droplets and continuous phase. Solid fats have a higher density than liquid oils; consequently the stability to gravitational stability depends on the solid fat content of the lipid droplets (McClements, 2005a). If the density of the lipid phase thus becomes close to that of the surrounding solvent phase, the stability of the solid lipid nanoparticles may be higher than that of their parent nanoemulsions. The density of SLN may be controlled by varying the ratio of solid to liquid oil within them.
- **Aggregation stability.** The presence of fat crystals within a lipid droplet makes it prone to partial coalescence (McClements, 2005a) or particle aggregation due to shape changes (Leal-Calderon et al., 2007). The susceptibility of an emulsion to this type of aggregation depends on the solid fat content, crystal distribution, interfacial properties of the lipid droplets, as well as the application of any shear forces (Sato and Garti, 1988; McClements et al., 1993; McClements and Dungan, 1997; Herrera et al., 1999; Kaneko et al., 1999; Hindle et al., 2000; Rousseau, 2000; Hartel, 2001; Katsuragi et al., 2001; Palanuwech and Coupland, 2003; Campbell et al., 2004; Himawan et al., 2006; Sonoda et al., 2006). The shape transformation can be a substantial problem for manufacturers of delivery systems, but may be somewhat counteracted by adding additional emulsifier to cover new surfaces formed during shape transformation processes.

- **Bioactive (antimicrobial) location.** The exposure of a bioactive component to the surrounding aqueous phase depends on its location within
  a lipid droplet, e.g. inner core, outer shell, or randomly distributed. For
  example, if the chemical degradation of a lipophilic component is catalyzed by an aqueous phase component (e.g. Fe^{2+}), then, having a greater
  fraction of the lipophilic component at the interface may be detrimental
  to its stability. On the other hand, if a rapid release and high interaction of
  the encapsulated lipophilic is desired (as might be the case for food
  antimicrobials) a preferential localization in the shell might be preferable.
- **Molecular diffusion.** The diffusion of molecules through a solid phase is
  slower than through a liquid phase (Leal-Calderon *et al.*, 2007). Consequently, a bioactive component may be protected against chemical degradation by being trapped within a solid matrix, as this would slow down
  movement of the bioactive compound to the droplet surface, as well as
  slow down penetration of any reactants (e.g. metals and oxygen) into the
  droplets. The crystallized matrix could thus also provide a convenient
  means to building a preservation system that is able to provide long-term
  protection by slowly releasing the antimicrobial into the food.

III.B.4. Manufacture of solid lipid nanoparticles

Antimicrobials carrying solid oil-in-water emulsions are typically prepared
via hot homogenization. In hot homogenization (Figure 24.5), the antimicrobial is mixed with a carrier triacylglyceride and heated to a minimum of
$5°C$ above melting of the highest melting component (T_m). The hot lipid
phase is then dispersed in an aqueous emulsifier solution, adjusted to the same
temperature as the lipid phase, typically using a thermostated high-speed
mixer. The hot pre-emulsion is then homogenized in a thermostatically
controlled microfluidizer with the interaction chamber kept at $T > T_m$.
Similar to manufacture of nanoemulsions described previously, the diameter
of droplets in emulsions can be controlled by varying the pressure that the
emulsion is subjected to while passing through the homogenizer, and the
number of passes. Emulsions are then cooled at specific cooling rates to
temperatures required to induce crystallization in the emulsion droplets.

III.B.5. Solid lipid nanoparticles as carriers of antimicrobials

The formation and study of solid lipid nanoparticles for food applications has
just begun. Thus, there are few research reports available in which the activity
of food antimicrobials in solid lipid nanoparticles has been described. However, poor solubility of functional lipophilic components has not only been an
issue in the food industry, but has been a substantial problem in the pharmaceutical industry as well since more and more newer drugs are lipid soluble.
SLN have been used to orally administer water-insoluble pharmaceutically

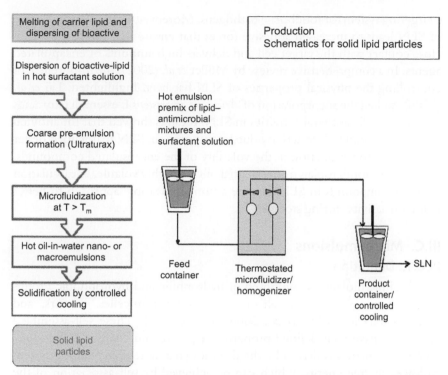

Figure 24.5 Schematics of formation of solid lipid nanoparticles using the hot-homogenization process. Hot lipid–antimicrobial mixes and surfactant are homogenized in a thermostated microfluidizer followed by cooling to induce crystallization.

active compounds to improve their stability and pharmacological activity. Upon oral administration, SLN have been shown to have much better performance compared to tablets, pellets, capsules, self-emulsifying drug–delivery systems (SEDDS) or powders. For example, gastric mucosa damage leading to gastric irritancy when using particles (20–30 μm) of naproxen was prevented by using SLN and was attributed to decrease in local high and prolonged concentration of naproxen attributed to reduced particle size (Liversidge and Conzentino, 1995). These improved release characteristics could be similarly useful for an antimicrobial delivery system.

In an interesting application, formation of stealth SLN has been reported by coating the solid lipid nanoparticles with a functional polymer that allows for specific interactions with organs or blood cells (to be designed). In the blood stream, this caused SLN to adhere to specific sites (Müller *et al.*, 1996b; Bocca *et al.*, 1998; Cavalli *et al.*, 1999; Peracchia *et al.*, 1999; Fundaro *et al.*, 2000; Gref *et al.*, 2000; Podio *et al.*, 2000; Müller *et al.*, 2001; Zara *et al.*, 2002; Peracchia, 2003; Goppert and Müller, 2005; Li *et al.*, 2005; Vonarbourg *et al.*, 2006; Wang and Wu, 2006). This approach might be transferable to produce a targeting system for foodborne pathogens. Moreover, the small particle size of SLN leads to increased adhesive forces that ensure that the particle is in close contact with the target cell and delivers high amounts of encapsulated agents. In a comprehensive review by Müller *et al.* (2002), the importance of controlling the physical properties of SLN has been highlighted. Lai *et al.* (2006) studied the incorporation of *Artemisia arborescens* L essential fatty acids with reportedly antiviral activities in SLN. Results showed that for antiviral carriers that tend to lose activity due to evaporation, SLN maintained better activity due to a reduction in the volatility of the encapsulated compounds. Since food antimicrobials from essential oils are highly volatile, encapsulation of these compounds in SLN may be a similarly feasible approach to better maintain activity during storage.

III.C. Microemulsions

III.C.1. Definition

Amphiphilic molecules composed of hydrophilic and hydrophobic groups dispersed in solvents can self-organize to form micelles, bilayers, and reversed micelles (Weiss *et al.*, 2006) (Figure 24.6). These self-assembled aggregates have well-defined properties, e.g. size and charge. Their spontaneous formation is driven by the thermodynamic tendency of the system to lower its free energy, which can be achieved by self-association of the hydrophobic groups of the surfactant tails, removing them from contact

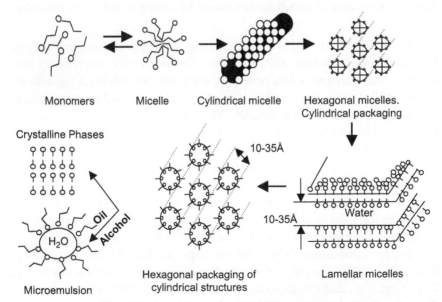

Figure 24.6 Schematic overview over type of self-assembled colloidal structures formed by surfactant monomers as a function of surfactant concentration.

with the polar solvent. Organic solvents of low solubility can be brought into solution by incorporation into micelles (Holmberg *et al.*, 2003). Micelles are spherical colloidal aggregates that form above a critical surfactant concentration, the so-called CMC (Figure 24.7). This increase in solubility in the presence of micelles is referred to as solubilization. McBain and Richards (1946) were the first to define solubilization as '. . . a particular mode of bringing into solution substances that are otherwise insoluble in a given medium, involving the presence of a colloidal solution whose particles take up and incorporate within or upon themselves the otherwise insoluble material.' As a consequence, *microemulsions* are formed. Microemulsions have been defined as 'the formation of a thermodynamically stable isotropic solution by solubilizing a compound normally insoluble or very slightly soluble in a given solvent by introducing an additional amphiphilic component or components' (Weiss *et al.*, 1996, 1997, 1999). Due to the fact the mechanism is based on an uptake and inclusion of the normally insoluble material in a micellar aggregate, the form and shape, as well as the temperature–concentration dependence, of the micelles are a major factor for solubilization.

Microemulsification has been of substantial interest to scientists in various fields of research such as cosmetics, pharmaceuticals, insecticides, and herbicides (Miller and Raney, 1992; Sjöblom, 1967). Solubilization, the underlying process in microemulsification, also has many biological aspects. Cell membranes can act as solubilizing agents (Sjöblom, 1967). The uptake of fat and cholesterol is closely associated with solubilization

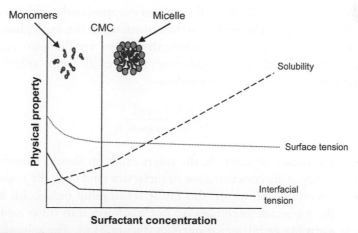

Figure 24.7   Change in surface tension, interfacial tension, and solubility as a function of surfactant concentration (Mulligan *et al.*, 2001). Micelles begin to form above the critical micellar concentration (CMC) causing interfacial and surface tension to remain constant and solubility to increase.

and consequently there is a high interest in understanding the mechanisms of solubilization and by what factors it is influenced. Bile acids, polar derivatives of cholesterol, are amphiphilic with detergent properties. Fat globules are emulsified into smaller droplets, thereby increasing the surface area accessible to lipid-hydrolyzing enzymes. The bile acids also solubilize lipid breakdown products such as mono- and diacylglycerols formed from hydrolysis of triacylglycerols. The solubilized structures can then be absorbed by the intestinal epithelial cells (Lehninger, 1993).

III.C.2. Functional properties of microemulsions

- *Maximum additive concentration (MAC) or solubilization capacity (SC).* The amount of a lipophilic antimicrobial that can be incorporated in a surfactant micelle to form a microemulsion is limited. The maximum amount of material that can be solubilized is referred to as the solubilization capacity (SC), the saturation concentration or maximum additive concentration (MAC; g solubilized compound per g of surfactant at a specific surfactant concentration) of the micelle. To theoretically derive the MAC, a partition coefficient P between the micellar system and the bulk phase can be established by treating micelles as a pseudophase:

$$P = \frac{x_{microemulsion}}{x_{bulk}} \qquad (24.1)$$

where $x_{microemulsion}$ is the mole fraction of encapsulated compound, the micelle, and x_{bulk} is the mole fraction of solute in the bulk phase (also known as monomer phase). Alternatively, an equilibrium constant K can be defined by considering the equilibrium reaction of *micelle + solute ↔ solute in microemulsion*

$$K = \frac{[Solute_{micelle}]}{[Solute][micelle]} \qquad (24.2)$$

The concentration of solute in the microemulsion should therefore increase linearly as the concentration of surfactant (and thus, the number of micelles) increases. However, this linear relationship only holds true as long as the surfactant assemblies do not transform into other structures such as wormlike or bilamellar micelles (Figure 24.6). The solubilization capacity is highly specific to the structure and composition of the system. Depending on the molecular nature of both the surfactant and the antimicrobial to be solubilized, the temperature, pressure, and presence of other surface-active materials or ionic compounds, more or less material can be incorporated (Weiss, 1999).

- **Solubilization kinetics.** Upon addition of a lipophilic antimicrobial to a micellar surfactant solution, the antimicrobial is transferred to the micellar phase forming the microemulsion. The solubilization kinetics is the speed with which this mass transport happens. The time it takes certain surfactant micelles to become fully saturated with non-polar molecules can range from a few seconds to several months. If the lipophilic antimicrobial is added in the form of coarse or fine emulsion droplets, a semi-empirical model can be used to describe the solubilization kinetics. If the process is assumed to have a first-order kinetics and depends on the concentration of non-solubilized material in the aqueous phase, as well as the MAC, then one can express the increase in micelle solubilized solute concentration with time as (Weiss *et al.*, 1996):

$$\frac{dc_{aq}}{dt} = k_i \left(\frac{A}{V}\right)(c_{sat} - c_{aq}) \qquad (24.3)$$

where c_{aq} is the concentration of solute in the micelle and c_{sat} is equal to the MAC, k_i is the rate of solubilization of a component i, A is the total surface area across which the mass transport occurs (i.e. total surface area of droplets), and V is the volume of oil in the system. Integration and introduction of the mean radius and volume fraction of the dispersed phase leads to the following equation:

$$\frac{1}{(c_0 - c_{sat})} ln\left[\frac{c_{sat}(c_0 - c_{aq})}{c_0(c_{sat} - c_{aq})}\right] = k_i \frac{6\rho_{total}}{d_{32}\rho_{oil}} t \qquad (24.4)$$

where c_0 is the concentration of solubilizate in the bulk, ρ_{total} is the density of the emulsion, d_{32} is the so-called Sauter-Diameter, and ρ_{oil} is the density of the oil. The solubilization kinetics can thus be readily assessed by measuring the decrease in turbidity, which corresponds to the decrease in droplet concentration after addition of the antimicrobial as a function of time.

- **Palisade layer composition.** The Palisade layer of a microemulsion is the outer shell of the microemulsion. Prior to the addition of the antimicrobial, the palisade layer is composed solely of surfactant monomers comprising the initial micelle. Upon addition of the lipophilic antimicrobial, the palisade layer composition and structure may be altered, since the antimicrobial may potentially insert into the Palisade layer. Very hydrophobic antimicrobials predominantly interact with the hydrophobic tails of the surfactants and thus have a negligible impact on the Palisade layer composition and structure, while partially hydrophilic antimicrobials (e.g. phenolic compounds) may have preferential interactions with both the surfactant head and tail groups. In this case, significant alterations in the composition of the Palisade layer can be observed. For example, the hydroxyl group of phenolic antimicrobials has been shown to be able to directly interact with the hydrophilic polyethyleneoxide group that forms the head of many synthetic surfactants (Gaysinsky et al., 2007). This phenomenon has important implications for the antimicrobial activity of the resulting microemulsion, since access to the encapsulated antimicrobial may differ for antimicrobials incorporated in the interior of the micelle rather than the Palisade layer.

- **Microemulsion charge.** The type of surfactant used for formulation of the micelle determines the charge of the micelle. For example, the charge may be neutral if non-ionic surfactants are used; or positive or negative, if cationic or anionic surfactants are used (Figure 24.8). In recent studies, it has been demonstrated that the ability of a microemulsion to carry a lipophilic compound can be significantly improved by not using a single surfactant system, but by using a binary or multiple surfactant system instead. By using combinations of surfactants (e.g. ionic and non-ionic), a fine-tuning of the surface charges of the self-assembled microemulsions is possible. For example, mixing a cationic and a non-ionic surfactant at various ratios results in micelles that carry charges ranging between that of

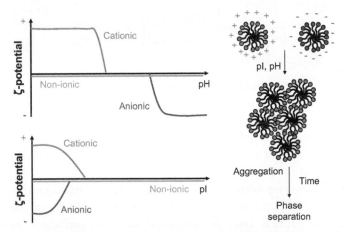

Figure 24.8 Comparison of pH and ionic strength stability of cationic, anionic, and non-ionic micelles. Cationic and anionic micelles may aggregate and breakdown above or below critical salt concentrations and pH.

the non-ionic surfactants (zero) or cationic surfactants (net positive), while mixing the anionic and non-ionic surfactants produces micelles with varying negative charges. The charge of the micelle is critical in determining the subsequent interaction with food pathogens or food ingredients. Since most microorganisms are negatively charged, positively charged micelles can be expected to be strongly attracted to the cells. If these micelles are loaded with an antimicrobial to form cationic antimicrobial microemulsions, the efficacy of the antimicrobial may be greatly increased. Conversely, if a slow release is desired, microemulsions may be designed to carry a slight negative charge to prevent accumulation at pathogen surfaces and burst release of the antimicrobial. Most importantly, mixed binary micelles composed of ionic and non-ionic surfactants not only have a much greater solubilization capacity, they are also typically more stable to pH changes and to addition of salt compared to purely ionic micelles (Figure 24.8).

- **Size of microemulsions.** The size of antimicrobial microemulsions may play an important role in their activity since the kinetics of the mass transport processes required to facilitate interaction with foodborne pathogens are influenced by the size of the nanocapsules. The size of microemulsions depends on the chemical structure of the component surfactants. A specific number of surfactant monomers, the so-called aggregation number, n, are required to form a single micelle. Depending on the chemical structure of the surfactant, the monomers assemble into a spherical vesicle with a thermodynamically optimal curvature, which governs the size of the final aggregate. Empty micelles have average diameters as small as 3–4 nm. Upon addition of the lipophilic antimicrobial the micelles begin to grow. The microemulsions formed may assume average diameters that are triple or quadruple that of the empty parent micelle. Microemulsions reach their maximal size when they are fully loaded, that is at the MAC. They grow proportionally smaller as the loading ratio is decreased (Gaysinsky et al., 2007a).

III.C.3. Manufacture of microemulsions

Formation of microemulsions is a very cheap and simple process. Two fundamental methods exist to produce a microemulsion. The first is to simply mix the lipophilic antimicrobial to be encapsulated with a surfactant solution and then mix the solution until all lipophilic material has been taken up. The process can be visually observed. The initial addition of antimicrobials introduces turbidity due to the presence of large droplets that scatter light. As these droplets are solubilized, the turbidity decreases and eventually the solution becomes transparent (Figure 24.9). Knowledge of the MAC is critical in this process, since addition of the antimicrobial above the

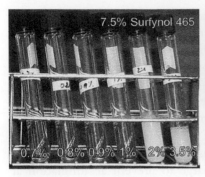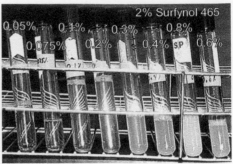

Figure 24.9 Visualization of formation of antimicrobial microemulsions through solubilization in micellar surfactant solutions. (Left) Test tubes containing 7.5 wt% Surfynol 465 and 0.7–3.5 wt% carvacrol after equilibration for 24 hours at room temperature. The maximum additive concentration (MAC) of carvacrol at 7.5 wt% surfynol is approximately 1 wt%. (Right) Test tubes containing 2% Surfynol 465 and 0.05–0.8 wt% carvacrol after equilibration for 24 hours.

concentration that can be loaded per number of micelles would lead to an excess non-dispersed antimicrobial. The unencapsulated antimicrobial would eventually phase separate and the preparation would thus be unsuitable for application. Alternatively, use of tertiary phase diagrams for mixtures of the three components, surfactant, antimicrobial, and solvent, can be made. Since the formation of microemulsions is thermodynamically driven, a dilution process using a mixture of surfactant and antimicrobial at appropriate mixing ratios can be used. Upon dilution, the solutions will spontaneously transfer into the appropriate phases that exist along the dilution line. It should be noted though that if a gel phase exists along the dilution line, formation of microemulsions may be impossible. In this case, addition of co-surfactants may be required to shift the location of the gel phase (Garti *et al.*, 2006; Spernath *et al.*, 2006; Kogan *et al.*, 2007).

To properly design a microemulsion composed of two or more surfactants, a thorough understanding of the physical processes involved in their formation is needed. As previously mentioned, knowledge of the critical micellar concentration is key, since uptake and encapsulation does not occur if no surfactant aggregates are present. Several models have been used to give a thermodynamic description of the formation of the parent colloidal aggregates prior to addition of the lipophilic functional components. The critical micellar concentration of mixed micelles, composed of two surfactants, can be calculated by using the so-called pseudophase formation model in which the mixed micellar phase is treated as a separate phase that is in thermodynamic equilibrium with the two surfactant monomers in the bulk phase (Rathmand and Scamehorn, 1989). This model is quite suitable above the critical micellar concentration, which represents the monomer saturation and the beginning of the micelle formation. The second model is the so-called mass action model that assumes that micelles and surfactant monomers exist in an association–dissociation equilibrium (Nishikido, 1993). Either model is capable of predicting the concentrations of each of the surfactant monomers in the binary mixture required to form micelles.

III.C.4. Microemulsions as antimicrobial delivery systems

Over the past 5 years, a series of investigations on the use of microemulsions to deliver food antimicrobials and to inhibit the growth of foodborne pathogens such as *E. coli* O157:H7 and *L. monocytogenes* were conducted (Gaysinsky *et al.*, 2005a, b, 2006, 2007a, b, 2008). Carvacrol- and eugenol-containing microemulsions were prepared by dispersing the antimicrobials in micellar solutions of Surfynol® 485W and Surfynol® 465. UV-visible spectroscopy was used to obtain pH and temperature stability phase diagrams; particle size was determined by dynamic light scattering, and structural information about microemulsions was obtained by nuclear magnetic resonance spectroscopy (NMR). In this study, carvacrol and eugenol could both be rapidly incorporated (<10 min) in surfactant micelles. Depending on the surfactant–antimicrobial combination, the particles produced varied in size between 5 and 20 nm. All microemulsions were stable over a wide range of pHs. However, temperature and essential oil component concentration influenced the stability of the microemulsions, and the nanocapsules broke down at lower temperatures during heating with increasing encapsulated essential oil component concentration. For example, 0.9% of eugenol encapsulated in Surfynol® 485W exhibited turbidity (cloud point) at 55°C, while at 0.5%, a temperature of 70°C was needed to reach the cloud point. Results of this study were then used to formulate a stable antimicrobial microemulsion, which was used in subsequent antimicrobial activity tests.

Growth inhibition of four strains of *E. coli* O157:H7 (H1730, F4546, 932, and E0019) and *L. monocytogenes* (Scott A, 101, 108, and 310) by essential oil (EO) components (carvacrol and eugenol) solubilized in microemulsions (Surfynol® 465 and 485W) was determined. Concentrations of phytophenols encapsulated were varied between 0.02 and 0.9 wt% depending on the compound, surfactant type, and surfactant concentration (0.5–5%). Both the antimicrobial and the surfactant were found to influence the antimicrobial efficacy of the microemulsions. Generally, the antimicrobial activity of Surfynol® 485 in combination with eugenol was higher than the combination with carvacrol. In a comparison of surfactants, Surfynol® 485 also displayed a higher activity when compared to Surfynol® 465.

Subsequently, growth inhibition of four strains of *E. coli* O157:H7 (H1730, F4546, 932, and E0019) and *L. monocytogenes* (Scott A, 101, 108, and 310) by eugenol in microemulsions (Surfynol® 485W) adjusted to pH 5, 6, and 7 and incubated at 10, 22, and 32°C was determined. Eugenol encapsulated in surfactant micelles inhibited both microorganisms at pH 5, 6, and 7. At pH 5, some inhibition occurred in the absence of eugenol, i.e. by the surfactant itself, but addition of > 0.2% eugenol led to complete inhibition of both microorganisms. Not surprisingly, inhibition decreased with increasing pH, i.e. the MIC was 0.2, 0.5, and 0.5% of micellar encapsulated eugenol solutions at pH 5, 6, and 7, respectively. Nevertheless, the antimicrobial activity of the microemulsions was generally maintained over a broad range of pH and temperatures typical for many food products.

Lastly, the antimicrobial activity of eugenol in microemulsions added to ultra-high-temperature (UHT) pasteurized milk containing different percentages of fat (0, 2, 4%) was evaluated. Addition of eugenol microemulsions led to complete inactivation of both strains of *E. coli* O157:H7 in less than 1 hour but only reduced both strains of *L. monocytogenes* by 1 log. Pure eugenol was less inhibitory against both strains of *L. monocytogenes*. The fat content affected the antimicrobial efficacy of microemulsions, i.e. microemulsions completely inhibited *L. monocytogenes* after 24 h in skim and 2% fat milk but resulted in just 1 log reduction after 24 h in full-fat milk. The results of this study showed that microemulsions are effective delivery systems, but that they should be engineered to exclude interaction with interfering lipid phases and promote interaction with target organisms such as *E. coli* O157:H7.

In conclusion, much work has gone into the design of antimicrobial microemulsions to combat foodborne diseases. Particular progress is currently being made by designing mixed microemulsion systems, i.e. microemulsions that are composed of combinations of a variety of surfactants. Our initial research has shown that this not only improves loading of the antimicrobial but targeting of microbial surfaces as well.

III.D. Liposomes
III.D.1. Definition

Liposomes are spherical particles formed from polar lipids (e.g. phosphatidylcholine or phosphatidylethanolamine) or mixtures of polar lipids with cholesterol or ergosterol; liposomes are also components available in abundance in nature. Their name is derived from a combination of the words lipid and the Greek word 'soma' meaning body. The size of liposomes can range from tens of nanometers to tens of micrometers depending on the method of manufacturing (Taylor *et al.*, 2005a). In polar solvents such as water, polar lipids generally have a tendency to self-assemble in the form of bilayer membranes. These bilayer membranes are relatively flexible and under shear can be forced to assume a curvature that leads to the generation of particles composed of a thin shell of a polar lipid bilayer; this bilayer surrounds an interior compartment that consists of the initial solvent in which the polar lipids were dispersed prior to the application of shear. Depending on the degree of shear applied during the manufacturing process, liposomes may assume a variety of different structures, such as unilamellar liposomes (which have a single bilayer membrane shell), multilamellar vesicles (liposomes with multiple bilayer membranes stacked in a concentric configuration similar to the skins of an onion), and multivesicular vesicles (liposomes that may contain other randomly sized liposomes in their interior (Figure 24.10).

Liposomes can serve as a separate microenvironment in which food antimicrobials and other functional components can be incorporated and their activity maintained despite changes in the surrounding aqueous phase (Bolsman *et al.*, 1988; Bouwstra, 1996). Interestingly, liposomes are able to encapsulate both lipophilic and hydrophilic functional components. In the case of lipophilic compounds, the bilayer membrane acts as the host environment to the compounds. The compounds are solubilized in the interior of the bilayer, which consists of the self-assembled fatty acid tails of the polar lipids via a process also known as ad-solubilization (a combination of adsorption/solubilization processes). Alternatively, hydrophilic compounds may be encapsulated in the interior of the liposomes. In this case, the liposomal membrane must provide a barrier to diffusion of compounds from

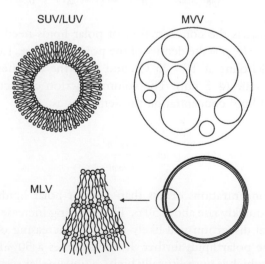

Figure 24.10 Schematics of liposomal structures that can be manufactured. SUV/LUV: small and large unilamellar vesicles; MVV: multiple vesicular vesicles; MLV: multiple lamellar vesicles.

the interior to the exterior. This mass transport may be initiated when a concentration difference between the interior and exterior of the liposomes exists. For example, after manufacturing, liposomes are typically added to a food or beverage system where the aqueous phase environment differs from that of the interior of the liposomes.

III.D.2. Functional properties of liposomes

- *Loading capacity.* Loading capacity of liposomes varies significantly depending on whether water-soluble or lipid-soluble antimicrobials are encapsulated. If water-soluble antimicrobials are encapsulated, very high loading ratios are feasible. For unilamellar liposomes, the shell is typically 4–5-nm thick. Thus for a 100-nm-sized liposome, the internal volume is approximately seven times larger than the volume occupied by the shell. This ratio decreases when the liposomes become smaller. In multilamellar liposomes, the internal volume may also be smaller since more of the volume will be occupied by polar lipids. Conversely, if lipid-soluble materials are encapsulated in the shell material, only relatively small concentrations of active material can be encapsulated. This is because the particle number n quickly rises with the concentration of polar lipids. If r is the average radius of liposomes, n can be calculated from the concentration of polar lipids in the system $c_{pl,total}$ as:

$$n = \frac{c_{pl,total}}{c_{pl.liposome}} = \frac{c_{pl,total}}{\frac{4}{3}\pi\left(r^3 - (r - \Delta r)^3\right) \cdot \rho_l} \tag{24.5}$$

where $c_{pl,liposome}$ is the concentration of polar lipids needed to form an individual vesicle, ρ is the density of the polar lipid (e.g. 1.015 g/cm^3 for phosphatidylcholine at T $= 25°$C), and Δr is the thickness of the liposomal membrane (4–5 nm). The volume fraction of occupied space as a function of lipid concentration thus emerges as:

$$\phi = \frac{\frac{4}{3}\pi r^3 n}{V_{total}} \tag{24.6}$$

Thus, for concentrations of less than 10% of polar lipids, the volume fraction can quickly rise above 50%. The resulting increase in viscosity of the liposomal dispersion will likely limit the increasing of the concentration of the polar lipids further. If one assumes a 50% loading of the membrane (which is unrealistically high), a unilamellar system occupying more than 50% of the entire volume of the food system would contain less than 5% of active ingredient. The loading ratio can however be significantly increased in multilamellar vesicles where concentric bilayers are formed.

- **Retention efficiency.** Retention efficiency depends on the properties of the encapsulated material and that of the liposomal shell material, as well as environmental conditions (pH, temperature). If hydrophilic compounds are encapsulated in the interior of the capsule, their retention may be improved if they carry a similar charge as the polar lipids comprising the shell due to induced electrostatic interactions. Using this principle, pH gradients can in effect be established, which is a unique functionality of liposomes. On the other hand, the membrane is flexible and undergoes natural undulations that can lead to a release of the material over time. In the case of incorporated lipid compounds, the situation is more complex and depends on the interaction of the lipid material with the polar lipids that comprise the shell of the liposomes.

- **Stability.** Liposomes are much more stable to gravitational separation than emulsions since the density difference between the two phases (particle and solvent) is much smaller than the density difference driving the separation of oil-in-water emulsions. Nevertheless, as opposed to microemulsions, liposomes are not thermodynamically stable. They do have a tendency to merge over time since electrostatic charges on the polar lipids are often small and insufficient to prevent liposomal collisions.

- **Optical properties.** The optical properties of liposomes differ from that of emulsions and microemulsions. In the case of microemulsions, particles are too small to scatter light and the systems are thus transparent. In the case of emulsions, systems may be transparent if the particle diameters are below approximately 90–100 nm and strongly turbid (milky-white) at average particle diameters above 120 nm. Liposomes are only composed of a thin shell of polar lipids and thus do not strongly scatter light. Even at average particle diameters of above 150–200 nm, liposomes are only slightly opaque and are completely transparent below 100 nm. Encapsulation of functional components can somewhat change the optical appearance due to the fact that the refractive index at the interface between solvent and internal phase changes and the size of liposomes may be altered.

- **Rheology.** As mentioned above, the effective volume occupied by the particles in a liposome suspension is often much greater than that of the polar lipid used to create the dispersion. For a dilute system:

$$\eta = \eta_0(1 + 2.5\phi_{off}) \tag{24.7}$$

For an emulsion, ϕ_{eff} is given as:

$$\phi_{eff} = \frac{V_{droplets}}{V_{emulsion}} \tag{24.8}$$

where $V_{droplets}$ can be simply calculated from the density ρ_{Lipid} and mass m_{Lipid} of the lipid as:

$$V_{droplets} = \frac{m_{Lipid}}{\rho_{Lipid}} \tag{24.9}$$

In a liposomal suspension, only a small portion of the lipid V_{Lipid} occupies the total volume of a liposome $V_{Liposome}$:

$$V_{Liposomes} = \frac{4}{3}n\pi r^3 \tag{24.10}$$

$$V_{Lipid} = \frac{4}{3}\pi n(r^3 - (r - \delta)^3) = \frac{m_{Lipid}}{\rho_{Lipid}} \tag{24.11}$$

Thus, the viscosity of a liposomal system may be much larger than that of an equivalent system with the same amount of lipid dispersed in the form of droplets:

$$\eta = \eta_0(1 + 2.5\phi_{eff}R) \tag{24.12}$$

where

$$R = \frac{V_{Liposome}}{V_{Lipid}} = \frac{\frac{4}{3}\eta\pi r^3}{\frac{m_{Lipid}}{\rho_{Lipid}}} \tag{24.13}$$

III.D.3. Manufacture of liposomes

Liposomes may be manufactured using a variety of different manufacturing procedures including thin-film rehydration, freeze-drying rehydration, reverse-phase evaporation, detergent depletion, membrane extrusion, high-pressure homogenization or microfluidization, and ultrasonication by probe or bath. For large-scale manufacture of liposomes with sizes below 100 nm, microfluidization has emerged as the predominant method. An in-depth review of manufacturing methods can be found elsewhere (Taylor et al., 2005a).

III.D.4. Liposomes as antimicrobial delivery systems

While liposomes have been used for delivery of various bioactive compounds, the application of liposomes for the delivery of food antimicrobials is still in its early stages. Thapon and Brule (1986) used liposomes to encapsulate lysozyme and nisin to prevent spoilage of cheeses. Degnan and Luchansky (1992) reported anti-listerial effects of the bacteriocin, pediocin AcH, in beef tallow and muscle slurries using PC liposomes. Bacteriocin activity increased by 28% following encapsulation compared to non-encapsulated pediocin. The authors reported that pediocin AcH activity decreased rapidly when unencapsulated bacteriocin was added to slurries or tallow, which was likely a function of cross-reactions with the food component leading to loss of activity. Benech et al. (2002a, b) encapsulated nisin Z (asparagine-containing variant of nisin) in hydrogenated phosphatidylcholine liposomes and added nisin Z containing liposomes or unencapsulated nisin Z to cheese at a final concentration of 300 IU/g. *Listeria innocua* counts decreased by 3 log CFU/g over a 6-month ripening period (Benech et al., 2002a, b). Nisin Z activity in liposomes at the end of the 6-month ripening period was ~90% of the initial activity, with a final concentration of *L. innocua* below 10 CFU/g cheese. In recent years, we carried out subsequent experiments that focused on encapsulation of water-soluble antimicrobials such as nisin and lysozyme (Were et al., 2004a, b; Taylor et al., 2005a, b, 2007a, b, 2008).

- *Entrapment efficiency, leakage and stability of liposomes containing nisin and lysozyme.* Effect of lipid composition [phosphatidylcholine (PC), phosphatidylglycerol (PG), and cholesterol] on size, thermal stability, and entrapment efficiency of polypeptide antimicrobials in liposomal nanocapsules was investigated. PC, PC:cholesterol (70:30), and PC:PG:cholesterol (50:20:30) liposomes containing nisin resulted in entrapment efficiencies of 63, 54, and 59% with particle sizes of 144, 223, and 167 nm for PC, PC:cholesterol (70:30), and PC:PG:cholesterol (50:20:30), respectively. Encapsulation of lysozyme yielded entrapment efficiencies of 61, 60, and 61%, with particle sizes of 161, 162, and 174 nm, respectively. The highest concentration of antimicrobials was encapsulated in 100% PC liposomes. Nisin induced slightly more leakage compared to lysozyme. Based on these results, PC liposome mixtures with varying acyl chain lengths (C16:0–C18:0), with or without entrapped nisin, were then used for temperature stability measurements. While liposomes, with or without nisin, maintained a stable size when heated to 90°C, phase transitions were measured in the bilayer membranes at 44.8, 54.3, and 54.9°C with increasing chain length of the polar lipids, suggesting that formulations have to be pasteurized prior to addition to the product. The results of this study showed that nisin and lysozyme can be retained at ~60% in PC or PC:PG liposomes and that those formulations can be pasteurized prior to addition to foods.

- *Antimicrobial efficacy of nisin and lysozyme in liposomes against* L. **monocytogenes.** Nisin- and lysozyme-containing liposomes were prepared by hydrating dried lipids with buffer containing nisin, the fluorescence probe calcein and nisin, or calcein and lysozyme, followed by centrifugation, sonication, and collection of encapsulated liposomes by size-exclusion chromatography. Antimicrobial concentration in liposomes was determined by bicinchonic acid assay prior to determination of antimicrobial activity against strains of L. *monocytogenes*. When nisin was encapsulated in liposomes, protein concentrations of 0.39, 0.27, and 0.23 mg/ml for phosphatidylcholine (PC), PC:cholesterol (7:3), and PC:phosphatidylglycerol (PG):cholesterol (5:2:3), respectively, were obtained. Encapsulation of nisin with calcein yielded protein concentrations of 0.35, 0.39, and 0.28 mg/ml for PC, PC:cholesterol (7:3), and PC:PG:cholesterol (5:2:3), respectively. Encapsulation of calcein with lysozyme resulted in protein concentrations of 0.43, 0.26, and 0.19 mg/ml for PC, PC:cholesterol (7:3), and PC:PG:cholesterol (5:2:3) respectively. Encapsulated nisin in 100% PC and PC:cholesterol liposomes inhibited bacterial growth by >2 log CFU/ml compared to free nisin. Growth inhibition with liposomal lysozyme was strain dependent, with greater inhibition observed with strains 310 and Scott A with PC:cholesterol (7:3), and PC:PG:cholesterol (5:2:3) liposomes. The results of this study demonstrated enhanced antimicrobial activity of nisin and lysozyme when encapsulated in liposomes with better results obtained for liposomes containing nisin.

- *Antimicrobial activity of liposome-entrapped nisin in milk.* Phosphatidylcholine and PG were used to encapsulate powdered nisin at a final concentration of 5 IU/ml. Liposomes were composed of PC (100%), PC:PG 8:2 (mol fraction), and PC:PG 6:4 (mol fraction). Liposomes were added to UHT-processed skim, 1%, and whole milk at a final concentration of 2 mM of active ingredient (nisin). Strains of Listeria mono-cytogenes were added at a final concentration of 4 log CFU/ml. Samples were homogenized and incubated aerobically at 25°C to simulate temperature abuse. Milk samples were incubated for 48 h and survivors enumerated on non-selective media. Table 24.2 shows the survivor populations in log CFU/ml after 48 h incubation at 25°C for 1% low-fat milk. Unencapsulated nisin at 5 IU/ml was able to exert inhibitory effects initially, but bacterial populations were able to eventually recover to populations approximating that of the positive control at the end of experimentation. Liposomal nisin was able to effectively inhibit the Gram-positive pathogen strains throughout experiments; significant bacteriostatic inhibition was observed for both *L. monocytogenes* strains, regardless of milk fat concentration. The results showed significant improvements in activity in milk upon inclusion in phospholipid liposomes.

IV. EMERGING NANOENCAPSULATION SYSTEMS

IV.A. Double-layered nanocapsules

Recently, it has been shown that the stability and functional performance of conventional (nano) particles can often be greatly improved by coating the particles with polymers to form monomolecular layers, using the so-called layer-by-layer (LbL) electrostatic deposition method (Ogawa *et al.*, 2003a, b; Guzey *et al.*, 2004; Gu *et al.*, 2005; Klinkesorn *et al.*, 2005; McClements, 2005a; Harnsilawat *et al.*, 2006a, b; Klinkesorn *et al.*, 2006; Guzey and McClements, 2007). This LbL electrostatic deposition method is based on the electrostatic attraction that a charged polymer may experience in the presence of an oppositely charged surface or interface. The subsequent adsorption of the polymer onto the surface or interface will typically lead to a reversal of the charges since the total number of charges on the adsorbed polymer is greater than the number of charges on the surface. Once the charge has reversed, additional adsorption of polymers onto the surface is no longer possible since additional polymer molecules will be repelled from the surfaces. This allows for (1) fine control over the dimension of the adsorbed polymer layer since only a single polymer layer can be adsorbed, and (2) build up of multiple layers by sequential addition of other polymers, each having an opposite charge to the previously deposited polymer (Decher and Schlenoff, 2003). In the case of emulsions, this technique has been shown to have a number of benefits. Under certain circumstances, emulsions containing oil droplets surrounded by multilayer interfaces have been found to have better stability to environmental stresses than conventional oil-in-water emulsions with single-layer interfaces (Moreau *et al.*, 2003; Ogawa *et al.*, 2003a, b; Gu *et al.*, 2004; Guzey *et al.*, 2004; Aoki *et al.*, 2005; Guzey and McClements, 2006; Harnsilawat *et al.*, 2006a, b). For example, stability of emulsion droplets coated with an additional layer of polymer prevented

Table 24.2 Concentrations of foodborne pathogens inhibited by 50 IU/ml nisin in 1% low-fat milk after 48 h incubation at 25°C, as a function of liposome entrapment and liposome formulation (Taylor et al., 2007)

Antimicrobial treatment	L. monocytogenes Scott A (log CFU/ml)				L. monocytogenes 310 (log CFU/ml)			
	0 h		48 h		0 h		48 h	
	Mean	SD	Mean	SD	Mean	SD	Mean	SD
Control	4.42	0.38	8.00	0.27	4.52	0.1	8.03	0.04
Free nisin	4.43	0.01	7.57	0.08	4.52	0.24	7.22	0.22
PC–nisin	4.23	0.09	4.93	0.33	4.09	0.16	4.83	0.12
PC:PG 8:2-nisin	4.28	0.01	4.50	0.06	4.23	0.21	4.84	0.06
PC:PG 6:4-nisin	4.21	0.01	4.73	0.09	4.31	0.2	4.33	0.2

PC, phosphatidyl choline; PG, phosphatidyl glycerol.

coalescence and improved stability to freezing and thawing, and drying and rehydration.

Recently, the formation of double-layered particles has been expanded to a variety of other particles including liposomes (Laye *et al.*, 2008). For example, a variety of polymers such as chitosan, β-lactoglobulin, fish gelatin, and casein were successfully adsorbed on the surface of liposomes with dramatic improvements in the long-term stability of the particles. The engineering of the properties of the nano-structured shells around the particles also allows for a modification of the colloidal interactions that may have an influence on antimicrobial activity, which is strongly influenced by capsule–microorganism interactions (Caruso and Mohwald, 1999a, b; Caruso and Schuler, 2000; Decher and Schlenoff, 2003). However, one of the current issues with this approach is that the structures are principally formed by electrostatic interactions and structures may thus be disrupted if ionic strength, pH, and temperature of the system changes. Current research focuses on preventing this structure breakdown by covalently cross-linking the adsorbed layers after they have been formed around a particle. Multi-layered nanoparticles may have additional functionalities when compared to their simpler parent systems. The formation of an additional shell around the particles may allow for additional functional components to be entrapped within the oil droplet *core* or within the laminated *shell* surrounding the droplets. For example, a lipophilic functional component could be incorporated into the oil phase prior to homogenization, whereas a hydrophilic ionic functional component could be used to make up one of the layers surrounding the oil droplets. The release of a functional component trapped in the *core* of the delivery system could be controlled by designing the response of the *shell* to the environment.

IV.B. Nanofibers

Nanofibers, another nanoscalar structure, are currently being developed for future food applications. Nanofibers are fibers with average diameters below 100 nm and may be used as food-packaging materials, ingredients, sensors, and processing aids (Figure 24.11a). These fibers have been shown to possess unique properties that distinguish them from larger non-woven fibers, such as a high orientation of polymers within the fibers that leads to mechanically superior properties, e.g. increased tensile strength and large surface-to-mass ratio, high porosity (Kim and Reneker, 1999; Frenot and Chronakis, 2003). To manufacture these fibers, a technique called electrospinning is used. In electrospinning, a high-voltage electric field is applied to a (bio)polymer solution that is pumped through a small orifice. The applied electrical field forces the polymers to be ejected from the tip of the orifice in the form of a fiber jet. The high velocity and high surface area of the jet causes the

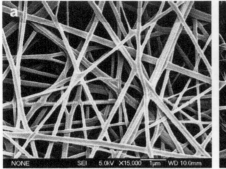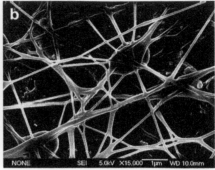

Figure 24.11 Scanning electron microscopy images of polyvinylacetate (PVA) nanofibers spun from 8 wt% aqueous PVA solution in (a) the absence of and (b) the presence of hexadecane-Tween 20 stabilized nanoemulsions. Image (b) indicates that some of the emulsion droplet coalesced to form a large emulsion droplet at fiber junctions surrounded by PVA.

solvent to rapidly evaporate, leading to the deposition of a solidified nanofiber on a grounded target. These fibers have already found wide applications in tissue engineering, wound healing, drug delivery, medical implants, dental applications, biosensors (Li and Xia, 2004; Zhang et al., 2005, 2006a, b), military protective clothing, filtration media, and other industrial applications (Doshi and Reneker, 1995; Huang et al., 2003; Subbiah et al., 2005) and are thus likely to find future applications in the food industry.

IV.C. Multi-assembled nanocapsule aggregates

An exciting and novel development in the area of nanoencapsulation is the directed assembly and architectural design of aggregate structures (Figure 24.12). Fundamentally, the relatively well-developed nano-encapsulation systems described previously could be combined to build more complex structures by thoroughly understanding interactions that govern the assembly process. The process can almost be pictured as that of a child having a set of building blocks with different shapes that can now be put together to build much more complex structures. In the case of the food scientist and with respect to food antimicrobials, the fundamental building blocks are the antimicrobial nanoencapsules; the shapes are the antimicrobial efficacies and the spectrum of activity of the assembled aggregates. For example, liposomes could be loaded with antimicrobial nanoemulsions or solid lipid nanoparticles instead of a simple water-soluble antimicrobial, thereby allowing for finer control over the release and overcoming issues of pH and salt stability in antimicrobial-loaded solid lipid nanoparticles and nanoemulsions. Moreover, this could allow for a combination of

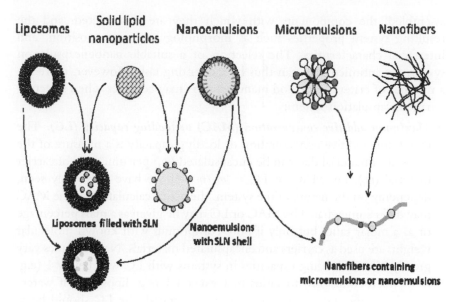

Liposomes   Solid lipid nanoparticles   Nanoemulsions   Microemulsions   Nanofibers

Liposomes filled with SLN

Nanoemulsions with SLN shell

Nanofibers containing microemulsions or nanoemulsions

Liposomes filled with nanoemulsions

Figure 24.12 Examples of assembly of aggregate structures from basic nano-encapsulation systems, including liposomes, solid-lipid nanoparticles, nanoemulsions, microemulsions, and nanofibers. A combination of structures can lead to production of liposomes filled with solid lipid nanoparticles or nanoemulsions, nanoemulsions stabilized by solid lipid nanoparticles, and nanofibers containing microemulsions or nanoemulsions.

a nanocapsule containing both water- and lipid-soluble antimicrobials. Alternatively, nanofibers have already been manufactured in our laboratories that contain nanoemulsions or microemulsions (Figure 24.11b). It should be noted though that the assembly processes are becoming increasingly more complex compared to those of simpler nanoencapsulation systems and costs may thus be increased. On the other hand, these systems would combine the advantages of a number of nanoencapsulation systems, thereby leading to novel functionalities previously not seen.

V. SELECTION AND EVALUATION CRITERIA FOR NANOENCAPSULATED ANTIMICROBIALS

The number of available nanoscalar delivery systems that may be used to deliver not only antimicrobials but other functional food components such as antioxidants, flavors, nutraceuticals, and colors is steadily increasing. We discussed above how these systems differ: in the way they are fabricated and assembled, the components with which they are constructed, and the resulting system properties such as morphology, size, composition, and interfacial characteristics. The selection of a suitable nanoencapsulation system for antimicrobials can thus be a daunting task. However, there are a number of criteria that food manufacturers may consider when selecting a nanoencapsulation system:

- **Maximum additive concentration (MAC) or loading capacity (LC).** The maximum additive concentration or loading capacity is a measure of the amount of material that can be encapsulated (m_e) per unit mass of carrier material (m_c), i.e. MAC or LC $= m_e/m_c$. As we have previously seen, depending on the nature of the system, the actual calculation of the MAC may differ somewhat. The MAC or LC may be given as a mass percentage or as a molar ratio, but only if pure compounds with known molecular weights are used as carriers and encapsulated materials. Nanocapsules vary greatly in their loading capacities in systems with LC well below 1 (e.g. most microemulsions) to ratios that exceed 1 (e.g. liposomes if water-soluble compounds are encapsulated). In general, the LC should be as high as possible to minimize the costs of the encapsulation system. Alternatively, the encapsulated compounds need to be increasingly active to offset reduced loading ratios.

- **Minimum inhibitory concentration (MIC).** The MIC is a measure of the efficacy of an antimicrobial and from an application point of view is a key property that needs to be determined. Unfortunately, there is still considerable discussion going on in the scientific community as to which microbial protocol is best suited to determine the efficacy. Protocols range from *in vitro* tests such as agar diffusion, agar/broth dilution, gradient, and spiral plating to actual application tests that involve recovery and enumeration of microorganisms from the food system. During the development stage, food manufacturers may use *in vitro* tests for screening of suitable candidate systems, but validation of candidate system in their actual food product is an essential prerequisite to ensure that their encapsulated antimicrobial has the required activity. In terms of activity, it is important to note that not only the efficacy but also the *specificity* of the antimicrobial, i.e. the spectrum of activity, may be altered upon nanoencapsulation. For example, unencapsulated antimicrobials may not have any strain specificity, but the encapsulated system may show increased (or decreased) activity against specific strains due to the altered interaction with microbial surfaces.

- **Mechanism of delivery and action.** An antimicrobial delivery system must be designed so that it carries the functional component to a particular site-of-action (microbial surface or microbial cell interior) and then releases it. The release may have to be at a controlled rate or in response to a particular environmental trigger (e.g. pH, ionic strength, enzyme activity, or temperature). For example, unencapsulated antimicrobials may in fact initially be more active and even lead to a reduction of the viable population, but after a short period, after the antimicrobial has been used up, the surviving microorganisms may again begin to replicate and rapidly grow back beyond initial contamination levels. In contrast, upon encapsulation, the initial activity may in fact be lower than that of the unencapsulated antimicrobial, resulting only in growth suppression rather than kill; the bacteriostatic activity may be maintained over a significantly longer period of time since the active compound will continue to be released at a steady rate from the nanocapsules. Moreover, nanocapsules may be designed to more strongly interact with bacterial surfaces by, for example, changing the interfacial properties of the nanocapsules. Introduction of a positive charge can lead to a stronger interaction with microbial surfaces, causing the local concentration of the antimicrobial in the vicinity of the microorganisms to increase. In other cases, the nanocapsule may be small enough to pass the microbial cell barrier and facilitate a penetration and release of compounds into the intracellular space. In all these cases, it is the nanoencapsulation system that governs the activity of the antimicrobial rather than the antimicrobial. It is thus imperative for food manufacturers to understand the mechanism of delivery and action of the nanoencapsulation system.

- **Delivery efficiency (DE).** The delivery efficiency is a measure of the ability of the nanoscalar delivery system to transport the encapsulated compound to the required site of action, e.g. $DE = 100 \times (m_{e,i} - m_{e,f})/m_{e,i}$, where $m_{e,i}$ and $m_{e,f}$ are the initial and final masses of encapsulated compounds at the target site, respectively. Ideally, one would like the delivery efficiency to be as high as 100%, i.e. with all of compound delivered to the site of action. In practice, this is highly unlikely since interactions with other food ingredients may be minimized but not completely prevented. In the case of antimicrobials, it may be difficult to measure the DE as the measurement requires recovery of microorganisms through a separation process, followed by identification of the concentration of the antimicrobial in the cells. However, during screening studies using microbial model systems to select antimicrobial carrier candidate systems, determination of the *DE* may be feasible and offer a means to engineer the nanoscalar carrier to better deliver compounds to the site of action.

- **Other selection criteria.** In addition to the above-mentioned evaluation criteria, food manufacturers may consider a number of other factors such as, for example, the economics of the nanoscalar system. Questions that processors may consider include: (1) can the antimicrobial system be manufactured at a reasonable cost at large scale, (2) can it be manufactured from food-grade (GRAS) ingredients, and (3) is the system physically stable and able to retain the compound throughout the shelf life of the product in which the nanocapsules have been applied? In terms of retention, one would ideally like to retain 100% of the encapsulated antimicrobial in the capsules, but as demonstrated in the case of liposomes, some leakage may invariably occur.

VI. CONCLUSIONS

In this chapter we attempted to give the reader a glimpse of the rapidly progressing field of nanoencapsulation, by focusing on encapsulation of food antimicrobials and demonstrating the potential of these novel systems to improve the quality and safety of foods. Because of the pace of development in this field, this chapter can provide no more than a snapshot in time. Nevertheless, we have illustrated that nanotechnology is truly an enabling science that allows food scientists to move beyond the crude concept of top-to-bottom manufacture of food systems, and instead use a bottom-to-top approach so as to more intelligently assemble functional structures. Nanoscience as such has led the groundwork for an emerging discipline that may be better termed 'Food Architecture' or 'Food Material Engineering' that in years to come may have a significant impact on the development of new foods and food ingredients. In the context of food safety, the technology clearly has the potential to improve the utilization of food preservatives to combat foodborne diseases and overcome activity losses upon introduction in complex food systems.

ACKNOWLEDGMENTS

We gratefully acknowledge the contributions of the many students and colleagues who have over the years contributed to our development of antimicrobial nanoencapsulation systems, including Barry D. Bruce, Lilian Were, T. Matthew Taylor, Dustin Carnahan, Sarisa Suriyarak, Christina Kriegel, Alessandra Arecchi, Celine Laye, and others. Much of the work presented above was supported by grants from the United States Department of Agriculture and the United States Environmental Protection Agency, as well as the State of Massachusetts and the State of Tennessee.

REFERENCES

Acamovic, T., & Brooker, J. D. (2005). Biochemistry of plant secondary metabolites and their effects in animals. *Proceed. Nutrition Soc.*, *64*, 403–412.

Allos, B. M., Moore, M. R., Griffin, P. M., & Tauxe, R. V. (2004). Surveillance for sporadic foodborne disease in the 21st century: the FoodNet perspective. *Crit. Infect. Dis.*, *38*(Supplement 3), S115–S120.

Anonymous. (2006a). FoodNet Surveillance Report for 2004 (Final Report). Centers for Disease Control and Prevention.

Anonymous. (2006b). Preliminary FoodNet data on the incidence of infection with pathogens transmitted commonly through food – 10 states, United States, 2005. Morbidity and Mortality Weekly 55(14), 395.

Aoki, T., Decker, E. A., & McClements, D. J. (2005). Influence of environmental stresses on stability of O/W emulsions containing droplets stabilized by multilayered membranes produced by a layer-by-layer electrostatic deposition technique. *Food Hydrocolloids*, *19*(2), 209–220.

Awad, T., Helgason, T., Kristbergsson, K., Decker, E. A., Weiss, J., & McClements, D. J. (2008). Effect of cooling and heating rates on polymorphic transformation inducing gelation of solid lipid nanoparticles (SLN) of tripalmitin. *Food Biophys.*, *3*(2), 155–162.

Aziz, N. H., Farag, S. E., Mousa, L. A. A., & Abo-Zaid, M. A. (1998). Comparative antibacterial and antifungal effects of some phenolic compounds. *Microbios*, *93*, 43–54.

Bagamboula, C. F., Uyttendaele, M., & Debevere, J. (2003). Antimicrobial effect of spices and herbs on *Shigella sonnei* and *Shigella flexneri*. *J. Food Protect.*, *66*(4), 668–673.

Benech, R.-O., Kheadr, E. E., Lacroix, C., & Fliss, I. (2002a). Antibacterial activities of nisin Z encapsulated in liposomes or produced in situ by mixed culture during Cheddar cheese ripening. *Appl. Environ. Microbiol.*, *68*(11), 5607–5619.

Benech, R.-O., Kheadr, E. E., Laridi, R., Lacroix, C., & Fliss, I. (2002b). Inhibition of *Listeria innocua* in Cheddar cheese by addition of nisin Z in liposomes or by in situ production in mixed culture. *Appl. Environ. Microbiol.*, *68*(8), 3683–3690.

Biatwright, J. (1976). Antimicrobial activity of hop oil emulsion. *J. Institute Brewing*, *82*, 334–335.

Bjorkroth, J. (2005). Microbiological ecology of marinated meat products. *Meat Sci.*, *70*(3), 477–480.

Bocca, C., Caputo, O., Cavalli, R., Gabriel, L., Miglietta, A., & Gasco, M. (1998). Phagocytic uptake of fluorescent stealth and non-stealth solid lipid nanoparticles. *Intern. J. Pharmaceut.*, *175*(2), 185–193.

Bolsman, T. A. B. M., Veltmaat, T. G. F., & van Os, N. M. (1988). The effect of surfactant structure on the rate of oil solubilization into aqueous surfactant solutions. *J. Am. Oil Chem. Soc.*, *65*, 280–283.

Bouwstra, J. A. (1996). Transport of model drugs across the skin applied in vesicles in-vitro and in-vivo. *Euro. J. Pharmaceut. Sci.*, *4*, S42.

Branen, J. K., & Davidson, P. M. (2000). Activity of hydrolyzed lactoferrin against food-borne pathogenic bacteria in growth media: The effect of EDTA. *Letters Appl. Microbiol.*, *30*, 233–237.

Brul, S., & Coote, P. (1999). Preservative agents in foods: Mode of action and microbial resistance mechanisms. *Intern. J. Food Microbiol.*, *50*(1–2), 1–17.

Brul, S., Coote, P., Oomes, S., Mensonides, F., Hellingwerf, K., & Klis, F. (2002). Physiological actions of preservative agents: prospective of use of modern microbiological techniques in assessing microbial behaviour in food preservation. *Intern. J. Food Microbiol.*, *79*(1–2), 55–64.

Buege, D. R., Searls, G., & Ingham, S. C. (2006). Lethality of commercial whole-muscle jerky manufacturing processes against *Salmonella* serovars and *Escherichia coli* O157:H7. *J. Food Protect.*, *69*, 2091–2099.

Burt, S. (2004). Essential oils: Their antimicrobial properties and potential applications in food – A review. *Intern. J. Food Microbiol., 94*, 223–253.

Busquets, R., Puignou, L., Galceran, M. T., & Skog, K. (2006). Effect of red wine marinades on the formation of heterocyclic amines in fried chicken breast. *J. Agricultural Food Chem., 54*(21), 8376–8384.

Campbell, S. D., Goff, H. D., & Rousseau, D. (2004). Modeling the nucleation and crystallization kinetics of a palm stearin/canola oil blend and lard in bulk and emulsified form. *J. Am. Oil Chem. Soc., 81*(3), 213–219.

Carlos, A. M. A., & Harrison, M. A. (1999). Inhibition of selected microorganisms in marinated chicken by pimento leaf oil and clove oleoresin. *Appl. Poultry Sci., 8*, 100–109.

Caruso, F., & Mohwald, H. (1999a). Preparation and characterization of ordered nano-particle and polymer composite multilayers on colloids. *Langmuir, 15*(23), 8276–8281.

Caruso, F., & Mohwald, H. (1999b). Protein multilayer formation on colloids through a stepwise self-assembly technique. *J. Am. Chem. Soc., 121*(25), 6039–6046.

Caruso, F., & Schuler, C. (2000). Enzyme multilayers on colloid particles: Assembly, stability, and enzymatic activity. *Langmuir, 16*(24), 9595–9603.

Cavalli, R., Bocca, C., Miglietta, A., Caputo, O., & Gasco, M. R. (1999). Albumin adsorption on stealth and non-stealth solid lipid nanoparticles. *Stp Pharma Sci., 9*(2), 183–189.

CDC (Center for Disease Control and Prevention). (2006a). Ongoing multistate outbreak of *Escherichia coli* serotype O157:H7 infections associated with consumption of fresh spinach – United States, September 2006. Mortality Morbidity Weekly Reports 55(38), 1045–1046.

CDC. (2006b). Botulism associated with commercial carrot juice – Georgia and Florida, September 2006. Mortality and Morbidity Weekly Reports 55, 1098–1099.

Chao, S. C., & Young, D. G. (2000). Screening for inhibitory activity of essential oils on selected bacteria, fungi and viruses. *J. Essential Oil Res., 12*, 630–649.

Conner, D. E., & Beuchat, L. R. (1984). Effects of essential oils from plants on growth of food spoilage yeast. *J. Food Sci., 49*, 429–434.

Cooperative State Research, Education and Extension Service. (2007). National Research Initiative (NRI) Competitive Grants Program - Request for Application. U.S. Dept. of Agriculture.

Cowan, M. M. (1999). Plant products as antimicrobial agents. *Clin. Microbiol. Rev., 12*(4), 564–582.

Davidson, P. M., Sofos, J. N., & Branen, A. L. (2005). *Antimicrobials in Food*. Boca Raton, FL: CRC Press.

Deans, S. G., & Svoboda, K. P. (1989). Antibacterial activity of summer savory (*Satureja hortensis* L) essential oil and its constituents. *J. Horticult. Sci., 64*(2), 205–210.

Decher, G., & Schlenoff, J.B. (eds.) (2003). *Multilayer thin films: sequential assembly of nanocomposite materials*. Weinheim: Wiley-VCH.

Degnan, A. J., & Luchansky, J. B. (1992). Influence of beef tallow and muscle on the antilisterial activity of pediocin AcH and liposome-encapsulated pediocin AcH. *J. Food Protect., 55*, 552–554.

Dias, M. L. N., Carvalho, J. P., Rodrigues, D. G., Graziani, S. R., & Maranhao, R. C. (2007). Pharmacokinetics and tumor uptake of a derivatized form of paclitaxel associated to a cholesterol-rich nanoemulsion (LDE) in patients with gynecologic cancers. Cancer Chemother. *Pharmacol., 59*(1), 105–111.

Dickinson, E. (2003). Hydrocolloids at interfaces and the influence on the properties of dispersed systems. *Food Hydrocolloids, 17*, 25–39.

Doores, S. (1999). *Food safety: Current status and future needs*. Washington, D.C.: American Academy of Microbiology.

Doshi, J., & Reneker, D. H. (1995). Electrospinning process and applications of electrospun fibers. *J. Electrost., 35*(2–3), 151–160.

Frenot, A., & Chronakis, I. S. (2003). Polymer nanofibers assembled by electrospinning. *Curr. Opin. Colloid Interface Sci., 8*(1), 64–75.

Friberg, S., Larsson, K., & Sjoblom, J. (eds.) (2004). *Food Emulsions.* New York: Marcel Dekker.

Friedman, M., Kozukue, N., & Harden, L. A. (2000). Cinnamaldehyde content in foods determined by gas chromatography-mass spectrophotometry. *J. Agricultural Food Chem., 48*, 5702–5709.

Friedman, M., Henika, P. R., & Mandrell, R. E. (2002). Bactericidal activities of plant essential oils and some of their isolated constituents against *Campylobacter jejuni, Escherichia coli, Listeria monocytogenes,* and *Salmonella enterica. J. Food Protect., 65*(10), 1545–1560.

Fundaro, A., Cavalli, R., Bargoni, A., Vighetto, D., Zara, G. P., & Gasco, M. R. (2000). Non-stealth and stealth solid lipid nanoparticles (SLN) carrying doxorubicin: Pharmacokinetics and tissue distribution after i.v. administration to rats. *Pharmacol. Res., 42*(4), 337–343.

Garti, N., Avrahami, M., & Aserin, A. (2006). Improved solubilization of Celecoxib in U-type non-ionic microemulsions and their structural transitions with progressive aqueous dilution. *J. Colloid Interface Sci., 299*(1), 352–365.

Gaysinsky, S., Davidson, P. M., Bruce, B. D., & Weiss, J. (2005a). Growth inhibition of *Escherichia coli* O157:H7 and *Listeria monocytogenes* by carvacrol and eugenol encapsulated in surfactant micelles. *J. Food Protect., 68*(12), 2559–2566.

Gaysinsky, S., Davidson, P. M., & Bruce, B. D., and Weiss, J. (2005b). Stability and antimicrobial efficiency of eugenol encapsulated in surfactant micelles as affected by temperature and pH. *J. Food Protect., 68*(7), 1359–1366.

Gaysinsky, S., McClements, D. J., & Weiss, J. (2006). Emulsions as antimicrobial delivery systems: Influence of essential oil concentration on emulsion stability. Annual Meeting of the Institute of Food Technologists, Orlando, FL.

Gaysinsky, S., Davidson, P. M., McClements, D. J., & Weiss, J. (2008). Formulation and characterization of phytophenol-carrying microemulsions. *Food Biophys., 3*(1), 54–65.

Gaysinsky, S., Davidson, P. M., & Weiss, J. (2007a). Emulsions and Microemulsions as Antimicrobial Delivery Systems. Annual Meeting of the American Oil Chemist' Society, Quebec City, Canada.

Gaysinsky, S., Taylor, T. M., Davidson, P. M., Bruce, B. D., & Weiss, J. (2007b). Antimicrobial efficacy of eugenol microemulsions in milk against *Listeria monocytogenes* and *Escherichia coli* O157:H7. *J. Food Protect., 70*(11), 2631–2637.

Gaysinsky, S., & Weiss, J. (2007). Aromatic and spice plants: Uses in food safety. *Stewart Postharvest Solutions, 4*, 9–16.

Goppert, T. M., & Müller, R. H. (2005). Adsorption kinetics of plasma proteins on solid lipid nanoparticles for drug targeting. *Intern. J. Pharmaceut., 302*(1–2), 172–186.

Gref, R., Lück, M., Quellec, P., Marchand, M., Dellacherie, E., Harnisch, S., Blunk, T., & Müller, R. H. (2000). Stealth' corona-core nanoparticles surface modified by polyethylene glycol (PEG): influences of the corona (PEG chain length and surface density) and of the core composition on phagocytic uptake and plasma protein adsorption. *Colloids Surf. B Biointerfaces, 18*(3–4), 301–313.

Gu, Y. S., Decker, E. A., & McClements, D. J. (2004). Influence of pH and iota-carrageenan concentration on physicochemical properties and stability of beta-lactoglobulin-stabilized oil-in-water emulsions. *J. Agricultural Food Chem., 52*(11), 3626–3632.

Gu, Y. S., Decker, A. E., & McClements, D. J. (2005). Production and characterization of oil-in-water emulsions containing droplets stabilized by multilayer membranes consisting of beta-lactoglobulin, iota-carrageenan and gelatin. *Langmuir, 21*(13), 5752–5760.

Guzey, D., Kim, H. J., & McClements, D. J. (2004). Factors influencing the production of O/W emulsions stabilized by beta-lactoglobulin-pectin membranes. *Food Hydrocolloids, 18*(6), 967–975.

Guzey, D., & McClements, D. J. (2006). Influence of environmental stresses on O/W emulsions stabilized by β-lactoglobulin–pectin and β-lactoglobulin–pectin–chitosan membranes produced by the electrostatic layer-by-layer deposition technique. *Food Biophys., 1*(1), 30–40.

Guzey, D., & McClements, D. J. (2007). Impact of electrostatic interactions on formation and stability of emulsions containing oil droplets coated by beta-lactoglobulin-pectin complexes. *J. Agricultural Food Chem., 55*(2), 475–485.

Harnsilawat, T., Pongsawatmanit, R., & McClements, D. J. (2006a). Influence of pH and ionic strength on formation and stability of emulsions containing oil droplets coated by beta-lactoglobulin-alginate interfaces. *Biomacromolecules, 7*, 2052–2058.

Harnsilawat, T., Pongsawatmanit, R., & McClements, D. J. (2006b). Stabilization of model beverage cloud emulsions using protein-polysaccharide electrostatic complexes formed at the oil-water interface. *J. Agricultural Food Chem., 54*, 5540–5547.

Hartel, R. (2001). *Crystallization in Foods.* Gaithersburg: Aspen.

Hazan, R., Levine, A., & Abeliovich, H. (2004). Benzoic acid, a weak organic acid food preservative, exerts specific effects on intracellular membrane trafficking pathways in *Saccharomyces cerevisiae. Appl. Environ. Microbiol., 70*(8), 4449–4457.

Helgason, T., Awad, T., Decker, E., Kristbergsson, K., McClements, D. J., & Weiss, J. (2008). Influence of polymorphic transformations on gelation of tripalmitin solid lipid nanoparticle suspensions. *J. Am. Oil Chem. Soc., 85*(6), 501–511.

Herrera, M. L., de León Gatti, M., & Hartel, R. W. (1999). A kinetic analysis of crystallization of a milk fat model system. *Food Res. Intern., 32*(4), 289–298.

Hili, P., Evans, C. S., & Veness, R. G. (1997). Antimicrobial action of essential oils: the effect of dimethylsulphoxide on the activity of cinnamon oil. *Letters Appl. Microbiol., 24*, 269–275.

Himawan, C., Starov, V. M., & Stapley, A. G. F. (2006). Thermodynamic and kinetic aspects of fat crystallization. *Adv. Colloid Interface Sci., 122*(1–3), 3–33.

Hindle, S., Povey, M. J. W., & Smith, K. (2000). Kinetics of crystallization in n-hexadecane and cocoa butter oil-in-water emulsions accounting for droplet collision-mediated nucleation. *J. Colloid Interface Sci., 232*(2), 370–380.

Hirasa, K., & Takemasa, M. (1998). *Spice Science and Technology.* New York: Marcel Dekker.

Holmberg, K. B., Jönsson, B., Kronberg, B., & Lindman, B. (2003). *Surfactants and Polymers in Aqueous Solutions.* New York: Wiley.

Huang, Z.-M., Zhang, Y.-Z., Kotaki, M., & Ramakrishna, S. (2003). A review on polymer nanofibers by electrospinning and their applications in nanocomposites. *Composites Sci. Technol., 63*(15), 2223–2253.

Institute of Food Technologists. (2006). Antimicrobial resistance: Implications for the food system. *Comprehensive Rev. Food Sci. Food Safety, 5*, 71–137.

Jenning, V., Schafer-Korting, M., & Gohla, S. (2000). Vitamin A-loaded solid lipid nanoparticles for topical use drug release properties. *J. Controlled Release, 66*(2–3), 115–126.

Jumaa, M., Furkert, F. H., & Müller, B. W. (2002). A new lipid emulsion with high antimicrobial efficacy using chitosan. *Euro. J. Pharmaceutics Biopharmaceutics, 63*, 115–123.

Kaneko, N., Horie, T., Ueno, S., Yano, J., Katsuragi, T., & Sato, K. (1999). Impurity effects on crystallization rates of n-hexadecane in oil-in-water emulsions. *J. Crystal Growth, 197*(1–2), 263–270.

Katsuragi, T., Kaneko, N., & Sato, K. (2001). Effects of addition of hydrophobic sucrose fatty acid oligoesters on crystallization rates of n-hexadecane in oil-in-water emulsions. *Colloids and Surfaces B: Biointerfaces, 20*(3), 229–237.

Kim, J. S., & Reneker, D. H. (1999). Mechanical properties of composites using ultrafine electrospun fibers. *Polymer Composites, 20*(1), 124–131.

Klinkesorn, U., Sophanodora, P., Chinachoti, P., McClements, D. J., & Decker, E. A. (2005). Increasing the oxidative stability of liquid and dried tuna oil-in-water emulsions with electrostatic layer-by-layer deposition technology. *J. Agricultural Food Chem., 53*(11), 4561–4566.

Klinkesorn, U., Sophanodora, P., Chinachoti, P., Decker, E. A., & McClements, D. J. (2006). Characterization of spray-dried tuna oil emulsified in two-layered interfacial membranes prepared using electrostatic layer-by-layer deposition. *Food Res. Intern., 39*(4), 449–457.

Kogan, A., Aserin, A., & Garti, N. (2007). Improved solubilization of carbamazepine and structural transitions in non-ionic microemulsions upon aqueous phase dilution. *J. Colloid Interface Sci., 315*(2), 637–647.

Kong, B., Wang, J., & Xiong, Y. L. (2007). Antimicrobial activity of several herb and spice extracts in culture medium and in vacuum packaged pork. *J. Food Protect., 70,* 641–647.

Lai, F., Wissing, S. A., Müller, R. H., & Fadda, A. M. (2006). *Artemisia arborescens* L essential oil-loaded solid lipid nanoparticles for potential agricultural application: Preparation and characterization. *AAPS PharmSciTech., 7*(1), E2.

Lambert, R. J., & Stratford, M. (2004). Weak-acid preservatives: modelling microbial inhibition and response. *J. Appl. Microbiol., 42,* 157–164.

Laye, C., McClements, D. J., & Weiss, J. (2008). Formation of biopolymer-coated liposomes by electrostatic deposition of chitosan. *J. Food Sci., 73*(5), N7–N15.

Leal-Calderon, F., Thivilliers, F., & Schmitt, V. (2007). Structured emulsions. *Curr. Opin. Colloid Interface Sci., 12*(4–5), 206–212.

Lehninger, A. L., Nelson, P. L., & Cox, M. M. (1993). *Principles of Biochemistry.* New York: Worth Publishers.

Li, D., & Xia, Y. N. (2004). Electrospinning of nanofibers: Reinventing the wheel? *Adv Materials, 16*(14), 1151–1170.

Li, J. C., Sha, X. Y., & Fang, X. L. (2005). 9-nitrocamptothecin nanostructured lipid carrier system: in vitro releasing characteristics, uptake by cells, and tissue distribution in vivo. *Yao Xue Xue Bao, 40*(11), 970–975.

Liversidge, G. G., & Conzentino, P. (1995). Drug particle-size reduction for decreasing gastric irritancy and enhancing absorption of naproxen in rats. *Intern. J. Pharmaceutics, 125*(2), 309–313.

Lombardi Borgia, S., Regehly, M., Sivaramakrishnan, R., Mehnert, W., Korting, H. C., Danker, K., Röder, B., Kramer, K. D., & Schäfer-Korting, M. (2005). Lipid nanoparticles for skin penetration enhancement – correlation to drug localization within the particle matrix as determined by fluorescence and parelectric spectroscopy. *J. Controlled Release, 110*(1), 151–163.

López-Malo, A., Alzamora, S. M., & Guerrero, S. (2000). Natural antimicrobials from plants. In S. M. Alzamora, M. S. Tapia, & and A. López-Malo (Eds.), *Minimally Processed Fruits and Vegetables* (pp. 237–258). Gaithersburg, Maryland: Aspen.

McBain, J. W., & Richards, P. H. (1946). Solubilization of insoluble organic liquids by detergents. *Industrial Engineer. Chem., 38*(6), 642–646.

McClements, D. J. (1999). *Food Emulsions: Principles, Practice, and Techniques.* Boca Raton: CRC Press.

McClements, D. J. (2002a). Colloidal basis of emulsion color. *Curr. Opin. Colloid Interface Sci., 7*(5–6), 451–455.

McClements, D. J. (2002b). Theoretical prediction of emulsion color. *Adv. Colloid Interface Sci., 97*(1–3), 63–89.

McClements, D. J. (2005a). *Food Emulsions: Principles, Practice, and Techniques.* Boca Raton: CRC Press.

McClements, D. J. (2005b). Theoretical analysis of factors affecting the formation and stability of multilayered colloidal dispersions. *Langmuir, 21*(21), 9777–9785.

McClements, D. J., Dickinson, E., Dungan, S. R., Kinsella, J. E., Ma, J. G., & Povey, M. J. W. (1993). Effect of emulsifier type on the crystallization kinetics of oil-in-water emulsions containing a mixture of solid and liquid droplets. *J. Colloid Interface Sci.,* 160(2), 293–297.

McClements, D. J., & Dungan, S. R. (1997). Effect of colloidal interactions on the rate of interdroplet heterogeneous nucleation in oil-in-water emulsions. *J. Colloid Interface Sci.,* 186(1), 17–28.

Mehnert, W., & Mader, K. (2001). Solid lipid nanoparticles: Production, characterization and applications. *Adv. Drug Delivery Rev.,* 47(2–3), 165–196.

Menon, K. V., & Garg, S. R. (2001). Inhibitory effect of clove oil on *Listeria monocytogenes* in meat and cheese. *J. Food Microbiol.,* 18, 647–650.

Miller, A. M., & Raney, K. H. (1993). Solubilization & emulsification: Mechanisms of detergency. *Colloids Surfaces A,* 74, 169–215.

Moleyar, V., & Narasimham, P. (1992). Antimicrobial activity of essential oil components. *Intern. J. Food Microbiol.,* 16, 337–342.

Moreau, L., Kim, H. J., Decker, E. A., & McClements, D. J. (2003). Production and characterization of oil-in-water emulsions containing droplets stabilized by beta-lactoglobulin-pectin membranes. *J. Agricultural Food Chem.,* 51(22), 6612–6617.

Müller, R. H., Freitas, C., zur Mühlen, A., & Mehnert, W. (1996a). Solid lipid nano-particles (SLN) for controlled drug delivery. *Euro. J. Pharmaceutical Sci.,* 4(Supplement 1), S75.

Müller, R. H., Maassen, S., Weyhers, H., & Mehnert, W. (1996b). Phagocytic uptake and cytotoxicity of solid lipid nanoparticles (SLN) sterically stabilized with poloxamine 908 and poloxamer 407. *J. Drug Target.,* 4(3), 161–170.

Müller, R. H., Maassen, S., Schwarz, C., & Mehnert, W. (1997). Solid lipid nanoparticles (SLN) as potential carrier for human use: interaction with human granulocytes. *J. Control. Release,* 47(3), 261–269.

Müller, R. H., Jacobs, C., & Kayser, O. (2001). Nanosuspensions as particulate drug for-mulations in therapy. Rationale for development and what we can expect for the future. *Adv. Drug Deliv. Rev.,* 47(1), 3–19.

Müller, R. H., Radtke, M., & Wissing, S. A. (2002). Solid lipid nanoparticles (SLN) and nanostructured lipid carriers (NLC) in cosmetic and dermatological preparations. *Adv. Drug Deliv. Rev.,* 54, S131–S155.

Mulligan, C. N., Yong, R. N., & Gibbs, B. F. (2001). Surfactant-enhanced remediation of contaminated soil: a review. *Engineer. Geol.,* 60, 371–380.

Nishikido, N. (1993). Thermodynamic models for mixed micellar systems. In K. Ogino, & M. and Abe (Eds.), *Mixed Surfactant Systems.* New York: Marcel Dekker.

NNI (National Nanotechnology Initiative). (2008). About the NNI. Accessed website 04/08. (). http://www.nano.gov/.

Ogawa, S., Decker, E. A., & McClements, D. J. (2003a). Influence of environmental conditions on the stability of oil in water emulsions containing droplets stabilized by lecithin-chitosan membranes. *J. Agricultural Food Chem.,* 51(18), 5522–5527.

Ogawa, S., Decker, E. A., & McClements, D. J. (2003b). Production and characterization of O/W emulsions containing cationic droplets stabilized by lecithin-chitosan membranes. *J. Agricultural Food Chem.,* 51(9), 2806–2812.

Oka, S. (1964). Mechanism of antimicrobial effect of various food preservatives. *Intern. Symposium Food Microbiol.,* 4, 3–16.

Palanuwech, J., & Coupland, J. N. (2003). Effect of surfactant type on the stability of oil-in-water emulsions to dispersed phase crystallization. *Colloids and Surfaces A – Physicochemical and Engineering Aspects,* 223(1–3), 251–262.

Paster, N., Juven, B. J., Shaaya, E., Menasherov, M., Nitzan, R., Weisslowicz, H., & Ravid, U. (1990). Inhibitory effect of oregano and thyme essential oils on moulds and foodborne bacteria. *Lett. Appl. Microbiol.,* 11, 33–37.

Peracchia, M. T. (2003). Stealth nanoparticles for intravenous administration. *Stp Pharma Sci.,* 13(3), 155–161.

Peracchia, M. T., Harnisch, S., Pinto-Alphandary, H., Gulik, A., Dedieu, J. C., Desmaele, D., d'Angelo, J., Müller, R. H., & Couvreur, P. (1999). Visualization of in vitro protein-rejecting properties of PEGylated stealth polycyanoacrylate nanoparticles. *Biomaterials, 20*(14), 1269–1275.

Podio, V., Zara, G. P., Carazzone, M., Cavalli, R., & Gasco, M. R. (2000). Biodistribution of stealth and non-stealth solid lipid nanospheres after intravenous administration to rats. *J. Pharm. Pharmacol., 52*(9), 1057–1063.

Rathmand, J. F., & Scamehorn, J. F. (1989). Electrostatic model to describe mixed ionic/nonionic micellar nonidealities. *Langmuir, 2*(3), 354–361.

Rico-Muñoz, E., & Davidson, P. M. (1983). The effect of corn oil and casein on the antimicrobial activity of phenolic antioxidants. *J. Food Sci., 48*, 1284–1288.

Rota, C., Carramiñana, J. J., Burillo, J., & Herrera, A. (2004). In vitro antimicrobial activity of essential oil from aromatic plants against selected foodborne pathogens. *J. Food Protect, 67*(6), 1252–1256.

Rousseau, D. (2000). Fat crystals and emulsion stability – a review. *Food Res. Intern., 33*, 3–11.

Sato, K., & Garti, N. (1988). *Crystallization and polymorphism of fats and fatty acids.* New York: Marcel Dekker.

Schramm, L. L. (1992). *Emulsions: Fundamentals and Applications in the Petroleum Industry.* Washington: Am. Chem. Soc.

Shelef, L. A. (1983). Antimicrobial effect of spices. *J. Food Safety, 6*, 29–44.

Siekmann, B., & Westesen, K. (1994). Thermoanalysis of the recrystallization process of the melt-homogenized glyceride nanoparticle. *Colloids Surf., 3*, 159–175.

Sikkema, J., de Bont, J. A., & Poolman, B. (1994). Interactions of cyclic hydrocarbons with biological membranes. *J. Biol. Chem., 269*, 8022–8028.

Sivaramakrishnan, R., Nakamura, C., Mehnert, W., Korting, H. C., Kramer, K. D., & Schäfer-Korting, M. (2004). Glucocorticoid entrapment into lipid carriers – characterisation by parelectric spectroscopy and influence on dermal uptake. *J. Control. Release, 97*(3), 493–502.

Sjoeblom, L. (1967). Pharmaceutical applications and physiological aspects of solubilization. In K. Shinoda (ed.), *Solvent Properties of Surfactant Solutions.* New York: Marcel Dekker.

Smith-Palmer, A., Stewart, J., & Fyfe, L. (1998). Antimicrobial properties of plant essential oils and essences against important food-borne pathogens. *Lett. Appl. Microbiol., 26*, 118–122.

Sonoda, T., Takata, Y., Ueno, S., & Sato, K. (2006). Effects of emulsifiers on crystallization behavior of lipid crystals in nanometer-size oil-in-water emulsion droplets. *Crystal Growth Design, 6*(1), 306–312.

Spernath, A., Aserin, A., & Garti, N. (2006). Fully dilutable microemulsions embedded with phospholipids and stabilized by short-chain organic acids and polyols. *J. Colloid Interface Sci., 299*(2), 900–909.

Subbiah, T., Bhat, G. S., Tock, R. W., Parameswaran, S., & Ramkumar, S. S. (2005). Electrospinning of nanofibers. *J. Appl. Polymer Sci., 96*(2), 557–569.

Sznitowska, M., Janicki, S., Dabrowska, E. A., & Gajewska, M. (2002). Physicochemical screening of antimicrobial agents as potential preservatives for submicron emulsions. *Euro. J. Pharmaceutical Sci., 15*, 489–495.

Tagne, J.-B., Kakumanu, S., Ortiz, D., Shea, T., & Nicolosi, R. J. (2008). A nanoemulsion formulation of tamoxifen increases its efficacy in a breast cancer cell line. *Mol. Pharmaceutics, 5*(2), 280–286.

Taylor, T. M., Davidson, P. M., Bruce, B., & Weiss, J. (2005a). Liposomal nanocapsules in food science and agriculture. *Crit. Rev. Food Sci. Technol., 45*, 587–605.

Taylor, T. M., Davidson, P. M., Bruce, B. D., & Weiss, J. (2005b). Ultrasonic spectroscopy and differential scanning calorimetry of liposomal encapsulated nisin. *J. Agricultural Food Chem., 53*, 8722–8728.

Taylor, T. M., Gaysinksy, S., Davidson, P. M., Bruce, B. D., & Weiss, J. (2007). Characterization of antimicrobial bearing liposomes by zeta-potential, vesicle size and encapsulation efficiency. *Food Biophys., 4,* 1–9.

Taylor, T. M., Bruce, B. D., Weiss, J., & Davidson, P. M. (2008). *Listeria monocytogenes* and *Escherichia coli* O157:H7 inhibition in vitro by liposome-encapsulated nisin and ethylene diaminetetraacetic acid. *J. Food Safety, 28*(2), 183–197.

Thapon, J. L., & Brule, G. (1986). Effets du pH et de la forme ionize sur l'affinit lysozymes-caseines. *Le Lait, 66,* 19–30.

Ultee, A., & Smith, E. J. (2002). The phenolic hydroxyl group of carvacrol is essential for action against the food-borne pathogen *Bacillus cereus. J. Appl. Environ. Microbiol., 63,* 620–624.

Velluti, A., Sanchis, V., Ramos, A. J., Egido, J., & Marin, S. (2003). Inhibitory effect of cinnamon, clove, lemongrass, oregano and palmarose essential oils on growth and fumonisin B1 production by *Fusarium proliferatum* in maize grain. *Intern. J. Food Microbiol., 89*(2–3), 145–154.

Videira, M. A., Botelho, M. F., Santos, A. C., Gouveia, L. F., Pedroso de Lima, J. J., & Almeida, A. J. (2002). Lymphatic uptake of pulmonary delivered radiolabelled solid lipid nanoparticles. *J. Drug Target., 10*(8), 607–613.

Vonarbourg, A., Passirani, C., Saulnier, P., & Benoit, J.-P. (2006). Parameters influencing the stealthiness of colloidal drug delivery systems. *Biomaterials, 27*(24), 4356–4373.

Walstra, P. (1993). Principles of emulsion formation. *Chem. Engineer. Sci., 48,* 333.

Walstra, P. (2003). *Physical Chemistry of Foods.* New York: Marcel Dekker.

Wang, Y., & Wu, W. (2006). In situ evading of phagocytic uptake of stealth solid lipid nanoparticles by mouse peritoneal macrophages. *Drug Deliv., 13*(3), 189–192.

Wang, X. Y., Jiang, Y., Wang, Y.-W., Huang, M. T., Ho, C. T., & Huang, Q. R. (2008). Enhancing anti-inflammation activity of curcumin through O/W nanoemulsions. *Food Chem., 108*(2), 419–424.

Weiss, J. (1999). *Mass Transport Phenomena in Oil-in-Water Emulsions Containing Surfactant Micelles. Department of Food Science.* Amherst: University of Massachusetts, Ph.D. Thesis.

Weiss, J., Coupland, J. N., & McClements, D. J. (1996). Solubilization of hydrocarbon emulsion droplets suspended in non-ionic surfactant micelle solutions. *J. Phys. Chem., 100,* 1066–1071.

Weiss, J., Coupland, J. N., Brathwaite, D., & McClements, D. J. (1997). Influence of olecular structure of hydrocarbon emulsion droplets on their solubilization in non-ionic surfactants. *Colloids Surfaces A, 121,* 53–60.

Weiss, J., & McClements, D. J. (2000a). Influence of Ostwald ripening on rheology of oil-in-water emulsions containing electrostatically stabilized droplets. *Langmuir, 16*(5), 2145–2150.

Weiss, J., & McClements, D. J. (2000b). Mass transport phenomena in oil-in-water emulsions containing surfactant micelles: Solubilization. *Langmuir, 16*(14), 5879–5883.

Weiss, J., Takhistov, P., & Weiss, J. (2006). Functional materials in food nanotechnology. *J. Food Sci., 71*(9), R107–R116.

Were, L. M., Bruce, B. D., Davidson, P. M., & Weiss, J. (2004a). Encapsulation of nisin and lysozyme in liposomes enhances efficacy against *Listeria monocytogenes. J. Food Protect., 67*(5), 922–927.

Were, L. M., Bruce, B. D., Davidson, P. M., & Weiss, J. (2004b). Size, stability and entrapment efficiency of phospholipid nanocapsules containing polypeptide antimicrobials. *J. Agricultural Food Chem., 51*(27), 8073–8079.

Westesen, K., Siekmann, B., & Koch, M. H. J. (1993). Investigations on the physical state of lipid nanoparticles by synchrotron-radiation X-ray-diffraction. *Intern. J. Pharmaceutics, 93*(1–3), 189–199.

Wissing, S. A., Kayser, O., & Müller, R. H. (2004). Solid lipid nanoparticles for parenteral drug delivery. *Adv. Drug Deliv. Rev., 56*(9), 1257–1272.

Yang, Y., Feng, J.-F., Zhang, H., & Luo, J.-Y. (2006). Optimization preparation of chan-su-loaded solid lipid nanoparticles by central composite design and response surface method. *Zhongguo Zhong Yao Za Zhi, 31*(8), 650–653.

Zara, G. P., Cavalli, R., Bargoni, A., Fundarò, A., Vighetto, D., & Gasco, M. R. (2002). Intravenous administration to rabbits of non-stealth and stealth doxor-ubicin-loaded solid lipid nanoparticles at increasing concentrations of stealth agent: Pharmacokinetics and distribution of doxorubicin in brain and other tissues. *J. Drug Target., 10*(4), 327–335.

Zhang, L., Hou, S., Maob, S., Weic, D., Song, X., & Lu, Y. (2004). Uptake of folat-econjugated albumin nanoparticles to the SKOV3 cells. *Intern. J. Pharmaceutics, 287*(1–2), 155–162.

Zhang, Y., Lim, C. T., Ramakrishna, S., & Huang, Z.-M. (2005). Recent development of polymer nanofibers for biomedical and biotechnological applications. *J. Mater. Sci. Mater. Med., 16*(10), 933–946.

Zhang, Y. Z., Feng, Y., Huang, Z.-M., Ramakrishna, S., & Lim, C. T. (2006a). Fabrication of porous electrospun nanofibres. *Nanotechnology, 17*(3), 901–908.

Zhang, Y. Z., Wang, X., Feng, Y., & Ramakrishna, S. (2006b). Coaxial electrospinning of (fluo-rescein isothiocyanate-conjugated bovine serum albumin)-encapsulated poly (epsilon-caprolactone) nanofibers for sustained release. *Biomacromolecules, 7*(4), 1049–1057.

zur Mühlen, A., Schwarz, C., & Mehnert, W. (1998). Solid lipid nanoparticles (SLN) for controlled drug delivery – Drug release and release mechanism. *Euro. J. Pharmaceutics Biopharmaceutics, 45*(2), 149–155.

INDEX

Note: Page numbers followed by 'f' indicate figures, 't' indicate tables, and 'n' indicate footnotes.

Printed and bound by CPI Group (UK) Ltd, Croydon, CR0 4YY

03/10/2024

01040410-0014